T0398627

A Practical Guide to Geometric Regulation for Distributed Parameter Systems

MONOGRAPHS AND RESEARCH NOTES IN MATHEMATICS

Series Editors

John A. Burns
Thomas J. Tucker
Miklos Bona
Michael Ruzhansky

Published Titles

Application of Fuzzy Logic to Social Choice Theory, John N. Mordeson, Davender S. Malik and Terry D. Clark

Blow-up Patterns for Higher-Order: Nonlinear Parabolic, Hyperbolic Dispersion and Schrödinger Equations, Victor A. Galaktionov, Enzo L. Mitidieri, and Stanislav Pohozaev

Difference Equations: Theory, Applications and Advanced Topics, Third Edition, Ronald E. Mickens

Dictionary of Inequalities, Second Edition, Peter Bullen

Iterative Optimization in Inverse Problems, Charles L. Byrne

Modeling and Inverse Problems in the Presence of Uncertainty, H. T. Banks, Shuhua Hu, and W. Clayton Thompson

Monomial Algebras, Second Edition, Rafael H. Villarreal

Partial Differential Equations with Variable Exponents: Variational Methods and Qualitative Analysis, Vicenţiu D. Rădulescu and Dušan D. Repovš

A Practical Guide to Geometric Regulation for Distributed Parameter Systems, Eugenio Aulisa and David Gilliam

Signal Processing: A Mathematical Approach, Second Edition, Charles L. Byrne

Sinusoids: Theory and Technological Applications, Prem K. Kythe

Special Integrals of Gradshetyn and Ryzhik: the Proofs – Volume I, Victor H. Moll

Forthcoming Titles

Actions and Invariants of Algebraic Groups, Second Edition, Walter Ferrer Santos and Alvaro Rittatore

Analytical Methods for Kolmogorov Equations, Second Edition, Luca Lorenzi

Complex Analysis: Conformal Inequalities and the Bierbach Conjecture, Prem K. Kythe

Computational Aspects of Polynomial Identities: Volume l, Kemer's Theorems, 2nd Edition Belov Alexey, Yaakov Karasik, Louis Halle Rowen

Cremona Groups and Icosahedron, Ivan Cheltsov and Constantin Shramov

Geometric Modeling and Mesh Generation from Scanned Images, Yongjie Zhang

Groups, Designs, and Linear Algebra, Donald L. Kreher

Handbook of the Tutte Polynomial, Joanna Anthony Ellis-Monaghan and Iain Moffat

Forthcoming Titles (continued)

Lineability: The Search for Linearity in Mathematics, Juan B. Seoane Sepulveda, Richard W. Aron, Luis Bernal-Gonzalez, and Daniel M. Pellegrinao

Line Integral Methods and Their Applications, Luigi Brugnano and Felice Iaverno

Microlocal Analysis on R^n and on NonCompact Manifolds, Sandro Coriasco

Nonlinear Functional Analysis in Banach Spaces and Banach Algebras: Fixed Point Theory Under Weak Topology for Nonlinear Operators and Block Operators with Applications Aref Jeribi and Bilel Krichen

Practical Guide to Geometric Regulation for Distributed Parameter Systems, Eugenio Aulisa and David S. Gilliam

Reconstructions from the Data of Integrals, Victor Palamodov

Special Integrals of Gradshetyn and Ryzhik: the Proofs – Volume II, Victor H. Moll

Stochastic Cauchy Problems in Infinite Dimensions: Generalized and Regularized Solutions, Irina V. Melnikova and Alexei Filinkov

Symmetry and Quantum Mechanics, Scott Corry

MONOGRAPHS AND RESEARCH NOTES IN MATHEMATICS

A Practical Guide to Geometric Regulation for Distributed Parameter Systems

Eugenio Aulisa

Texas Tech University

Lubbock, TX, USA

David Gilliam

Texas Tech University

Lubbock, TX, USA

CRC Press

Taylor & Francis Group

Boca Raton   London   New York

CRC Press is an imprint of the
Taylor & Francis Group, an **informa** business

A CHAPMAN & HALL BOOK

CRC Press
Taylor & Francis Group
6000 Broken Sound Parkway NW, Suite 300
Boca Raton, FL 33487-2742

© 2016 by Taylor & Francis Group, LLC
CRC Press is an imprint of Taylor & Francis Group, an Informa business

Printed on acid-free paper
Version Date: 20150417

International Standard Book Number-13: 978-1-4822-4013-9 (Hardback)

Library of Congress Cataloging-in-Publication Data

Aulisa, Eugenio.
  A Practical Guide to Geometric Regulation for Distributed Parameter Systems / Eugenio Aulisa and David Gilliam, Texas Tech University, Lubbock, Texas.
    pages cm. -- (Monographs and research notes in mathematics)
  "A Chaphman & Hall book."
  Includes bibliographical references and index.
  ISBN 978-1-4822-4013-9 (alk. paper)
  1. Regulators (Mathematics) 2. Control theory. 3. Geometry, Algebraic. I. Gilliam, D. S. (David S.), 1946- II. Title.

QA247.A95 2015
003'.78--dc23   2015010900

Visit the Taylor & Francis Web site at
http://www.taylorandfrancis.com

and the CRC Press Web site at
http://www.crcpress.com

Dedication

In memory of Chris Byrnes,
a pioneer in the development of geometric regulation,
a powerful mathematician and a great friend.
His presence can be found throughout this book.
The book is also dedicated to John Burns in thanks for
his constant encouragement in our research efforts.

David Gilliam would also like to dedicate the book to
his wonderful and extremely patient wife.

Eugenio Aulisa also dedicates this book to his
wonderful wife and son.

Contents

Acknowledgments

David Gilliam would like to acknowledge the many years of support from the Air Force Office of Scientific Research through grants to Christopher I. Byrnes. He would also like to acknowledge the nearly 30 years of constant encouragement and research interest by H.T. Banks and John A. Burns. These special individuals have had a profound effect on my professional life and work. Their commitment to mathematical research has impacted and inspired countless other mathematicians.

Eugenio Aulisa would like to thank the Department of Mathematics and Statistics at Texas Tech University, and in particular all his colleagues and collaborators in the Applied Mathematics group for their constant support and encouragement. With special thanks to Akif Ibragimov who in the last ten years has been both a mentor and a friend. Eugenio Aulisa also would like to acknowledge the National Science Foundation Division of Mathematical Sciences for its continuous support and funding.

We both want to thank the editors Bob Stern and Sarfraz Khan for their optimistic encouragement and professionalism. We also want to thank the reviewers for the careful reading and the many useful suggestions.

Lastly we would like to thank our families for their support and understanding as we committed many hours working on the book.

Preface

In this work we consider a central class of problems in control theory, referred to as regulator problems, which generally corresponds to controlling the output of a system to achieve asymptotic tracking of prescribed trajectories and/or asymptotic rejection of disturbances.

In one of the many possible versions usually called the output regulation problem the objective is to design a feedback control which internally stabilizes a given linear or nonlinear control system so that the output of the resulting closed-loop system asymptotically converges to, or tracks, a given reference signal in the presence of external disturbances. In the literature it is generally assumed that the reference signals and external disturbances are modeled by a so-called exosystem (or exogenous) system.

Regulator theory has a long history in both the finite- and infinite-dimensional cases and there is a vast literature on these subjects, some of which can be found in the references. In this book we are primarily interested in regulation in the case in which the state space of the fixed control system is infinite-dimensional, in which case we say the control system is governed by a distributed parameter system. This is in contrast to the case in which the state space is finite-dimensional and the system is said to be lumped.

There are many versions of the output regulation problem and many different approaches to solving them. The central focus of this book will be the so-called State Feedback Regulator Problem (SFRP), in which the controller is provided with full information of the state of the control system and exosystem. In this work we do not cover the important second main problem referred to as the Error Feedback Regulator Problem in which only the components of the error are available for measurement.

We also assume that the uncontrolled control system is exponentially stable in which case we only need to assume that the output of the composite system is available for measurement. If the control system is not stable we would assume that the system is exponentially stabilizable, in which case one would first find a stabilizing state feedback to render the system exponentially stable. Then, with a stable system in hand, we would return to our milder assumption that only the measured output is available for processing in the controller for solving the problem of asymptotic regulation.

The SFRP has a rich history in the scientific literature. We restrict our historical discussion to the literature that has lead to the particular geometric approach considered in the book. In particular our methodology follows the pioneering work for linear finite-dimensional systems carried out by numerous authors during the 1970's and 1980's (cf., Davison [35], Francis and Wonham [40], Francis [41], Wonham [91]. In particular, Francis [41] showed that the solvability of a multivariable linear regulator problem corresponds to the solvability of a system of two linear matrix equations, called the *regulator* or *Francis* equations. Hautus [46] gave necessary and sufficient conditions for solvability of the *Hautus* equations which contain the regulator equations as a special case. Indeed, for finite-dimensional linear systems, the Hautus conditions given in [58] state that no eigenvalue of the exosystem is an invariant zero of the control system. In 1990 [56], C.I. Byrnes and A. Isidori extended the results of Francis to finite-dimensional nonlinear systems, in case the control system is exponentially stabilizable and the exosystem has bounded trajectories that do not trivially converge to zero. In particular, they give necessary and sufficient conditions for solvability of the regulator problem in terms of solvability of a pair of nonlinear regulator equations. The main mathematical tool employed by C.I. Byrnes and A. Isidori in [56] is the center manifold theorem (see also [31]) which forms the basis of the geometric method in the nonlinear case.

Several authors extended the pioneering geometric methods for linear finite-dimensional systems to linear distributed parameter systems beginning in the early 1980's, [76, 74, 81, 79] and 1990's [30]. Of these works the most closely related is that of J.M. Schumacher [81, 79] and C.I. Byrnes, D. Gilliam and I. Lauko [30]. Schumacher considered control systems whose dynamics are governed by a discrete spectral operator whose generalized eigenvectors form a complete set. In particular, the proof employed by the author requires that the state operator satisfies the spectrum decomposition property (cf., [57, 34]), the spectrum determined growth condition and a controllability condition that implies the stabilizability of the control system with a finite-dimensional controller. In the later work [30], the reference signals and disturbances considered in [81, 79] are assumed to be generated by a finite-dimensional exosystem.

Our main applications are concerned with distributed parameter control systems. To be more precise we are mostly interested in systems governed by partial differential equations. The theoretical mathematical analysis of such systems has been the source of numerous research articles and books. We do not attempt to give an exhaustive list of references to this subject but throughout the text there will be some references to relevant literature. To provide the reader with a starting point we list just a few monographs for control of distributed parameter systems [4, 5, 6, 7, 34, 62, 84]. We also want to make clear that this book is not intended as a theoretical research monograph in the theory of partial differential equations. While almost all of our examples concern dynamical systems governed by partial differential equations we do not devote detailed discussions of the important topics such as existence, uniqueness, regularity and long time behavior of solutions of equations since any attempt to do so would make the book far too long and would detract from the main purposes of the book. For detailed discussions of issues related to the theory of distributions, Sobolev spaces and the theory of partial differential equations we direct the reader to one of the excellent books available in the literature, for example, [37, 36, 39, 47, 61, 60, 90, 92].

In the simplest case of bounded input and output operators, it is shown that sufficient conditions for the design of a finite-dimensional controller that solves the error feedback regulator problem is that there exists a solution to a certain set of operator equations that we call the regulator equations. Construction of a finite-dimensional controller, as opposed to an infinite-dimensional controller, is simply a matter of applying all the assumptions imposed on the system to obtain finite-dimensional approximations to infinite-dimensional operators using eigenfunction approximation. See for example [34] problems 5.22, 5.23 page 261 which are based on the work in [79, 80]. We also mention the many works of S. Pohjolainen including [43] in which the author provides sufficient conditions for the existence of finite-dimensional controllers solving problems of tracking and disturbance rejection for stable parabolic systems (in the case the system generates an analytic semigroup) and in the case that the exosystem is a finite-dimensional linear system with a complete set of eigenfunctions (i.e., a matrix representation is diagonalizable). As pointed out in [43] this does not account for disturbances or signals to be tracked of the form $t^p \sin(\alpha t)$ for $p \geq 1$. Not surprising, hidden in this work is also a solution based on the solvability of the regulator equations. Consider for example page 486 of [43], and the discussion from formula (15) up to (17). But more recent work by S. Pohjolainen, his students and co-authors have lead to a remarkable extension of the geometric theory of output regulation in a number of different directions (cf. [70, 45, 44, 69, 53, 52]).

The paper [30] contains an extension of the geometric methods and provides a characterization analogous to the results found in [56] nonlinear lumped systems. In this work the authors consider infinite-dimensional linear control systems, assuming that the control and observation operators are bounded on the Hilbert state space. The systematic approach presented in [30] for the design of feedback control schemes is described, once again, in terms of the solution of the so-called regulator equations. Indeed, solvability of state feedback regulator problem is characterized in terms of solvability of the regulator equations. For systems described by partial differential equations the regulator equations typically reduce to elliptic boundary value problems that can be solved off-line. The results also apply to retarded functional differential equations in which case the regulator equations reduce to a finite-dimensional linear system of equations.

Extending the above results to the case of systems for which the input and outputs are given in terms of unbounded operators on the Hilbert state space is somewhat more challenging. Such controller design has to take advantage of truly distributed parameter effects which have no lumped counterpart, and therefore could not be designed on the basis of lumped approximations to the distributed parameter models. Thus there are numerous technical obstacles that need to be overcome en route to extending the geometric theory of regulation to the case of unbounded input and output operators. For example, for unbounded B, even if A generates an analytic semigroup it may happen that $(A + B)$ does not generate a C_0 semigroup. Further, for unbounded B and C expressions such as CB may not make sense in the Hilbert state space. On the other hand, there is considerable interest in the case of unbounded inputs and outputs that arise, for example, in the study of boundary control systems governed by partial differential equations. Typical applications include actuators and sensors supported at isolated points or on lower-dimensional hyper-surfaces in, or on the boundary of, a spatial domain.

As the title of the monograph suggests, we hope to provide a practical guide to the use of geometric methods in solving problems of regulation. Over the course of the book we will cover linear and nonlinear examples and the case of bounded and unbounded actuators and sensors. We start with the simplest cases, linear problems with bounded inputs and outputs. Then we consider linear examples with unbounded inputs and outputs. Next we consider nonlinear problems beginning with bounded inputs and outputs and finally consider nonlinear problems with unbounded inputs and outputs.

The book is organized as follows: In the first part of the book we consider regulation for linear systems. In Chapter 1 we consider the case of regulation for Bounded Input and Output Operators. This case most closely resembles the finite-dimensional linear case. In this chapter we present the mathematical setup including the description of the general control problem, exosystem and statement of the problem closely following the development in [30]. In particular, we use a version of the main solvability theorem found in [30] along with a detailed outline of the proof. Several simple examples are given to motivate the results.

In Chapter 2 we again consider linear regulation but now with unbounded control and sensing. This material is more complicated to describe and general theorems in this area have only very recently become available. For example, the state feedback regulator problem for the class of stable regular linear systems can be found in [66]. Once again throughout this chapter several simple examples are given to motivate the results.

In Chapter 3 we present a collection of more challenging and interesting examples of regulation for linear distributed parameter systems with unbounded inputs and outputs.

The second part of the book is devoted to regulation for nonlinear systems. In Chapter 4 we present some theoretical results for nonlinear systems along with some motivating examples. Just as was done in the first part of the book, we begin the discussion in the case of bounded input and outputs for which we can state and prove a result characterizing solvability of a local version of the state feedback regulator problem in terms of solvability of nonlinear regulator equations. Our approach for more general nonlinear systems and input and output operators begins with set-point regulation problems with feedback control obtained by solving the nonlinear regulator equations. After this we quickly progress to problems for which the geometric theory based on center manifold does not directly apply. Nevertheless, the idea of attractive invariance of a dynamic version of the regulator equations still does persist and we show how it can be used to solve a series of increasingly complex regulation problems.

Finally in the final Chapter 5 we present the most complex and interesting applications of the book in a series of examples that we hope will motivate more interest in the geometric design method. In contrast to the 1D examples given in Chapter 4 most of the examples in Chapter 5 are presented for nonlinear partial differential equations in bounded domains in 2D Euclidean space. Examples include control of a Navier-Stokes flow in a forked domain, several examples of control for a Non-Isothermal Navier-Stokes flow in a box domain, control of 2D Burgers' and Chatee-Infante equations and, finally, an example for a vibrating nonlinear beam.

A general theme throughout the book is the emphasis on numerical solvability of the regulator equations using off the shelf software. In particular one does not need to generate a new specialized software to solve each regulation problem. For consistency we have chosen to use the finite element software package Comsol to carry out our numerical simulations. But with our presentation of the necessary equations, written in a software-independent format, one can adapt the solution to available software. We hope that the many examples presented in the book will generate some interest among scientists and control engineers in geometric design methods as an alternative to the much more widely studied optimal design methods.

Software Notes

In this work we use the software MATLAB®

```
The MathWorks, Inc.
3 Apple Hill Drive
Natick, MA 01760-2098 USA
Tel: 508-647-7000, Fax: 508-647-7001
E-mail: info@mathworks.com Web: www.mathworks.com
```

We also use COMSOL Multiphysics® simulation software.

```
COMSOL, Inc.
744 Cowper Street
Palo Alto, CA 94301 USA
Tel: +1-650-324-9935, Fax: +1-650-324-9936
info@comsol.com
```

Part I

Regulation for Linear Systems

Chapter 1

Regulation: Bounded Input and Output Operators

1.1   Setup and Statement of Problem

1.1.1   The Control System

In the first part of this book we focus on regulation problems for linear distributed parameters systems with bounded input and output operators. As we have already mentioned in the introduction our main interest is in applications to systems governed by partial differential equations. For such systems the case of bounded input and output operators does not contain the most interesting applications. However, many unbounded operators have bounded approximations that are widely considered in the literature and since the historical development begins with this case that's where we begin. Moreover, this case most closely parallels the classical results for finite-dimensional linear systems. Although we are interested in applications to control systems governed by PDEs, we choose to present our basic model as an abstract control system

$$z_t = Az + B_{\mathrm{d}}d + B_{\mathrm{in}}u, \tag{1.1}$$

$$z(0) = z_0, \quad z_0 \in \mathcal{Z}, \tag{1.2}$$

$$y(t) = Cz(t), \tag{1.3}$$

where $z \in \mathcal{Z}$ is the state of the system evolving in $\mathcal{Z}$, a complex separable Hilbert space (state space) with inner product denoted by $\langle \cdot, \cdot \rangle$ and norm $\|\varphi\| = \langle \varphi, \varphi \rangle^{1/2}$. We use z_0 to denote the initial condition, $y(t)$ the measured output, and $u(t)$ is the control input. Here $u \in \mathcal{U}$ the input space and $y \in \mathcal{Y}$ the output space where $\mathcal{U}$ and $\mathcal{Y}$ are finite- or infinite-dimensional Hilbert input and output spaces, respectively. A is an unbounded densely defined operator, with domain $\mathcal{D}(A)$ in $\mathcal{Z}$, and it is assumed to be the infinitesimal generator of a strongly continuous semigroup (see, for example, [34, 72]) on $\mathcal{Z}$, $B_{\mathrm{d}} \in \mathcal{L}(\mathcal{U}, \mathcal{Z})$ and $C \in \mathcal{L}(\mathcal{Z}, \mathcal{Y})$. The symbol $\mathcal{L}(W_1, W_2)$ denotes the set of all bounded linear operators from a Hilbert space W_1 to a Hilbert space W_2. When $W = W_1 = W_2$ we write $\mathcal{L}(W)$. The term $d(t)$ represents a disturbance. In general B_{d} refers to a *disturbance input* operator and B_{in} refers to a *control input* operator.

1.1.2   Exosystem: Reference and Disturbance Signals

We will assume that there is a finite-dimensional linear system, referred to as the exogenous system (or exosystem), that produces a reference output $y_r(t)$ (1.5) and which is also used to model the disturbance $d(t)$ (1.6)

$$\frac{dw}{dt}(t) = Sw(t), \tag{1.4}$$

$$y_r(t) = Qw(t), \tag{1.5}$$

$$d(t) = Pw(t), \tag{1.6}$$
$$w(0) = w_0. \tag{1.7}$$

Here $\mathcal{W}$ is the finite-dimensional exosystem state space, $S \in \mathcal{L}(\mathcal{W})$, $Q \in \mathcal{L}(\mathcal{W}, \mathcal{Y})$ and $P \in \mathcal{L}(\mathcal{W}, \mathcal{Z})$.

1.1.3 Some Important Assumptions

Remark 1.1. Stability of the origin for the uncontrolled problem, i.e., the problem with all $u = 0$ and $d = 0$, is a critical component of the theoretical development (see [30, 55]) of the geometric approach to regulation. There are many different versions of stability but here we mean exponential stability which can be described as follows. A system

$$z_t(x, t) = Az(x, t),$$
$$z(x, 0) = z_0(x),$$

is said to be exponentially stable if there are positive constants M and α such that for all initial data $z_0 \in \mathcal{Z}$ the solution z satisfies

$$\|z(\cdot, t)\| = \left\| e^{At} z_0 \right\| \leq M e^{-\alpha t} \|z_0\| \quad \text{for all } t \geq 0.$$

In this case we say the operator A is stable, meaning that it generates an exponentially stable semigroup.

If the operator A is not stable, (does not generate a stable semigroup) then we must first introduce a feedback mechanism that stabilizes the control system. The problem of finding such a feedback law is the stabilization problem and is not the same as the regulator problem considered here. Since our interest is with tracking and disturbance rejection and not stabilization we assume that the control system in question is already stable in order to avoid the extra layer of complication.

Therefore, in order to keep the scope of this book reasonable we do not consider the most general version of the Linear State Feedback Regulator Problem for systems with bounded control and observation as considered in [30]. In particular, in [30] a standing assumption is that the pair (A, B) is stabilizable, i.e., there exists $K \in \mathcal{L}(\mathcal{Z}, \mathcal{U})$ such that $(A + BK)$ is stable. So once and for all, in this work we assume that the operator A is already stable. This assumption allows us to focus more on the problems of interest (tracking and disturbance rejection) and not on the important but separate problem of stabilization. We make this assumption more explicit with the following assumption.

Assumption 1.1. *In this book we will assume that the operator A generates an exponentially stable C_0 semigroup in $\mathcal{Z}$.*

We also impose the following assumption on the exosystem.

Assumption 1.2. *The exosystem is neutrally stable, as in [56]. In the linear case, this is equivalent to the origin being Lyapunov stable forward and backward in time. This implies that $\sigma(S) \subset i\mathbb{R}$ (the imaginary axis, here $i = \sqrt{-1}$) and S in (1.4) has no nontrivial Jordan blocks. Here and below we use the notation $\sigma(T)$ for the spectrum of an operator T. Also, by $\rho(T)$ we will denote the resolvent set of T.*

Remark 1.2. Assumption 1.2 is not the most general that could be given. It is the one given in [30] and covers most common examples, including harmonic and set-point tracking. One possible generalization is that we could actually allow S to have eigenvalues on the imaginary axis corresponding to nontrivial Jordan blocks. This allows us to track and reject polynomial signals.

1.1.4   State Operators & Bounded Actuators and Sensors

1.1.4.1   Examples of State Operators

In many of the examples of systems governed by partial differential equations considered in this book the state operator A is defined in terms of a uniformly elliptic, formally self-adjoint, linear partial differential operator L in the infinite-dimensional Hilbert state space $\mathcal{Z} = L^2(\Omega)$ where Ω is a bounded domain in $\mathbb{R}^n$ with piecewise smooth boundary. For example, in the second-order case,

$$L = \sum_{i,j=1}^{n} \frac{\partial}{\partial x_j}\left(a_{ij}(x)\frac{\partial}{\partial x_i}\right) - a(x), \qquad (1.8)$$

where $a, a_{ij} \in C^\infty(\overline{\Omega})$ ($\overline{\Omega}$ is the closure of Ω), $a_{ij}(x) = a_{ji}(x)$, and $a(x) \geq 0$ for all $x \in \Omega$. By uniform ellipticity we mean that there exist constants $0 < c_1 < c_2 < \infty$ such that

$$c_1|\xi|^2 \leq \sum_{i,j=1}^{n} a_{ij}(x)\xi_i\xi_j \leq c_2|\xi|^2 \quad \forall\, \xi \in \mathbb{R}^n \quad \text{and} \quad x \in \overline{\Omega}.$$

(Here $|\cdot|$ denotes the Euclidean norm in $\mathbb{R}^n$).

The operator A is usually defined as a restriction of the operator L to a dense domain $\mathcal{D}(A) \subset \mathcal{Z}$ in terms of homogeneous boundary conditions of Dirichlet or generalized Neumann or Robin type. The generalized Neumann or Robin boundary conditions are given in terms of the co-normal derivative operator as follows: Let $\nu(x) = (\nu_1(x),\, \nu_2(x),\, \cdots,\, \nu_n(x))$ denote the exterior unit normal vector at the point $x \in \partial\Omega$ and define the vector

$$\mu(x) = (\mu_1(x),\, \mu_2(x),\, \cdots,\, \mu_n(x)) \quad \text{where} \quad \mu_j(x) = \sum_{i=1}^{n} a_{ij}(x)\nu_i(x).$$

Then the co-normal derivative is defined by $\partial\varphi/\partial\mu = \mu \cdot \nabla\varphi$, where the right hand side is the dot product of μ with the gradient of φ. In the special case in which $L = \Delta$, the Laplace operator, the co-normal derivative simply becomes the exterior normal derivative.

1.1.4.2 Examples of Bounded Input and Output Operators

In our intended applications in this chapter the output operator C is defined in terms of a set of bounded operators C_i as weighted integral of the solution $z(x, t)$ in some part Ω_j of the domain Ω, i.e.,

$$y_i(t) = C_i z = \frac{1}{|\Omega_i|} \int_{\Omega_i} z(x, t)\, dx, \qquad (1.9)$$

where

$$|\Omega_i| = \int_{\Omega_i} dx > 0,$$

denotes the Lebesgue measure of the set $\Omega_j \subset \Omega$.

More generally in the same setting

$$y_i(t) = C_i z = \langle z, \Psi_i \rangle = \int_{\Omega} z(x, t)\Psi_i(x)\, dx,$$

for some $\Psi_i \in L^2(\Omega)$. For example in (1.9) we have

$$\Psi_i(x) = \frac{1}{|\Omega_i|}\mathbf{1}_{\Omega_i}(x),$$

where

$$\mathbf{1}_{\Omega_i}(x) = \begin{cases} 0, & \text{if } x \notin \Omega_i, \\ 1, & \text{if } x \in \Omega_i, \end{cases} \qquad (1.10)$$

denotes the characteristic (or indicator) function. The output y is then given by

$$y = Cz = \begin{bmatrix} C_1(z) & C_2(z) & \cdots & C_{n_c}(z) \end{bmatrix}^{\mathsf{T}}.$$

The input operators B_{d} and B_{in} are similarly given as

$$B_{\mathrm{d}}d(t) = \sum_{j=1}^{n_{\mathrm{d}}} B_{\mathrm{d}}^j\, d_j(t), \quad B_{\mathrm{in}}u(t) = \sum_{j=1}^{n_{\mathrm{in}}} B_{\mathrm{in}}^j\, u_j(t), \qquad (1.11)$$

where B_{d}^j and B_{in}^j are in $\mathcal{Z}$, $d_j(t)$ and $u_j(t)$ are scalar disturbances and control inputs, respectively. Very often we will have B_{in}^j and $B_{\mathrm{d}}^j(x)$ defined as the characteristic function of a bounded subset of Ω. More specifically we often have expressions of the form

$$B_{\mathrm{in}}^j(x) = \frac{1}{|\Omega_j|}\mathbf{1}_{\Omega_j}(x),$$

where, in order to ensure $B_{\mathrm{in}}^j \in \mathcal{Z}$, we assume that the set Ω_j has positive (Lebesgue) measure, i.e., $|\Omega_j| > 0$.

Let us consider our first example that should help to clarify the notation introduced above in a very simple context.

Example 1.1 (Periodic Tracking for the Heat Equation). Consider a controlled one-dimensional heat equation on the interval $[0,1]$ with homogeneous Dirichlet (or Robin) boundary condition at $x = 0$ and a Neumann boundary condition at $x = 1$ (cf. Curtain-Zwart [34], Example 2.1.1, page 13):

$$z_t(x,t) = z_{xx}(x,t) + B_\mathrm{d}d + B_\mathrm{in}u(t), \tag{1.12}$$
$$z(0,t) = 0, \qquad\qquad\qquad \text{(Dirichlet BC)} \tag{1.13}$$
$$z_x(1,t) = 0, \qquad\qquad\qquad \text{(Neumann BC)} \tag{1.14}$$
$$z(x,0) = \phi(x),$$
$$y(t) = Cz(t). \tag{1.15}$$

We formulate the system (1.12)–(1.15) in the form (1.1), (1.2) by choosing the Hilbert state space $\mathcal{Z} = L^2(0,1)$. In this case the maximal elliptic operator defined by $L = d^2/dx^2$ in (1.8) with domain is $\mathcal{D}(L) = H^2(0,1)$, the usual Sobolev space of functions in $\mathcal{Z}$ with square integrable second derivative, or more explicitly, the functions $\varphi \in \mathcal{Z}$ which together with $d\varphi/dx$ are absolutely continuous on $(0,1)$ and for which $d^2\varphi/dx^2 \in \mathcal{Z}$. The operator $A = d^2/dx^2$ is then defined as a restriction of L given in terms of the boundary conditions in (1.13) and (1.14). In this way we obtain A as an unbounded densely defined self-adjoint operator in $\mathcal{Z}$, namely,

$$A\varphi = \varphi'', \ \ \mathcal{D}(A) = \{\varphi \in H^2(0,1) : \varphi(0) = 0, \ \ \varphi'(1) = 0\}.$$

As it is well-known the spectrum of A is purely discrete

$$\sigma(A) = \{\lambda_k\}_{k=0}^\infty, \ \ \lambda_k = -\mu_k^2, \ \mu_k = \left(k - \frac{1}{2}\right)\pi,$$

with a corresponding complete set of orthonormal eigenvectors

$$\varphi_k(x) = \sqrt{2}\sin(\mu_k x), \ \ k = 1, 2, \cdots.$$

The operator A generates a strongly continuous (in fact, analytic) semigroup on $\mathcal{Z}$ given in terms of the eigenfunction expansion

$$e^{At}\varphi = \sum_{j=0}^\infty e^{\lambda_j t}\langle\varphi, \varphi_j\rangle\,\varphi_j.$$

In this example, we consider a single-input single-output system with bounded disturbance operator B_d and scalar input and output operators B_in and C so that $\mathcal{Y} = \mathcal{U} = \mathbb{R}$.

1. *Input Operator:* The temperature input is spatially uniform over a small interval about a fixed point $x_{in} = x_0 \in (0,1)$.

$$B_\mathrm{in}u = b(x)u, \ \ b(x) = \frac{1}{2\nu_0}\mathbf{1}_{[x_0 - \nu_0, x_0 + \nu_0]}(x). \tag{1.16}$$

Here the characteristic function, $\mathbf{1}_{[x_1,x_2]}(x)$, is defined in (1.10). Clearly B_{in} is a bounded linear operator defined by multiplication with the function $b \in \mathcal{Z}$.

2. *Output Operator:* The measured output is the average temperature over a small interval about a point $x_{out} = x_1 \in (0,1)$.

$$C\phi = \int_0^1 c(x)\phi(x)\, dx,$$

$$c(x) = \frac{1}{2\nu_1}\mathbf{1}_{[x_1-\nu_1,x_1+\nu_1]}(x). \tag{1.17}$$

Since $C\varphi = \langle \varphi, c \rangle$ with $c \in \mathcal{Z}$ we see that C is a bounded linear observation functional on $\mathcal{Z}$.

3. *Disturbance Operator:* For this specific example we consider a constant disturbance $d(t) = M_d \in \mathbb{R}$ that enters across the entire interval so that $B_{\text{d}} = 1$. We could just as easily consider a disturbance that only enters through a portion of the domain as for the control input B_{in}.

The design objective is to construct a control $u(t)$ that will force the output $y(t)$ to asymptotically track a periodic reference trajectory of the form $y_r(t) = A_r \sin(\alpha t)$ as $t \to +\infty$. In this case we can take the exogenous system in (1.4) to be governed by the 3×3 system

$$\frac{dw}{dt} = Sw, \quad S = \begin{bmatrix} 0 & \alpha & 0 \\ -\alpha & 0 & 0 \\ 0 & 0 & 0 \end{bmatrix}, \quad w(0) = \begin{bmatrix} 0 \\ A_r \\ M_d \end{bmatrix},$$

where $w \in \mathcal{W} = \mathbb{R}^3$ and $S \in \mathcal{L}(\mathcal{W})$. The solution of the previous system is

$$w = \begin{bmatrix} w_1 \\ w_2 \\ w_3 \end{bmatrix} = \begin{bmatrix} A_r \sin(\alpha t) \\ A_r \cos(\alpha t) \\ M_d \end{bmatrix}.$$

In terms of our earlier notation we have $P = [0,0,1]$ in (1.4) and $Q = [1,0,0]$ in (1.5) so that $y_r = Qw$ and $d = Pw$.

1.1.5 Statement of the Regulation Problem

We denote the error between the measured and reference outputs by

$$e(t) = y(t) - y_r(t) = Cz(t) - Qw(t).$$

The main objective of regulation is to force the measured output to track the reference signal while rejecting the disturbance $d(t)$. In other words we want the error $e(t)$ to go to 0 as t goes to ∞.

Recall that throughout this book we will impose the assumption of exponential stability on the uncontrolled system as discussed in Remark 1.1. If this were not the case then we would need to modify the statement of Problem 1.1 below to include a state feedback law in order to obtain the necessary stability for the regulatory theory. As we have already pointed out, the stabilization problem is important but it is a separate question from the tracking and disturbance rejection problem considered here. Therefore in our tracking problems the feedback control law will always be a function of the state of the exosystem and not the state of the control system. For a discussion of the more general situation we refer the reader to the paper [30]. With this we can state the main problem considered in this chapter as follows.

Problem 1.1 (State Feedback Regulator Problem). *Find a feedback control law in the form*

$$u(t) = \Gamma w(t)$$

with $\Gamma \in \mathcal{L}(\mathcal{W}, \mathcal{U})$, such that, for the closed loop system (consisting of the interconnection of the control system (1.1)–(1.3) and the exosystem (1.4)–(1.7) with feedback control $u = \Gamma w$)

$$\frac{dz}{dt}(t) = Az(t) + (B_{in}\Gamma + B_d P)w(t),$$

$$\frac{dw}{dt}(t) = Sw(t),$$

the norm of the error satisfies

$$\|e(t)\| = \|Cz(t) - Qw(t)\| = \|y(t) - y_r(t)\| \longrightarrow 0 \quad as \ t \to \infty, \qquad (1.18)$$

for any initial conditions $z_0 \in \mathcal{Z}$ in (1.3) and $w_0 \in \mathcal{W}$ in (1.7).

In general $e(t)$ is a finite-dimensional vector, thus all l_p norms in (1.18) are equivalent.

1.2 Main Theoretical Result

One of the truly impressive aspects of the geometric theory of regulation is that solvability of the State Feedback Regulator Problem (SFRP) can be characterized in terms of the solvability of a pair of operator equations referred to as the Regulator Equations. The main result in the present setting is the following (cf. Theorem IV.1 in [30]).

Theorem 1.1. *Under the Assumptions 1.1, 1.2, the linear SFRP is solvable if and only if there exist mappings $\Pi \in \mathcal{L}(\mathcal{W}, \mathcal{Z})$ with $\mathrm{Ran}(\Pi) \subset \mathcal{D}(A)$ and $\Gamma \in \mathcal{L}(\mathcal{W}, \mathcal{U})$ satisfying the "Regulator Equations,"*

$$\Pi S = A\Pi + B_{in}\Gamma + B_d P, \tag{1.19}$$

$$C\Pi = Q. \tag{1.20}$$

In addition, if the equations (1.19) and (1.20) are solvable, a feedback law solving the SFRP is given by

$$u(t) = \Gamma w(t).$$

Proof. The Theorem follows from the special form of the closed loop composite system in $\mathfrak{X} = \mathcal{Z} \times \mathcal{W}$. Namely, if we introduce the feedback $u = \Gamma w$ then the system obtained by combining the control system (1.1)–(1.3) and the exosystem (1.4)–(1.7) produces the so-called closed loop system:

$$\frac{d}{dt}X = \mathcal{A}X, \quad \text{with } \mathcal{A} = \begin{bmatrix} A & (B_{in}\Gamma + P) \\ 0 & S \end{bmatrix}, \tag{1.21}$$

$$\mathcal{Y} = [C, -Q]X, \quad X = \begin{bmatrix} z \\ w \end{bmatrix} \in \mathfrak{X}.$$

Here $\mathcal{A}$ generates a C_0 semigroup

$$\mathcal{T}_\mathcal{A}(t) = \exp{(\mathcal{A}t)} = \begin{bmatrix} e^{At} & * \\ 0 & e^{St} \end{bmatrix},$$

in the Hilbert space $\mathfrak{X} = \mathcal{Z} \times \mathcal{W}$ and the spectrum of $\mathcal{A}$ (in $\mathfrak{X}$) decomposes as

$$\sigma(\mathcal{A}) = \sigma(A) \cup \sigma(S) \text{ and } \sigma(A) \cap \sigma(S) = \emptyset.$$

Furthermore, $\mathcal{A}$ satisfies the spectrum decomposition condition at $\beta = 0$, as defined in [57] (see also [34], pages 71 and 232). Namely, we have

$$\sigma(A) \subset \mathbb{C}_\alpha^- \text{ for some } \alpha < 0 \text{ and } \sigma(S) \subset \overline{\mathbb{C}_0^+} \text{ is finite.}$$

Here

$$\mathbb{C}_\beta^+ = \{s \in \mathbb{C} : \text{Re}\,(s) > \beta\} \quad \text{and} \quad \mathbb{C}_\alpha^- = \{s \in \mathbb{C} : \text{Re}\,(s) < -\alpha\}.$$

Therefore the composite state operator $\mathcal{A}$ satisfies the spectrum decomposition condition at $\beta = 0$. Thus we can conclude (as in [57]) that $\mathfrak{X}$ decomposes into the direct sum

$$\mathfrak{X} = \mathcal{V}^+ \oplus \mathcal{V}^-,$$

where $\mathcal{V}^\pm$ are invariant subspaces* under the corresponding C_0-semigroup $\mathcal{T}_\mathcal{A}(t)$ and also under $(sI - \mathcal{A})^{-1}$ for $s \in \rho(\mathcal{A})$. Also $\mathcal{V}^+ \subset \mathcal{D}(\mathcal{A})$, $\mathcal{A}\mathcal{V}^+ \subset \mathcal{V}^+$,

---

*A subspace X is invariant for an operator T if $TX \subset X$.

$\mathcal{A}(\mathcal{D}(\mathcal{A}) \cap \mathcal{V}^-) \subset \mathcal{V}^-$ and $\dim \mathcal{V}^+ = \dim(\mathcal{W})$. $\mathcal{A}$ restricted to $\mathcal{V}^+$ has all its eigenvalues in $\overline{\mathbb{C}_0^+}$ (i.e., they coincide with the eigenvalues of S), while $\mathcal{T}_\mathcal{A}$ restricted to $\mathcal{V}^-$ is exponentially stable. Therefore we can define a linear operator $\Pi \in \mathcal{L}(\mathcal{W}, \mathcal{Z})$ by the condition

$$\left\{ \begin{bmatrix} \Pi w \\ w \end{bmatrix} : w \in W \right\} = \mathcal{V}^+,$$

and we have $\mathrm{Ran}(\Pi) \subset \mathcal{D}(\mathcal{A})$. Here $\mathrm{Ran}(\Pi)$ denotes the range of the operator Π.

From the structure of $\mathcal{A}$ it is easy to see that

$$\left\{ \begin{bmatrix} z \\ 0 \end{bmatrix} : z \in Z \right\} = \mathcal{V}^-.$$

For every $w_0 \in W$, from the $\mathcal{A}$ invariance of $\mathcal{V}^+$ we can write

$$\begin{bmatrix} A & (B_{\mathrm{in}}\Gamma + P) \\ 0 & S \end{bmatrix} \begin{bmatrix} \Pi w_0 \\ w_0 \end{bmatrix} = \begin{bmatrix} \Pi S w_0 \\ S w_0 \end{bmatrix}.$$

This implies

$$\Pi S = A\Pi + B_{\mathrm{in}}\Gamma + B_{\mathrm{d}}P,$$

and therefore the first regulator equation (1.19) holds.

From the $\mathcal{T}_\mathcal{A}(t)$ invariance of $\mathcal{V}^+$

$$\mathcal{T}_\mathcal{A}(t) \begin{bmatrix} \Pi w_0 \\ w_0 \end{bmatrix} = \begin{bmatrix} \Pi e^{St} w_0 \\ e^{St} w_0 \end{bmatrix}, \quad \text{for all } w_0 \in W.$$

Thus for any initial condition $\begin{bmatrix} z_0 \\ w_0 \end{bmatrix} \in \mathfrak{X} = Z \times W$, the solution

$$\begin{bmatrix} z \\ w \end{bmatrix}(t) = \mathcal{T}_\mathcal{A}(t) \begin{bmatrix} z_0 \\ w_0 \end{bmatrix} = \mathcal{T}_\mathcal{A}(t) \begin{bmatrix} (z_0 - \Pi w_0) \\ 0 \end{bmatrix} + \mathcal{T}_\mathcal{A}(t) \begin{bmatrix} \Pi w_0 \\ w_0 \end{bmatrix}$$

$$= \mathcal{J} \exp\left(\begin{bmatrix} A & 0 \\ 0 & S \end{bmatrix} t \right) \mathcal{J}^{-1} \begin{bmatrix} (z_0 - \Pi w_0) \\ 0 \end{bmatrix} + \begin{bmatrix} \Pi e^{St} w_0 \\ e^{St} w_0 \end{bmatrix}$$

$$= \begin{bmatrix} e^{At}(z_0 - \Pi w_0) \\ 0 \end{bmatrix} + \begin{bmatrix} \Pi e^{St} w_0 \\ e^{St} w_0 \end{bmatrix}.$$

Applying $[C, -Q]$ we have

$$e(t) = Cz(t) - Qw(t) = Ce^{At}(z_0 - \Pi w_0) + [C\Pi - Q] e^{St} w_0.$$

From this equation, the Assumption 1.1 that $\exp(At)$ is an exponentially stable semigroup, and Assumption 1.2 which states that the exosystem is neutrally stable (and therefore trajectories must remain bounded away from 0), it is easy to see that

$$e(t) \xrightarrow{t \to \infty} 0 \quad \text{for all} \quad z_0 \in Z, \quad w_0 \in W \quad \text{if and only if} \quad \left[C\Pi - Q\right] = 0.$$

The theorem is proven. □

At this point there are (at least) two other different but related directions that we could have taken to prove the main Theorem 1.1: (1) Direct solution of the Sylvester equation; (2) Diagonalization of the composite operators in (1.21). Let us briefly describe what we mean by this and then give a simple proof of Theorem 1.1

Remark 1.3 (The First Regulator Equation).

- **Direct Solution of Sylvester Equation:** Recall that our assumptions require B_{in}, B_{d}, P, C be bounded, the exosystem be finite dimensional, and A to be stable. Therefore, as we have mentioned, the spectrum of $\mathcal{A}$ satisfies $\sigma(\mathcal{A}) = \sigma(A) \cup \sigma(S)$ and $\sigma(A) \cap \sigma(S) = \emptyset$. In this case it can be shown directly that, for each fixed $\Gamma \in \mathcal{L}(\mathcal{W}, \mathcal{U})$, the Sylvester equation $\Pi S - A\Pi = (B_{\text{in}}\Gamma + B_{\text{d}}P)$ has a unique solution Π given by

$$\Pi = \frac{1}{2\pi i} \int_\gamma (\lambda I - A)^{-1}(B_{\text{in}}\Gamma + B_{\text{d}}P)(\lambda I - S)^{-1} \, d\lambda,$$

  where $\gamma \subset \mathbb{C}$ is a simple closed positively oriented curve in $\rho(A)$ and $\sigma(S) \subset \text{Int}(\gamma)$. This formula is readily verified directly using the Cauchy integral formula and the fact that $(\lambda I - A)^{-1}$ is analytic on and inside the curve γ.

  Under our assumptions we can also express the solution for Π as in [30]

$$\Pi = \int_0^\infty e^{At}(B_{\text{in}}\Gamma + B_{\text{d}}P)e^{-St} \, dt.$$

  This formula is also useful for the case of infinite-dimensional exosystems.

- **Diagonalization of $\mathcal{A}$:** In our other alternative approach we could have shown that the first regulator equation is related to the diagonalization of the upper triangular block matrix operator $\mathcal{A}$. To this end let us define the operator

$$\mathcal{J} = \begin{bmatrix} I & -\Pi \\ 0 & I \end{bmatrix}, \quad \text{with inverse} \quad \mathcal{J}^{-1} = \begin{bmatrix} I & \Pi \\ 0 & I \end{bmatrix}.$$

  Our goal is to find conditions on an operator Π so that the operator $\mathcal{J}$ diagonalizes $\mathcal{A}$.

We have

$$
\mathcal{J}^{-1}\mathcal{A}\mathcal{J} = \left[\begin{array}{c|c} A & [A\Pi - \Pi S + (B_{\mathrm{in}}\Gamma + B_{\mathrm{d}}P)] \\ \hline 0 & S \end{array}\right] = \begin{bmatrix} A & 0 \\ 0 & S \end{bmatrix},
$$

where the last equality holds if and only if the $(1,2)$ position of the middle term is zero. Thus $\mathcal{J}$ diagonalizes $\mathcal{A}$ if and only if Π satisfies the first regulator equation.

1.3   The Transfer Function

In solving the regulator equations the transfer function from linear systems theory arises again and again. The transfer function of the linear triple (A, B_{in}, C) is obtained (formally) by applying the Laplace transform to the system

$$
z_t = Az + B_{\mathrm{in}}u, \quad z(0) = 0, \quad y = Cz,
$$

where the Laplace transform is defined for a function $f(t)$ on $[0, \infty)$ by

$$
\widehat{f}(s) = \int_0^\infty e^{-st} f(t)\, dt.
$$

We obtain

$$
s\widehat{z} = A\widehat{z} + B_{\mathrm{in}}\widehat{u}, \quad \widehat{y} = C\widehat{z},
$$

which can be written as

$$
(sI - A)\widehat{z} = B_{\mathrm{in}}\widehat{u}.
$$

For all $s \notin \sigma(A)$ we can solve for $\widehat{z}$

$$
\widehat{z} = (sI - A)^{-1} B_{\mathrm{in}}\widehat{u}
$$

and apply C to obtain

$$
\widehat{y}(s) = G(s)\widehat{u}(s),
$$

where $G(s)$ is the transfer function of the system governed by (A, B_{in}, C) defined by

$$
G(s) = C(sI - A)^{-1} B_{\mathrm{in}}, \quad \forall\, s \in \rho(A). \tag{1.22}
$$

1.3.1   Solvability Criteria for Regulator Equations

We have seen that the first regulator equation is solvable for every $\Gamma \in \mathcal{L}(\mathcal{W}, \mathcal{U})$. But one must be able to find a Γ so that the second regulator equation is also satisfied. We now investigate why the solvability of the second regulator equation is related to a frequency domain non-resonance condition for the transfer function.

For simplicity, in dealing with various properties of the transfer function, let us consider the case

$$\dim(\mathcal{U}) = \dim(\mathcal{Y}) = k < \infty.$$

In this case the transfer function is a $k \times k$ matrix. We shall also assume that $\det G(s)$ is not identically zero. In this case we can easily define the concept of transmission zero.

Definition 1.1. $s_0 \in \mathbb{C}$ is called a *transmission zero* if $\det G(s_0) = 0$.

Theorem 1.2. *Assume* C, B_{in}, B_d, P, Q *are bounded operators and all the hypotheses of Theorem 1.1 hold. Then the Regulator Equations are solvable provided no natural frequency of the exosystem is a transmission zero of the control system, i.e.,*

$$\lambda_j \in \sigma(S) \quad implies \quad \det G(\lambda_j) \neq 0.$$

Remark 1.4. Under an additional assumption of detectability it is shown in [30] that the non-resonance condition is both necessary and sufficient for solvability of the regulator equations. In this book we are mainly interested in conditions that imply solvability so we present only half of this more general result and avoid a discussion of detectability.

We present a short proof of Theorem 1.2 in the case S is diagonalizable. The result is valid for the more general case but the proof is more involved.

Remark 1.5. Under the additional assumption that S is diagonalizable in $\mathcal{W}$ with eigenvalues λ_j, eigenvectors Φ_j and a bi-orthogonal sequence Ψ_j, i.e., $\langle \Phi_i, \Psi_j \rangle = \delta_{ij}$ (the Kronecker delta function), of eigenvectors for S^* so that

$$S\Phi_j = \lambda_j \Phi_j, \quad S^* \Psi_i = \overline{\lambda_i} \Psi_i. \tag{1.23}$$

In this case the operator Π has the following representation

$$\Pi w = \sum_{j=1}^{k} \langle w, \Psi_j \rangle (\lambda_j I - A)^{-1} (B\Gamma + P) \Phi_j. \tag{1.24}$$

This formula follows immediately from the spectral theorem. We also note that this formula can also be used to extend our discussion to the case of infinite-dimensional exosystems governed by Riesz spectral operators.

Proof. Recall the regulator equations (1.19), (1.20). Let us assume that Π is given by (1.24) and consider the second regulator equation applied to a particular $w = \Phi_\ell$ for an arbitrary $\ell = 1, \cdots, k$, i.e., $C\Pi\Phi_\ell = Q\Phi_\ell$. Using the bi-orthogonality of (Φ_j, Ψ_j) in (1.23), we obtain

$$Q\Phi_\ell = C(\lambda_\ell I - A)^{-1}B_{\text{in}}\Gamma\Phi_\ell + C(\lambda_\ell I - A)^{-1}B_{\text{d}}P\Phi_\ell.$$

We can solve this expression for $G(\lambda_\ell)\Gamma\Phi_\ell$, for $\ell = 1, \cdots, k$ where $G(\lambda_\ell)$ is the transfer function evaluated at the eigenfrequency λ_ℓ of S. Namely we have

$$G(\lambda_\ell)\Gamma\Phi_\ell = Q\Phi_\ell - C(\lambda_\ell I - A)^{-1}B_{\text{d}}P\Phi_\ell,$$

where $G(\lambda_\ell)$ is the transfer function $G(s) = C(sI - A)^{-1}B_{\text{in}}$ evaluated at λ_ℓ.

Under the non-resonance assumption, $\det G(\lambda_j) \neq 0$ for all $\ell = 1, \cdots, k$, so we can solve for $\Gamma\Phi_\ell$ for all ℓ. Finally, using the representation of an arbitrary vector $w \in \mathcal{W}$ by

$$w = \sum_{\ell=1}^{k} \langle w, \Psi_\ell \rangle \Phi_\ell,$$

we have

$$\Gamma w = \sum_{\ell=1}^{k} \langle w, \Psi_\ell \rangle G(\lambda_\ell)^{-1} \left[Q\Phi_\ell - C(\lambda_\ell I - A)^{-1}P\Phi_\ell \right].$$

Notice that in the last formula it is critical that we can invert the transfer function evaluated at every eigenvalue of S. $\square$

1.4   SISO Examples with Bounded Control and Sensing

There are two important special cases of reference and disturbance signals that arise in applications: *Set Points* in which the reference or disturbance signal is time independent, and *Harmonic Signals* in which the reference or disturbance signal is a sinusoid $A_r \sin(\alpha t + \phi)$. We will now analyze these two special cases in a series of Single Input Single Output (SISO) examples.

1.4.1   Set-Point Tracking

For the set point control problem (without a disturbance) we want to track a constant $y_r = M_r \in \mathbb{R}$. In this case we have

$$w_t = Sw = 0, \quad w(0) = M_r \quad \Rightarrow \quad w(t) = M_r,$$

thus

$$S = 0, \quad Q = 1, \quad \text{and } P = 0,$$

and the regulator equations become

$$0 = A\Pi + B_{\text{in}}\Gamma, \quad C\Pi = 1.$$

The first of these equations implies

$$\Pi = -A^{-1}B_{\text{in}}\Gamma.$$

Applying the second regulator equation we have

$$1 = C\Pi = C(-A)^{-1}B_{\text{in}}\Gamma = G(0)\Gamma,$$

using the definition of the transfer function $G(s)$ in (1.22) for $s \in \rho(A)$. Finally under the assumption that $G(0) \neq 0$ we can write

$$\Gamma = G(0)^{-1}.$$

For the above calculations we needed $0 \in \rho(A)$ (which follows from Assumption 1.1). But we also need the more restrictive condition for $G(0)$ to be invertible. This is the non-resonance condition discussed in the previous session, and it requires that no eigenvalue of the exosystem is a "transmission zero" of the control system. In the present case this simply means that $G(0) \neq 0$.

1.4.2 Set-Point Disturbance Rejection

Here we want to reject a constant disturbance $d = M_d \in \mathbb{R}$. So we have

$$w_t = Sw = 0, \quad w(0) = M_d \quad \Rightarrow \quad w(t) = M_d,$$

thus

$$S = 0, \quad Q = 1 \quad \text{and } P = 1.$$

The regulator equations become

$$0 = A\Pi + B_{\text{d}} + B_{\text{in}}\Gamma, \quad C\Pi = 0.$$

The previous equation system is easily solved for Γ by

$$\Gamma = \frac{CA^{-1}B_{\text{d}}}{G(0)} = -\frac{G_{B_{\text{d}}}(0)}{G(0)}, \tag{1.25}$$

where we used the notation

$$G_{B_{\text{d}}}(s) = C(sI - A)^{-1}B_{\text{d}}.$$

Again we have to assume the non-resonance condition $G(0) \neq 0$.

1.4.3   Harmonic Tracking

For tracking a sinusoidal signal such as $y_r = A_r \sin(\alpha t)$, we choose an exosystem governed by a harmonic oscillator

$$w_t = Sw, \quad w(0) = \begin{bmatrix} 0 \\ A_r \end{bmatrix}, \quad S = \begin{bmatrix} 0 & \alpha \\ -\alpha & 0 \end{bmatrix},$$

whose solution is given by

$$w(t) = \begin{bmatrix} A_r \sin(\alpha t) \\ A_r \cos(\alpha t) \end{bmatrix}.$$

Thus we take

$$Q = [1, 0] \text{ and } P = [0, 0],$$

such that

$$y_r(t) = Qw = A_r \sin(\alpha t).$$

In this case $\mathcal{W} = \mathbb{R}^2$ so we look for

$$\Pi = [\Pi_1, \Pi_2], \quad \Pi_j \in \mathcal{Z}, \text{ and } \Gamma = [\Gamma_1, \Gamma_2] \in \mathbb{R}^2.$$

The two regulator equations applied to a general vector $w = \begin{bmatrix} w_1, & w_2 \end{bmatrix}^\mathsf{T} \in \mathcal{W}$ give the following system

$$\Pi Sw = A\Pi w + B_{\text{in}}\Gamma w, \quad C\Pi w = Qw.$$

When S corresponds to a harmonic oscillator there are many ways to solve these equations and each of them has its own merits. Here we have chosen a method that requires no skills beyond freshman algebra. We will exploit an additional reasonable assumption that the system is real, i.e.,

$$G(\bar{s}) = \overline{G(s)}, \quad \text{for all } s \notin \rho(A).$$

It is easy to see that the first regulator equation applied to a general vector $w = \begin{bmatrix} w_1, & w_2 \end{bmatrix}^\mathsf{T} \in \mathcal{W}$ can be written as

$$\alpha\Pi_1 w_2 - \alpha\Pi_2 w_1 = A\Pi_1 w_1 + A\Pi_2 w_2 + B_{\text{in}}\Gamma_1 w_1 + B_{\text{in}}\Gamma_2 w_2.$$

Since this equation has to hold for all w, we consider the special case $w_1 = 1$ and $w_2 = 0$ and then the case $w_1 = 0$ and $w_2 = 1$ which gives the two operator equations

$$-\alpha\Pi_2 - A\Pi_1 = B_{\text{in}}\Gamma_1,$$
$$\alpha\Pi_1 - A\Pi_2 = B_{\text{in}}\Gamma_2.$$

Noting that the eigenvalues of S are given by $\lambda = \pm i\alpha$, we are motivated to multiply the second of these two equations by $i = \sqrt{-1}$ and add the result to the first equation to get

$$(i\alpha I - A)\Pi_1 + i(i\alpha I - A)\Pi_2 = B_{\text{in}}i\Gamma_2 + B\Gamma_1.$$

Then since $i\alpha \notin \rho(A)$ we can apply $(i\alpha I - A)^{-1}$ to both sides and get

$$\Pi_1 + i\Pi_2 = (i\alpha I - A)^{-1}B_{\text{in}}\left(i\Gamma_2 + \Gamma_1\right).$$

If we apply C to both sides of the previous expression, and recall that the second regulator equation implies

$$C\Pi_1 = 1, \quad C\Pi_2 = 0,$$

we get

$$1 = C(i\alpha I - A)^{-1}B_{\text{in}}\left(i\Gamma_2 + \Gamma_1\right) = G(i\alpha)\left(i\Gamma_2 + \Gamma_1\right).$$

Rewriting $G(i\alpha)$ in terms of its real and imaginary parts, i.e.,

$$G(i\alpha) = \text{Re}\,(G(i\alpha)) + i\text{Im}\,(G(i\alpha)),$$

yields

$$1 = \big(\text{Re}\,(G(i\alpha)) + i\text{Im}\,(G(i\alpha))\big)\left(i\Gamma_2 + \Gamma_1\right).$$

Equating the real and imaginary parts it follows that

$$1 = \text{Re}\,(G(i\alpha))\Gamma_1 - \text{Im}\,(G(i\alpha))\Gamma_2,$$
$$0 = \text{Im}\,(G(i\alpha))\Gamma_1 + \text{Re}\,(G(i\alpha))\Gamma_2.$$

The solution to this system is given by

$$\Gamma_1 = \frac{\text{Re}\,(G(i\alpha))}{|G(i\alpha)|^2}, \quad \Gamma_2 = -\frac{\text{Im}\,(G(i\alpha))}{|G(i\alpha)|^2}.$$

Thus, to solve the problem of harmonic tracking the desired control is given by

$$\Gamma = [\Gamma_1, \Gamma_2] = \left[\text{Re}\,(G(i\alpha)^{-1}), \text{Im}\,(G(i\alpha)^{-1})\right]. \tag{1.26}$$

This last result follows directly from the fact that

$$G(i\alpha)^{-1} = \frac{1}{G(i\alpha)} = \frac{\overline{G(i\alpha)}}{|G(i\alpha)|^2} = \frac{\text{Re}\,(G(i\alpha)) - i\text{Im}\,(G(i\alpha))}{|G(i\alpha)|^2}.$$

Notice once again there is a non-resonance condition for solvability. Namely, we must have $G(i\alpha) \neq 0$.

1.4.4   Harmonic Disturbance Rejection

For rejecting a sinusoidal disturbance such as $d = A_d \sin(\beta t)$ once again we choose an exosystem governed by the harmonic oscillator

$$w_t = Sw, \quad w(0) = \begin{bmatrix} 0 \\ A_d \end{bmatrix}, \quad S = \begin{bmatrix} 0 & \beta \\ -\beta & 0 \end{bmatrix},$$

whose solution is given by

$$w(t) = \begin{bmatrix} A_d \sin(\beta t) \\ A_d \cos(\beta t) \end{bmatrix}.$$

Therefore we set

$$Q = [0, 0] \text{ and } P = [1, 0],$$

so that

$$y_r(t) = Qw = 0.$$

In this case $\mathcal{W} = \mathbb{R}^2$, so we look for

$$\Pi = [\Pi_1, \Pi_2], \quad \Pi_j \in \mathcal{Z}, \text{ and } \Gamma = [\Gamma_1, \Gamma_2] \in \mathbb{R}^2.$$

The two regulator equations applied to a general vector $w = \begin{bmatrix} w_1 & w_2 \end{bmatrix}^\mathsf{T} \in \mathcal{W}$ give the following system:

$$\Pi Sw = A\Pi w + B_d Pw + B_{in}\Gamma w, \quad C\Pi w = 0.$$

Writing out the above matrix multiplications we see that the first regulator equation becomes

$$\beta\Pi_1 w_2 - \beta\Pi_2 w_1 = A\Pi_1 w_1 + A\Pi_2 w_2 + B_d w_1 + B_{in}\Gamma_1 w_1 + B_{in}\Gamma_2 w_2.$$

We note that this equation has to hold for all w, so collecting the w_1 and w_2 terms together, similar to the previous example, we obtain

$$-\beta\Pi_2 - A\Pi_1 = B_d + B_{in}\Gamma_1,$$
$$\beta\Pi_1 - A\Pi_2 = B_{in}\Gamma_2.$$

Multiply the second of these two equations by $i = \sqrt{-1}$ and add the result to the first equation to get

$$(i\beta I - A)\Pi_1 + i(i\beta I - A)\Pi_2 = B_d + B_{in}i\Gamma_2 + B\Gamma_1.$$

Note that $i\beta \notin \rho(A)$ so we can apply $(i\beta I - A)^{-1}$ to both sides and get

$$\Pi_1 + i\Pi_2 = (i\beta I - A)^{-1} B_d + (i\beta I - A)^{-1} B_{in} (i\Gamma_2 + \Gamma_1).$$

If we apply C to both sides of the previous expression, and recall that the second regulator equation implies

$$C\Pi_1 = 0, \quad C\Pi_2 = 0,$$

we get

$$0 = C(i\beta I - A)^{-1}B_{\mathrm{d}} + C(i\beta I - A)^{-1}B_{\mathrm{in}}(i\Gamma_2 + \Gamma_1) = G_{B_{\mathrm{d}}}(i\beta) + G(i\beta)(i\Gamma_2 + \Gamma_1),$$

where again we have used the notation

$$G_{B_{\mathrm{d}}}(s) = C(sI - A)^{-1}B_{\mathrm{d}}.$$

Finally, solving for Γ, we get

$$\Gamma = [\Gamma_1, \Gamma_2] = \left[-\mathrm{Re}\ (G(i\beta)^{-1})\ \mathrm{Re}\ (G_{B_{\mathrm{d}}}(i\beta)) + \mathrm{Im}\ (G(i\beta)^{-1})\ \mathrm{Im}\ (G_{B_{\mathrm{d}}}(i\beta)), \right.$$
$$\left. -\mathrm{Re}\ (G(i\beta)^{-1})\ \mathrm{Im}\ (G_{B_{\mathrm{d}}}(i\beta)) - \mathrm{Im}\ (G(i\beta)^{-1}\ \mathrm{Re}\ (G_{B_{\mathrm{d}}}(i\beta))) \right].$$

Example 1.2 (Heat Equation). We now return to Example 1.1 for the one-dimensional heat equation. For this problem we have defined all the necessary ingredients in order to find the desired control $u = \Gamma w$. The regulator equations in this case take the form

$$\Pi S w = A\Pi w + B_{\mathrm{d}} P w + B_{\mathrm{in}}\Gamma w, \quad C\Pi w = Q w = w_1,$$

where

$$\Pi = [\Pi_1, \Pi_2, \Pi_3], \quad w = \begin{bmatrix} w_1 \\ w_2 \\ w_3 \end{bmatrix} = \begin{bmatrix} A_r \sin(\alpha t) \\ A_r \cos(\alpha t) \\ M_d \end{bmatrix},$$

and

$$S = \begin{bmatrix} 0 & \alpha & 0 \\ -\alpha & 0 & 0 \\ 0 & 0 & 0 \end{bmatrix}, \quad P = [0, 0, 1], \quad Q = [1, 0, 0].$$

The block diagonal structure of S then allows us to decouple the regulator equations into two separate parts. The first part, corresponding to the harmonic tracking, is given by

$$\Pi^\alpha S^\alpha w^\alpha = A\Pi^\alpha w^\alpha + B_{\mathrm{in}}\Gamma^\alpha w^\alpha, \quad C\Pi^\alpha w^\alpha = w_1$$

where

$$\Pi^\alpha = [\Pi_1, \Pi_2], \quad S^\alpha = \begin{bmatrix} 0 & \alpha \\ -\alpha & 0 \end{bmatrix}, \quad w^\alpha = [w_1, w_2]^\mathsf{T}, \quad Q^\alpha = [1, 0], \quad P^\alpha = [0, 0].$$

The second part, corresponding to the rejection of a constant disturbance, is given by

$$0 = A\Pi_3 w_3 + B_{\mathrm{d}} w_3 + B_{\mathrm{in}}\Gamma_3 w_3, \quad C\Pi_3 w_3 = Q_3 w_3 = 0$$

where

$$Q_3 = 0, \quad P_3 = 1.$$

Thus the regulator equations for the harmonic tracking become exactly those given in Section 1.4.3, whose solution for Γ^α is given in Equation (1.26), while the regulator equations for the set point disturbance rejection become exactly those given in Section 1.4.2 whose solution for Γ_3 is given by Equation (1.25). Combining these two solutions, we have

$$\Gamma = [\Gamma_1, \Gamma_2, \Gamma_3] = \left[\mathrm{Re}\,(G(i\alpha)^{-1}), \mathrm{Im}\,(G(i\alpha)^{-1}), G(0)^{-1}(CA^{-1}B_{\mathrm{d}}) \right].$$

The analytic computation of the control thus reduces to finding an explicit formula for the transfer function $G(s)$. For the present problem this is a simple task and it can be done by hand. Recall that $G(s) = C(sI - A)^{-1}B_{\mathrm{in}}$ so we first compute $\phi = (sI - A)^{-1}B_{\mathrm{in}}$ which amounts to solving the boundary value problem

$$s\phi(x) - \phi''(x) = b(x),$$

subject to the boundary conditions

$$\phi(0) = 0, \quad \phi'(1) = 0.$$

This is an elementary problem in ordinary differential equations, and for given values $x_0 = 0.75$, $x_1 = 0.25$ and $\nu_0 = \nu_1 = 0.25$ in Equations (1.16) and (1.17) it produces the following solution:

$$\phi(x) = \frac{1}{\sqrt{s}\cosh(\sqrt{s})} \left[\cosh(\sqrt{s}(1-x)) \int_0^x \sinh(\sqrt{s}\xi)\, b(\xi)\, d\xi \right.$$
$$\left. + \sinh(\sqrt{s}x) \int_0^x \cosh(\sqrt{s}(1-\xi))\, b(\xi)\, d\xi \right].$$

Applying C to this function gives us $G(s) = C(\phi) = C(sI - A)^{-1}B_{\mathrm{in}}$

$$G(s) = \frac{1}{\sqrt{s}\cosh(\sqrt{s})} \left(\frac{\cosh(\sqrt{s}/2) - 1}{\sqrt{s}/2} \right) \left(\frac{\sinh(\sqrt{s}/2)}{\sqrt{s}/2} \right). \qquad (1.27)$$

We note that

$$G(0) = \lim_{s \to 0} G(s) = \frac{1}{4}.$$

Furthermore we can easily compute $(CA^{-1}B_{\mathrm{d}})$ by solving

$$\phi''(x) = 1, \quad \phi(0) = 0, \quad \phi'(1) = 0,$$

which gives $\phi(x) = x^2/2 - x$. Applying C to this gives

$$(CA^{-1}B_{\mathrm{d}}) = -\frac{5}{24}.$$

This, in turn, gives

$$\Gamma_3 = \frac{(CA^{-1}B_{\mathrm{d}})}{G(0)} = -\frac{5}{6}.$$

We also readily compute Γ_1 and Γ_2 from (1.26) and (1.27).

In our numerical simulation we have set $M_d = 1$, $A_r = 1$, $\alpha = 2$ and as above taken $x_0 = 0.75$, $x_1 = 0.25$ and $\nu_0 = \nu_1 = 0.25$. We have chosen the initial condition $\varphi(x) = 4\cos(\pi x)$.

Fig. 1.1 depicts the reference signal $y_r(t) = \sin(\alpha t)$, and the controlled output $y(t) = Cz$ for the closed loop system. Fig. 1.2 depicts the error $e(t) = y(t) - y_r(t)$, and finally Fig. 1.3 contains the numerical solution for the temperature $z(x,t)$ for $x \in [0,1]$ and $t \in [0,6]$.

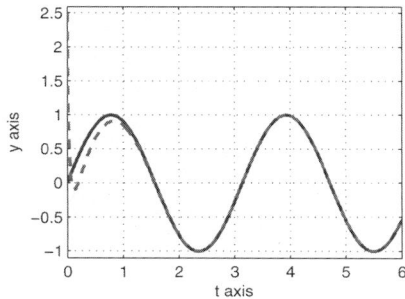

Fig. 1.1: $y(t)$ and $y_r(t)$.   **Fig. 1.2**: Plot of Error $e(t)$.

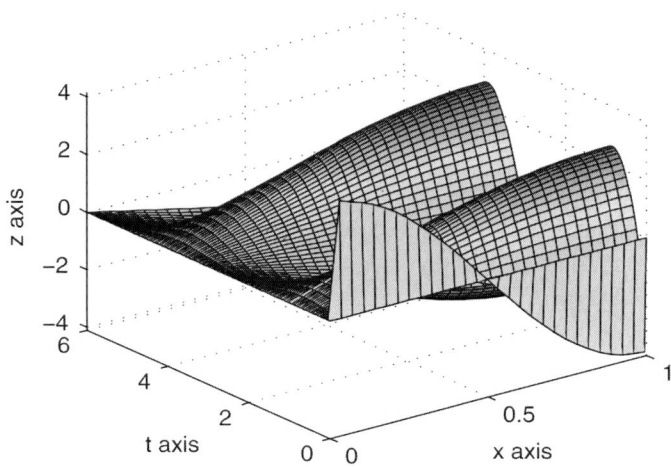

Fig. 1.3: Plot of Solution Surface.

Example 1.2 is a very simple example for which it is easy to find an explicit formula for the transfer function $G(s) = C(sI - A)^{-1}B_{\mathrm{in}}$. But in general this can be an extremely difficult task even for relatively simple distributed parameter systems governed by partial differential equations. One of the main goals of this work is to demonstrate how easily one can obtain numerical approximations to the necessary feedback controls, which, as we have already shown, essentially means computing the transfer function.

We now present an example (see [30]) of tracking for a one-dimensional damped wave equation. Due to the second-order time derivative, hyperbolic regulation problems require some extra work in order that they can be written in the form (1.1)–(1.3). In this example we explain the usual representation of the wave equation in standard dynamical systems form and proceed to solve the regulator problem using this formalism.

Example 1.3 (Damped Wave Equation). Consider a controlled one-dimensional damped wave equation governing the small vibrations of a finite string

$$z_{tt} = z_{xx} - 2\gamma z_t + B_{\mathrm{d}}d(t) + B_{\mathrm{in}}u(t), \tag{1.28}$$
$$z(0,t) = z(1,t) = 0,$$
$$y(t) = Cz(t),$$
$$z(0) = \phi \in H^2(0,1),$$
$$z_t(0) = \psi \in L^2(0,1).$$

Here $\gamma > 0$ is the damping coefficient. Our design objective in this example is to control the average motion of the string over a small fixed interval, centered at $x_{out} \in (0,1)$, to track a prescribed constant value given by $y_r(t) = M_r$. The output and the control input spaces are $Y = U = \mathbb{R}$. The input operator B_{in} is given in (1.16) and output operator C is given in (1.17). We also consider a disturbance that enters over a fixed interval centered at x_d. The disturbance operator B_{d} is given by formula (1.16), and the signal to be rejected is a periodic signal given by $d(t) = A_d \sin(\beta t)$.

In order to formulate this problem within the current framework we proceed in the usual way and define $A = d^2/dx^2$ in $\mathcal{Z} = L^2(0,1)$ as the self-adjoint operator defined by

$$\mathcal{D}(A) = \{\phi \in H^2(0,1) \,:\, \phi(0) = \phi(1) = 0\}$$
$$= H^2(0,1) \cap H_0^1(0,1).$$

Then we define the energy Hilbert space

$$\mathcal{H} = H_0^1(0,1) \oplus L^2(0,1) = \mathcal{D}((-A)^{1/2}) \oplus L^2(0,1),$$

with the norm

$$\|\Phi\|_{\mathcal{H}}^2 = \|\phi_1'\|^2 + \|\phi_2\|^2 \quad \text{for} \quad \Phi = \begin{pmatrix} \phi_1 \\ \phi_2 \end{pmatrix} \in \mathcal{H},$$

where $\| \cdot \|$ is the norm in $\mathcal{Z}$. Note that this norm is equivalent to the graph norm. Next we define the operator $\mathcal{A}$ with $\mathcal{D}(\mathcal{A}) = \mathcal{D}(A) \oplus H_0^1$ by

$$\mathcal{A} = \begin{pmatrix} 0 & I \\ A & -2\gamma I \end{pmatrix}.$$

With this we can write the system (1.28) in the standard state space form (1.1)–(1.3)

$$Z_t = \mathcal{A}Z + \mathcal{B}_d d + \mathcal{B}_{in} u, \quad Z(0) = \begin{bmatrix} \phi \\ \psi \end{bmatrix}, \quad y = \mathcal{C}Z, \qquad (1.29)$$

where

$$Z = \begin{bmatrix} z_1 \\ z_2 \end{bmatrix} = \begin{bmatrix} z \\ z_t \end{bmatrix},$$

$$\mathcal{B}_d d = \begin{bmatrix} 0 \\ B_d d \end{bmatrix}, \quad \mathcal{B}_{in} u = \begin{bmatrix} 0 \\ B_{in} u \end{bmatrix}, \quad \mathcal{C}Z = \begin{bmatrix} C & 0 \end{bmatrix} \begin{bmatrix} z_1 \\ z_2 \end{bmatrix} = C z_1.$$

It is straightforward to verify that $\mathcal{A}$ is maximal dissipative and therefore generates a contraction semigroup. In fact more is true, namely, $\mathcal{A}$ generates an exponentially stable semigroup. This can be seen by direct estimates or one can first show that $\mathcal{A}$ is a Riesz spectral operator satisfying the spectrum determined growth condition (see for example Theorem 2.3.5.c in Curtain-Zwart [34]), the spectrum lies along the line $\operatorname{Re}\lambda = -\gamma$ therefore the semigroup generated by $\mathcal{A}$ is exponentially stable. Therefore we can apply Theorem 1.1 and seek mappings Π and Γ solving the regulator equations to obtain the desired feedback law $u = \Gamma w$.

It is certainly possible to derive the regulator equations using (1.29) and proceeding as we did in the heat example. But due to the vector nature of that formula it turns out this approach is awkward and produces complicated terms that are completely unnecessary. Rather, it turns out that it is easier to derive the regulator equations from equation (1.28) by introducing a very useful simplifying idea for equations with a second-order time derivative term. Let us introduce the relation $z = \Pi w$ in (1.28) to get

$$z_{tt} = \Pi w_{tt} = \Pi S^2 w = \Pi \begin{bmatrix} 0 & 0 & 0 \\ 0 & 0 & \beta \\ 0 & -\beta & 0 \end{bmatrix}^2 w$$

$$= \Pi \begin{bmatrix} 0 & 0 & 0 \\ 0 & -\beta^2 & 0 \\ 0 & 0 & -\beta^2 \end{bmatrix} w = -\beta^2 (\Pi_2 w_2 + \Pi_3 w_3) = -\beta^2 \Pi^\beta w^\beta,$$

where

$$\Pi = [\Pi_1, \ \Pi^\beta] = [\Pi_1, [\Pi_2, \ \Pi_3]], \quad w = \begin{bmatrix} w_1 \\ w^\beta \end{bmatrix} = \begin{bmatrix} w_1 \\ w_2 \\ w_3 \end{bmatrix} = \begin{bmatrix} M_r \\ A_d \sin(\beta t) \\ A_d \cos(\beta t) \end{bmatrix}.$$

This leads to

$$-\beta^2 \Pi^\beta w^\beta = A\Pi w - 2\gamma S\Pi w + B_{\rm d}Pw + B_{\rm in}\Gamma w,$$
$$y_r = C\Pi w = C\Pi_1 w_1 + C\Pi_2 w_2 + C\Pi_3 w_3 = Qw = w_1,$$

where

$$Q = [Q_1, Q^\beta] = \begin{bmatrix} 1, & [0,0] \end{bmatrix} \text{ and } P = [P_1, P^\beta] = \begin{bmatrix} 0, & [1,0] \end{bmatrix}.$$

Grouping together the w_1 terms, one gets

$$0 = A\Pi_1 + B_{\rm in}\Gamma_1, \tag{1.30}$$
$$C\Pi_1 = 1, \tag{1.31}$$

while, grouping together the $w^\beta = [w_2, w_3]$ terms, one gets

$$-\beta^2 \Pi_2 = A\Pi_2 + 2\gamma\beta\Pi_3 + B_{\rm d} + B_{\rm in}\Gamma_2, \tag{1.32}$$
$$-\beta^2 \Pi_3 = A\Pi_3 - 2\gamma\beta\Pi_2 + B_{\rm in}\Gamma_3,$$
$$C\Pi_2 = 0, \quad C\Pi_3 = 0. \tag{1.33}$$

The regulator equations (1.30)–(1.31) for the set-point tracking are exactly the same as those described in Section 1.4.1, whose solution for Γ_1 is given by

$$\Gamma_1 = G(0)^{-1}.$$

To solve for Γ^β we solve the system (1.32)–(1.33) similarly to the procedure presented in Section 1.4.4, which produces the following solution:

$$\begin{aligned} \Gamma^\beta &= [\Gamma_2, \Gamma_3] \\ &= \big[-\operatorname{Re}\left(G(\delta)^{-1}\right)\operatorname{Re}\left(G_{B_{\rm d}}(\delta)\right) + \operatorname{Im}\left(G(\delta)^{-1}\right)\operatorname{Im}\left(G_{B_{\rm d}}(\delta)\right), \\ &\quad -\operatorname{Re}\left(G(\delta)^{-1}\right)\operatorname{Im}\left(G_{B_{\rm d}}(\delta)\right) - \operatorname{Im}\left(G(\delta)^{-1}\operatorname{Re}\left(G_{B_{\rm d}}(\delta)\right)\right)\big] \end{aligned} \tag{1.34}$$

where

$$\delta = i2\beta\gamma - \beta^2.$$

In Equation (1.34) once again we have the necessary non-resonance condition $\operatorname{Re}\left(G(\delta)\right) \neq 0$, and $\operatorname{Im}\left(G(\delta)\right) \neq 0$. We also note that $\delta \in \rho(A)$ since A is self-adjoint and therefore has real spectrum.

As we have already mentioned, since the original system is exponentially stable we may use the control $u = \Gamma w$. Thus the resulting closed loop system can be written as

$$z_{tt} = Az - 2\gamma z_t + B_{\rm d}d + B_{\rm in}\Gamma w. \tag{1.35}$$

In our numerical example we have set $x_{out} = 3/4$, $x_{in} = 1/4$, $x_d = 5/8$, $\nu_{out} = \nu_{in} = 1/4$, $\nu_d = 1/8$, $M_r = 1$, $A_d = 0.25$, $\beta = 0.5$, $\gamma = .5$. Finally we have chosen initial conditions $\phi(x) = 16x^2(1-x)^2$ and $\psi = 0.5\sin^2(\pi x)$. For these values the control gains turn out to be

$$\Gamma = [16, \ -2.148080, 0.03728473].$$

Fig. 1.4 depicts the reference signal $y_r(t) = 1$ and the controlled output $y(t) = Cz$ for the closed loop system (1.35). Fig. 1.5 depicts the error $e(t) = y(t) - y_r(t)$, and finally Fig. 1.6 contains the numerical solution for the displacement z for $x \in [0,1]$ and $t \in [0,30]$.

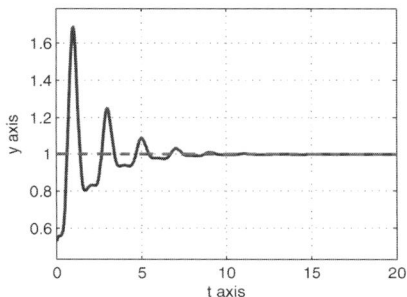

Fig. 1.4: $y(t)$ and $y_r(t)$.   **Fig. 1.5**: Plot of Error $e(t)$.

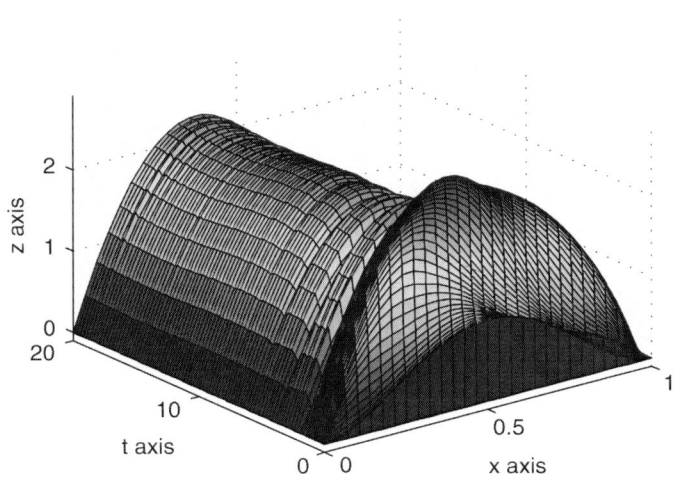

Fig. 1.6: Solution Surface $z(x,t)$.

1.5   The MIMO Case

1.5.1   Problem Setting and Notation

In order to simplify the notation for a Multi-Input and Multi-Output (MIMO) control problem let us introduce some notation

$$B_{\text{in}} = \begin{bmatrix} B_{\text{in}}^1, & B_{\text{in}}^2, & \cdots & B_{\text{in}}^{n_{\text{in}}} \end{bmatrix}, \quad B_{\text{in}}^j \in \mathcal{Z},$$

$$B_{\text{d}} = \begin{bmatrix} B_{\text{d}}^1, & B_{\text{d}}^2, & \cdots & B_{\text{d}}^{n_{\text{d}}} \end{bmatrix}, \quad B_{\text{d}}^j \in \mathcal{Z},$$

$$U = \begin{bmatrix} u_1 \\ u_2 \\ \vdots \\ u_{n_{\text{in}}} \end{bmatrix}, \quad D = \begin{bmatrix} d_1 \\ d_2 \\ \vdots \\ d_{n_{\text{d}}} \end{bmatrix}, \quad Y = \begin{bmatrix} y_1 \\ y_2 \\ \vdots \\ y_{n_{\text{c}}} \end{bmatrix}, \quad Y_r = \begin{bmatrix} y_{r,1} \\ y_{r,2} \\ \vdots \\ y_{r,n_{\text{c}}} \end{bmatrix}.$$

With this notation we can write our abstract control problem as

$$\frac{dz}{dt} = Az + B_{\text{d}}D + B_{\text{in}}U,$$

$$Y = Cz.$$

Here we have written the input, disturbance and output terms in matrix form as

$$B_{\text{d}}D = \sum_{j=1}^{n_{\text{d}}} B_{\text{d}}^j d_j(t), \quad B_{\text{in}}U = \sum_{j=1}^{n_{\text{in}}} B_{\text{in}}^j u_j(t), \quad Y = Cz = \begin{bmatrix} C_1(z) \\ C_2(z) \\ \vdots \\ C_{n_{\text{c}}}(z) \end{bmatrix},$$

where B_{d} and B_{in} are disturbance input and control input operators respectively, and C denotes the output operator. The transfer function has the form

$$G(s) = C(sI - A)^{-1}B_{\text{in}}.$$

In the case with n_{in} inputs and n_{c} outputs the transfer function is an $n_{\text{c}} \times n_{\text{in}}$ matrix with entries

$$g_{ij}(s) = C_i(sI - A)^{-1}B_{\text{in}}^j, \quad i = 1, \ldots, n_{\text{c}}, \ j = 1, \ldots, n_{\text{in}},$$

which we compute numerically for a fixed $s \in \rho(A)$ as follows: First we solve the n_{in} linear boundary value problems given by

$$(sI - A)X_j(x) = B_{\text{in}}^j(x), \quad j = 1, \ldots, n_{\text{in}} \ \text{and} \ x \in \Omega.$$

Notice that as long as the coefficients in these elliptic boundary value problems are sufficiently smooth the solution X_j will also be smooth by elliptic regularity so that $X_i \in \mathcal{D}(C_j)$.

With this we can assemble the $n_c \times n_{in}$ matrix $G(s)$, whose entries are

$$g_{ij}(s) = C_i X_j, \quad i = 1, \ldots, n_c, \ j = 1, \ldots, n_{in}.$$

While neutrally stable exosystems can generate rather complex periodic and quasi-periodic motions, two of the most important examples of regulation problems involve set point and harmonic tracking and disturbance rejection. For this reason we take extra care to illustrate a useful algorithm for solving Multi-Input, Multi-Output (MIMO) linear regulation problems when the disturbances and reference signals consist of a constant plus a sinusoid. In particular let us suppose that we want to track n_{in} reference signals of the form

$$y_{rj}(t) = M_{rj} + A_{rj} \sin(\alpha_j t) \text{ for } j = 1, \ \cdots n_{in}. \tag{1.36}$$

Also assume that we have n_d disturbance signals of the form

$$d_j(t) = M_{dj} + A_{dj} \sin(\beta_j t) \text{ for } j = 1, \ \cdots n_d. \tag{1.37}$$

In order to simplify matters we will assume that the number n_{in} of control inputs u_j equals the number n_{out} of measured outputs y_i and denote this common value by n_c. Then we have

$$y_i(t) = C_i(z)(t) \text{ for } i = 1, \cdots n_c,$$

and we have a vector of errors

$$e(t) = \begin{bmatrix} e_1(t) & e_2(t) & \cdots & e_{n_c}(t) \end{bmatrix},$$

where

$$e_j(t) = y_j(t) - y_{rj}(t).$$

Our objective is to find the control inputs u_j so that

$$\|e(t)\| \to 0, \quad \text{as } t \to \infty.$$

Here we assume there is a control system as in (1.1)–(1.3) in the form

$$z_t = Az + B_d D + B_{in} U,$$
$$z(0) = z_0, \quad z_0 \in \mathcal{Z},$$

$$y(t) = Cz(t),$$

where

$$B_d = \begin{bmatrix} B_d^1 & \cdots & B_d^{n_d} \end{bmatrix}, \quad D = \begin{bmatrix} d_1(t) & \cdots & d_{n_d}(t) \end{bmatrix}^\mathsf{T},$$
$$B_{in} = \begin{bmatrix} B_{in}^1 & \cdots & B_{in}^{n_c} \end{bmatrix}, \quad u = \begin{bmatrix} u_1(t) & \cdots & u_{n_c}(t) \end{bmatrix}^\mathsf{T}.$$

For reference signals and disturbances in the form (1.36) and (1.37) we seek a $k = 3(n_c + n_d)$-dimensional exosystem, i.e., $\mathcal{W} = \mathbb{R}^k$, in the form

$$\frac{dw}{dt}(t) = Sw(t),$$
$$Y_r(t) = Qw(t),$$
$$D(t) = Pw(t),$$
$$w(0) = w_0,$$

where for this example we have

$$S = \text{diag}\left[\boldsymbol{O}_{(n_c+n_d)}, S^{\alpha_1}, \cdots, S^{\alpha_{n_c}}, S^{\beta_1}, \cdots, S^{\beta_{n_d}}\right].$$

Here $\boldsymbol{O}_{(n_c+nd)}$ is the $(n_c + n_d) \times (n_c + n_d)$ zero matrix and

$$S^{\alpha_j} = \begin{bmatrix} 0 & \alpha_j \\ -\alpha_j & 0 \end{bmatrix}, \quad S^{\beta_j} = \begin{bmatrix} 0 & \beta_j \\ -\beta_j & 0 \end{bmatrix}.$$

Thus the exosystem has a state matrix with $(n_c + n_d)$ zeros which provides the constants M_{rj} and M_{dj}, and a set of $(n_c + n_d)$ harmonic oscillators that generate the sinusoids in the reference and disturbance signals.

The diagonal structure of the matrix S allows us to block divide the state w in the following form

$$w = \begin{bmatrix} w^0 \\ w^\alpha \\ w^\beta \end{bmatrix},$$

where

$$w^0 = \begin{bmatrix} w^{r0} \\ w^{d0} \end{bmatrix}, \quad w^\alpha = \begin{bmatrix} w^{\alpha_1} \\ \vdots \\ w^{\alpha_{n_c}} \end{bmatrix}, \quad w^\beta = \begin{bmatrix} w^{\beta_1} \\ \vdots \\ w^{\beta_{n_d}} \end{bmatrix},$$

and

$$w^{r0} \in \mathbb{R}^{n_c}, \quad w^{d0} \in \mathbb{R}^{n_d}, \quad w^{\alpha_j} \in \mathbb{R}^2, \quad w^{\beta_j} \in \mathbb{R}^2.$$

In particular we take for initial condition

$$w(0) = \begin{bmatrix} w^0(0) \\ w^\alpha(0) \\ w^\beta(0) \end{bmatrix},$$

where

$$w^{r0}(0) = \begin{bmatrix} M_{r1} \\ \vdots \\ M_{rn_c} \end{bmatrix}, \quad w^{d0}(0) = \begin{bmatrix} M_{d1} \\ \vdots \\ M_{dn_d} \end{bmatrix}, \quad w^{\alpha_j}(0) = \begin{bmatrix} 0 \\ A_{rj} \end{bmatrix}, \quad w^{\beta_j}(0) = \begin{bmatrix} 0 \\ A_{dj} \end{bmatrix}.$$

With this it is not difficult to see that the state w is given by

$$w^{r0}(t) = \begin{bmatrix} M_{r1} \\ \vdots \\ M_{rn_c} \end{bmatrix}, \quad w^{d0}(t) = \begin{bmatrix} M_{d1} \\ \vdots \\ M_{dn_d} \end{bmatrix},$$

$$w^{\alpha_j}(t) = \begin{bmatrix} A_{rj}\sin(\alpha_j t) \\ A_{rj}\cos(\alpha_j t) \end{bmatrix}, \quad w^{\beta_j}(t) = \begin{bmatrix} A_{dj}\sin(\beta_j t) \\ A_{dj}\cos(\beta_j t) \end{bmatrix}.$$

The controls are given by

$$u_j = \Gamma_j w = \begin{bmatrix} \Gamma^{r0}, & \Gamma^{d0}, & \Gamma_j^{\alpha_1}, & \cdots, & \Gamma_j^{\alpha_{n_c}}, & \Gamma_j^{\beta_1}, & \cdots, & \Gamma_j^{\beta_{n_d}} \end{bmatrix},$$

where

$$\Gamma^{r0} \in \mathbb{R}^{n_c}, \quad \Gamma^{d0} \in \mathbb{R}^{n_d}, \quad \Gamma^{\alpha_\ell} \in \mathbb{R}^2, \quad \Gamma_j^{\beta_\ell} \in \mathbb{R}^2 \quad \forall \ \ell.$$

The output operator is given in terms of n_c operators C_j by

$$Cz = \begin{bmatrix} C_1 z & C_2 z & \cdots & C_{n_c} z \end{bmatrix}^\mathsf{T}.$$

1.5.2 Solution Strategy

In order to keep the notation at a minimum we consider solving the regulator equation for this problem in the special case in which $n_c = 2$ and $n_d = 1$. We expect that it will be clear exactly how things go in any case with more measured outputs and more disturbance.

In this particular case $S^0 = \mathbf{0}_{3,3}$ and $S = \text{diag}[S^0, S^{\alpha_1}, S^{\alpha_2}, S^{\beta_1}]$. The exosystem state space is $\mathcal{W} = \mathbb{R}^9$ and the state variable has the natural decomposition given by

$$w = \begin{bmatrix} w^0 \\ w^\alpha \\ w^{\beta_1} \end{bmatrix},$$

where

$$w^0 = \begin{bmatrix} w^{r0} \\ w^{d0} \end{bmatrix}, \quad w^\alpha = \begin{bmatrix} w^{\alpha_1} \\ w^{\alpha_2} \end{bmatrix}, \quad w^\beta = \begin{bmatrix} w^{\beta_1} \end{bmatrix},$$

and

$$w^{r0} = \begin{bmatrix} w_1 \\ w_2 \end{bmatrix}, \quad w^{d0} = w_3, \quad w^{\alpha_1} = \begin{bmatrix} w_4 \\ w_5 \end{bmatrix}, \quad w^{\alpha_2} = \begin{bmatrix} w_6 \\ w_7 \end{bmatrix} \quad w^{\beta_1} = \begin{bmatrix} w_8 \\ w_9 \end{bmatrix}.$$

The initial conditions are

$$w(0) = \begin{bmatrix} w^0(0) \\ w^\alpha(0) \\ w^\beta(0) \end{bmatrix},$$

where

$$w^{r0}(0) = \begin{bmatrix} M_{r1} \\ M_{r2} \end{bmatrix}, \quad w^{d0}(0) = M_{d1},$$

$$w^{\alpha_1} = \begin{bmatrix} 0 \\ A_{r1} \end{bmatrix}, \quad w^{\alpha_2} = \begin{bmatrix} 0 \\ A_{r2} \end{bmatrix}, \quad w^{\beta_1} = \begin{bmatrix} 0 \\ A_{d1} \end{bmatrix}.$$

With this we have

$$w^0 = \begin{bmatrix} w_1 \\ w_2 \\ w_3 \end{bmatrix} = \begin{bmatrix} M_{r1} \\ M_{r2} \\ M_{d1} \end{bmatrix}, \quad w^\alpha = \begin{bmatrix} w_4 \\ w_5 \\ w_6 \\ w_7 \end{bmatrix} = \begin{bmatrix} A_{r1}\sin(\alpha_1 t) \\ A_{r1}\cos(\alpha_1 t) \\ A_{r2}\sin(\alpha_2 t) \\ A_{r2}\cos(\alpha_2 t) \end{bmatrix},$$

$$w^\beta = \begin{bmatrix} w_8 \\ w_9 \end{bmatrix} = \begin{bmatrix} A_{d1}\sin(\beta_1 t) \\ A_{d1}\cos(\beta_1 t) \end{bmatrix}.$$

The control system has the form

$$z_t = Az + B_{\mathrm{d}}Pw + B_{\mathrm{in}}^1 u_1 + B_{\mathrm{in}}^2 u_2$$

where $B_{\mathrm{in}} = [B_{\mathrm{in}}^1, B_{\mathrm{in}}^2]$. We will seek controls in the form

$$u_j = \Gamma_j w, \; j = 1, 2, \quad \Gamma_j = [\Gamma_j^0, \Gamma_j^\alpha, \Gamma_j^\beta] \in \mathbb{R}^9, \text{ with}$$

$$\Gamma_j^0 = [\Gamma_{j,1}^0, \; \Gamma_{j,2}^0, \; \Gamma_{j,3}^0],$$

$$\Gamma_j^\alpha = [\Gamma_j^{\alpha_1}, \; \Gamma_j^{\alpha_2}] = [[\Gamma_{j,1}^{\alpha_1}, \; \Gamma_{j,2}^{\alpha_1}], \; [\Gamma_{j,1}^{\alpha_2}, \; \Gamma_{j,2}^{\alpha_2}]], \text{ and}$$

$$\Gamma_j^\beta = [\Gamma_{j,1}^{\beta_1}, \; \Gamma_{j,2}^{\beta_1}].$$

The disturbance matrix P takes the form

$$P = [[0, 0, 1], [0, 0, 0, 0], [1, 0]] = [P^0, P^\alpha, P^\beta],$$

so that

$$Pw = w_3 + w_8 = M_{d1} + A_{d1}\sin(\beta_1 t) = D(t).$$

The reference signal matrix Q takes the form

$$Q = \begin{bmatrix} [1,0,0] & [1,0] & [0,0] & [0,0] \\ [0,1,0] & [0,0] & [1,0] & [0,0] \end{bmatrix} = \begin{bmatrix} Q_1^0 & Q_1^{\alpha_1} & Q_1^{\alpha_2} & Q_1^{\beta_1} \\ Q_2^0 & Q_2^{\alpha_1} & Q_2^{\alpha_2} & Q_2^{\beta_1} \end{bmatrix},$$

so that

$$Qw = \begin{bmatrix} w_1 + w_4 \\ w_2 + w_5 \end{bmatrix} = \begin{bmatrix} w_1 + w_4 \\ w_2 + w_5 \end{bmatrix} = \begin{bmatrix} M_{r1} + A_{r1}\sin(\alpha_1 t) \\ M_{r2} + A_{r2}\sin(\alpha_2 t) \end{bmatrix} = Y_r.$$

Thus the regulator equations for this problem have the form

$$\Pi S w = A \Pi w + B_{\mathrm{d}} P w + B_{\mathrm{in}}^1 \Gamma_1 w + B_{\mathrm{in}}^2 \Gamma_2 w,$$
$$C \Pi w = Q w.$$

Here Π has the following decomposition

$$\Pi = \left[\Pi^0, \Pi^{\alpha_1}, \Pi^{\alpha_2}, \Pi^{\beta_1}\right] = \left[[\Pi_1^0, \Pi_2^0, \Pi_3^0], [\Pi_1^{\alpha_1}, \Pi_2^{\alpha_1}], [\Pi_1^{\alpha_2}, \Pi_2^{\alpha_2}], [\Pi_1^{\beta_1}, \Pi_2^{\beta_1}]\right],$$

so that

$$\Pi w = \Pi^0 w^0 + \Pi^{\alpha_1} w^{\alpha_1} + \Pi^{\alpha_2} w^{\alpha_2} + \Pi^{\beta_1} w^{\beta_1}.$$

The diagonal structure of S gives us

$$\Pi S w = \Pi^0 S^0 w^0 + \Pi^{\alpha_1} S^{\alpha_1} w^{\alpha_1} + \Pi^{\alpha_2} S^{\alpha_2} w^{\alpha_2} + \Pi^{\beta_1} S^{\beta_1} w^{\beta_1}.$$

The output operator in this case is given by

$$C\Pi w = \begin{bmatrix} C_1 \Pi w \\ C_2 \Pi w \end{bmatrix},$$

and with the above notations the second regulator equation can be written as

$$\begin{bmatrix} C_1\Pi^0 & C_1\Pi^{\alpha_1} & C_1\Pi^{\alpha_2} & C_1\Pi^{\beta_1} \\ C_2\Pi^0 & C_2\Pi^{\alpha_1} & C_2\Pi^{\alpha_2} & C_2\Pi^{\beta_1} \end{bmatrix} = \begin{bmatrix} Q_1^0, Q_1^{\alpha_1}, Q_1^{\alpha_2}, Q_1^{\beta_1} \\ Q_2^0, Q_2^{\alpha_1}, Q_2^{\alpha_2}, Q_2^{\beta_1} \end{bmatrix}.$$

The block decomposition of all the various components allow us to systematically solve the regulator equations for the controls Γ_j. The following procedure should provide a guideline for solving more general problems. In particular, since the equations are required to hold for all $w \in \mathbb{R}^9$ we systematically choose some of the w_i values to be zero and some others not to be zero. This produces smaller versions of the regulator equations that can be more easily solved.

1. Let $w^0 \neq 0$ and $w^{\alpha_1} = w^{\alpha_2} = w^{\beta_1} = 0$ so that the regulator equations become

$$0 = \Pi^0 s^0 w^0 = A\Pi^0 w^0 + B_d P^0 w^0 + B_{in}^1 \Gamma_1^0 w^0 + B_{in}^2 \Gamma_2^0 w^0.$$

Recall that $w^0 = [w_1, w_2, w_3]^{\mathsf{T}}$. We successively set one component equal to one while setting the other two equal to zero. Recall also that $\Gamma_j^0 = [\Gamma_{j,1}^0, \Gamma_{j,2}^0, \Gamma_{j,3}^0]$.

(a) First set $w_1 = 1$, $w_2 = w_3 = 0$, which gives

$$0 = A\Pi_1^0 + B_{in}^1 \Gamma_{1,1}^0 + B_{in}^2 \Gamma_{2,1}^0.$$

We can easily solve this equation for Π_1^0

$$\Pi_1^0 = (-A^{-1})B_{in}^1 \Gamma_{1,1}^0 + (-A^{-1})B_{in}^2 \Gamma_{2,1}^0.$$

Applying C to Π_1^0, and using the second regulator equations we get

$$1 = C_1 \Pi_1^0 = g_{11}\Gamma_{1,1}^0 + g_{12}\Gamma_{2,1}^0,$$

$$0 = C_2 \Pi_1^0 = g_{21} \Gamma_{1,1}^0 + g_{22} \Gamma_{2,1}^0,$$

where we have introduced the notation g_{ij} for the components of the transfer function evaluated at $s = 0$

$$g_{ij} = C_i(-A^{-1})B_{\text{in}}^j, \quad \text{which gives } G = \begin{bmatrix} g_{11} & g_{12} \\ g_{21} & g_{22} \end{bmatrix}.$$

We then obtain

$$\begin{bmatrix} \Gamma_{1,1}^0 \\ \Gamma_{2,1}^0 \end{bmatrix} = G^{-1} \begin{bmatrix} 1 \\ 0 \end{bmatrix}.$$

(b) Next set $w_2 = 1$, $w_1 = w_3 = 0$, which gives

$$0 = A\Pi_2^0 + B_{\text{in}}^1 \Gamma_{1,2}^0 + B_{\text{in}}^2 \Gamma_{2,2}^0.$$

Once again we can easily solve this equation for Π_2^0. Applying C to Π_2^0, and using the second regulator equations we get

$$0 = C_1 \Pi_2^0 = g_{11} \Gamma_{1,2}^0 + g_{12} \Gamma_{2,2}^0,$$
$$1 = C_2 \Pi_2^0 = g_{21} \Gamma_{1,2}^0 + g_{22} \Gamma_{2,2}^0,$$

and find

$$\begin{bmatrix} \Gamma_{1,2}^0 \\ \Gamma_{2,2}^0 \end{bmatrix} = G^{-1} \begin{bmatrix} 0 \\ 1 \end{bmatrix}.$$

(c) Finally set $w_3 = 1$, $w_1 = w_2 = 0$, which gives

$$0 = A\Pi_3^0 + B_{\text{d}} + B_{\text{in}}^1 \Gamma_{1,3}^0 + B_{\text{in}}^2 \Gamma_{2,3}^0,$$

which can be easily solved for Π_3^0

$$\Pi_3^0 = (-A^{-1})B_{\text{d}} + (-A^{-1})B_{\text{in}}^1 \Gamma_{1,3}^0 + (-A^{-1})B_{\text{in}}^2 \Gamma_{2,3}^0.$$

Applying C to Π_3^0, and using the second regulator equations we get

$$0 = C_1 \Pi_3^0 = C_1(-A^{-1})B_{\text{d}} + g_{11} \Gamma_{1,3}^0 + g_{12} \Gamma_{2,3}^0,$$
$$0 = C_2 \Pi_3^0 = C_2(-A^{-1})B_{\text{d}} + g_{21} \Gamma_{1,3}^0 + g_{22} \Gamma_{2,3}^0.$$

We then obtain

$$\begin{bmatrix} \Gamma_{1,3}^0 \\ \Gamma_{2,3}^0 \end{bmatrix} = G^{-1} \begin{bmatrix} C_1(-A^{-1})B_{\text{d}} \\ C_2(-A^{-1})B_{\text{d}} \end{bmatrix}.$$

2. Next we turn our attention to calculate the controls corresponding to the harmonic terms in both the reference signals and the disturbance.

(a) First consider $w^{\alpha_1} \neq 0$ and $w^0 = w^{\alpha_2} = w^{\beta_1} = 0$. In this case the first regulator equation simplifies to

$$\Pi^{\alpha_1} S^{\alpha_1} w^{\alpha_1} = A\Pi^{\alpha_1} w^{\alpha_1} + B_{\text{in}}^1 \Gamma_1^{\alpha_1} w^{\alpha_1} + B_{\text{in}}^2 \Gamma_2^{\alpha_1} w^{\alpha_1}.$$

Notice that this equation has to hold for all w^{α_1} so we can suppress the variable and write this equation as an operator equation

$$\Pi^{\alpha_1} S^{\alpha_1} = A\Pi^{\alpha_1} + B_{\text{in}}^1 \Gamma_1^{\alpha_1} + B_{\text{in}}^2 \Gamma_2^{\alpha_1}.$$

We proceed by rewriting the equation in the form

$$-A\Pi^{\alpha_1} = -\Pi^{\alpha_1} S^{\alpha_1} + B_{\text{in}}^1 \Gamma_1^{\alpha_1} + B_{\text{in}}^2 \Gamma_2^{\alpha_1}.$$

Next we apply $(-A^{-1})$ to obtain

$$\Pi^{\alpha_1} = A^{-1} \Pi^{\alpha_1} S^{\alpha_1} + (-A^{-1}) B_{\text{in}}^1 \Gamma_1^{\alpha_1} + (-A^{-1}) B_{\text{in}}^2 \Gamma_2^{\alpha_1}.$$

If we now apply C_i and use the second regulator equation we have

$$Q_1^{\alpha_1} = [1, 0] = C_1 \Pi^{\alpha_1} = C_1 \left(A^{-1} \Pi^{\alpha_1} S^{\alpha_1} \right) + g_{11} \Gamma_1^{\alpha_1} + g_{12} \Gamma_2^{\alpha_1},$$

$$Q_2^{\alpha_1} = [0, 0] = C_2 \Pi^{\alpha_1} = C_2 \left(A^{-1} \Pi^{\alpha_1} S^{\alpha_1} \right) + g_{21} \Gamma_1^{\alpha_1} + g_{22} \Gamma_2^{\alpha_1}.$$

At this point we introduce an auxiliary variable $\widetilde{\Pi}^{\alpha_1} = [\widetilde{\Pi}_1^{\alpha_1}, \widetilde{\Pi}_2^{\alpha_1}]$ as the solution to the equation

$$A\widetilde{\Pi}^{\alpha_1} = \Pi^{\alpha_1} S^{\alpha_1}.$$

Therefore we can replace the above equations by

$$Q_1^{\alpha_1} = C_1 \left(\widetilde{\Pi}^{\alpha_1} \right) + g_{11} \Gamma_1^{\alpha_1} + g_{12} \Gamma_2^{\alpha_1},$$

$$Q_2^{\alpha_1} = C_2 \left(\widetilde{\Pi}^{\alpha_1} \right) + g_{21} \Gamma_1^{\alpha_1} + g_{22} \Gamma_2^{\alpha_1},$$

which we can write as a single equation as

$$\begin{bmatrix} Q_1^{\alpha_1} \\ Q_2^{\alpha_1} \end{bmatrix} = C\widetilde{\Pi}^{\alpha_1} + G \begin{bmatrix} \Gamma_1^{\alpha_1} \\ \Gamma_2^{\alpha_1} \end{bmatrix},$$

which provides the solution

$$\begin{bmatrix} \Gamma_1^{\alpha_1} \\ \Gamma_2^{\alpha_1} \end{bmatrix} = G^{-1} \left(\begin{bmatrix} Q_1^{\alpha_1} \\ Q_2^{\alpha_1} \end{bmatrix} - C\widetilde{\Pi}^{\alpha_1} \right).$$

Here the algebro-differential system[†] of equations

$$\Pi^{\alpha_1} S^{\alpha_1} = A\Pi^{\alpha_1} + B_{\text{in}}^1 \Gamma_1^{\alpha_1} + B_{\text{in}}^2 \Gamma_2^{\alpha_1}, \tag{1.38}$$

[†]By algebro-differential system we refer to a system of differential equations which are coupled to algebraic constraints. For example the pair of regulator equations for systems governed by PDEs provide algebro-differential systems.

$$A\widetilde{\Pi}^{\alpha_1} = \Pi^{\alpha_1} S^{\alpha_1}, \tag{1.39}$$

$$\begin{bmatrix} \Gamma_1^{\alpha_1} \\ \Gamma_2^{\alpha_1} \end{bmatrix} = G^{-1} \left(\begin{bmatrix} Q_1^{\alpha_1} \\ Q_2^{\alpha_1} \end{bmatrix} - C\widetilde{\Pi}^{\alpha_1} \right), \tag{1.40}$$

is fully coupled. By writing this system of equations in terms of the individual components, it is possible to solve the equations as done in Section 1.4.3. However, this is only practical for the simplest examples as the one-dimensional heat equation, and even then it requires a lot of work. For more sophisticated cases, a better solution is to employ a finite element package that can solve algebro-differential equations in the form of (1.38)–(1.40).

(b) Next we consider $w^{\alpha_2} \neq 0$ and $w^0 = w^{\alpha_1} = w^{\beta_1} = 0$. In this case the first regulator equation simplifies to

$$\Pi^{\alpha_2} S^{\alpha_2} w^{\alpha_2} = A\Pi^{\alpha_2} w^{\alpha_2} + B_{\text{in}}^1 \Gamma_1^{\alpha_2} w^{\alpha_2} + B_{\text{in}}^2 \Gamma_2^{\alpha_2} w^{\alpha_2}.$$

Notice that this equation has to hold for all w^{α_2} so we can suppress the variable and write the equation as an operator equation

$$\Pi^{\alpha_2} S^{\alpha_2} = A\Pi^{\alpha_2} + B_{\text{in}}^1 \Gamma_1^{\alpha_2} + B_{\text{in}}^2 \Gamma_2^{\alpha_2}.$$

We proceed by rewriting the equation in the form

$$-A\Pi^{\alpha_2} = -\Pi^{\alpha_2} S^{\alpha_2} + B_{\text{in}}^1 \Gamma_1^{\alpha_2} + B_{\text{in}}^2 \Gamma_2^{\alpha_2}.$$

Next we apply $(-A^{-1})$ to obtain

$$\Pi^{\alpha_2} = A^{-1} \Pi^{\alpha_2} S^{\alpha_2} + (-A^{-1}) B_{\text{in}}^1 \Gamma_1^{\alpha_2} + (-A^{-1}) B_{\text{in}}^2 \Gamma_2^{\alpha_2}.$$

If we now apply C_i and use the second regulator equation we have

$$Q_1^{\alpha_2} = [0,0] = C_1 \Pi^{\alpha_2} = C_1 \left(A^{-1} \Pi^{\alpha_2} S^{\alpha_2} \right) + g_{11} \Gamma_1^{\alpha_2} + g_{12} \Gamma_2^{\alpha_2},$$

$$Q_2^{\alpha_2} = [1,0] = C_2 \Pi^{\alpha_2} = C_2 \left(A^{-1} \Pi^{\alpha_2} S^{\alpha_2} \right) + g_{21} \Gamma_1^{\alpha_2} + g_{22} \Gamma_2^{\alpha_2}.$$

At this point we, once again, introduce an auxiliary variable $\widetilde{\Pi}^{\alpha_2} = [\widetilde{\Pi}_1^{\alpha_2}, \ \widetilde{\Pi}_2^{\alpha_2}]$ as the solution to the equation

$$A\widetilde{\Pi}^{\alpha_2} = \Pi^{\alpha_2} S^{\alpha_2}.$$

Therefore we can replace the above equations by

$$Q_1^{\alpha_2} = C_1 \left(\widetilde{\Pi}^{\alpha_2} \right) + g_{11} \Gamma_1^{\alpha_2} + g_{12} \Gamma_2^{\alpha_2},$$

$$Q_2^{\alpha_2} = C_2 \left(\widetilde{\Pi}^{\alpha_2} \right) + g_{21} \Gamma_1^{\alpha_2} + g_{22} \Gamma_2^{\alpha_2},$$

which we can write as a single equation as

$$\begin{bmatrix} Q_1^{\alpha_2} \\ Q_2^{\alpha_2} \end{bmatrix} = C\widetilde{\Pi}^{\alpha_2} + G \begin{bmatrix} \Gamma_1^{\alpha_2} \\ \Gamma_2^{\alpha_2} \end{bmatrix},$$

which provides the solution

$$\begin{bmatrix} \Gamma_1^{\alpha_2} \\ \Gamma_2^{\alpha_2} \end{bmatrix} = G^{-1} \left(\begin{bmatrix} Q_1^{\alpha_2} \\ Q_2^{\alpha_2} \end{bmatrix} - C\widetilde{\Pi}^{\alpha_2} \right).$$

Here again the algebro-differential system of equations

$$\Pi^{\alpha_2} S^{\alpha_2} = A\Pi^{\alpha_2} + B_{\text{in}}^1 \Gamma_1^{\alpha_2} + B_{\text{in}}^2 \Gamma_2^{\alpha_2}, \tag{1.41}$$

$$A\widetilde{\Pi}^{\alpha_2} = \Pi^{\alpha_2} S^{\alpha_2}, \tag{1.42}$$

$$\begin{bmatrix} \Gamma_1^{\alpha_2} \\ \Gamma_2^{\alpha_2} \end{bmatrix} = G^{-1} \left(\begin{bmatrix} Q_1^{\alpha_2} \\ Q_2^{\alpha_2} \end{bmatrix} - C\widetilde{\Pi}^{\alpha_2} \right), \tag{1.43}$$

is fully coupled. Note the similarity between the equation system (1.38)–(1.40) and the equation system (1.41)–(1.43).

(c) Finally we consider the case $w^\beta \neq 0$ and $w^0 = w^{\alpha_1} = w^{\alpha_2} = 0$. In this case the first regulator equation simplifies to

$$\Pi^{\beta_1} S^{\beta_1} w^{\beta_1} = A\Pi^{\beta_1} w^{\beta_1} + B_{\text{d}}[1,0]w^{\beta_1} + B_{\text{in}}^1 \Gamma_1^{\beta_1} w^{\beta_1} + B_{\text{in}}^2 \Gamma_2^{\beta_1} w^{\beta_1}.$$

Note that this equation has to hold for all w^{β_1} so we can suppress the variable and write the equation as an operator equation

$$\Pi^{\beta_1} S^{\beta_1} = A\Pi^{\beta_1} + B_{\text{d}} P^{\beta_1} + B_{\text{in}}^1 \Gamma_1^{\beta_1} + B_{\text{in}}^2 \Gamma_2^{\beta_1},$$

where $P^{\beta_1} = [1,0]$. We proceed by rewriting the equation in the form

$$-A\Pi^{\beta_1} = -\Pi^{\beta_1} S^{\beta_1} + B_{\text{d}} P^{\beta_1} + B_{\text{in}}^1 \Gamma_1^{\beta_1} + B_{\text{in}}^2 \Gamma_2^{\beta_1}.$$

Next we apply $(-A^{-1})$ to obtain

$$\Pi^{\beta_1} = A^{-1}\Pi^{\beta_1} S^{\beta_1} + (-A^{-1})B_{\text{d}} P^{\beta_1} + (-A^{-1})B_{\text{in}}^1 \Gamma_1^{\beta_1} + (-A^{-1})B_{\text{in}}^2 \Gamma_2^{\beta_1}.$$

If we now apply C_i and use the second regulator equation we have

$$Q_1^{\beta_1} = [0,0] = C_1\Pi^{\beta_1} = C_1\left(A^{-1}\Pi^{\beta_1} S^{\beta_1}\right) + C_1\left[(-A^{-1})B_{\text{d}} P^{\beta_1}\right] + g_{11}\Gamma_1^{\beta_1} + g_{12}\Gamma_2^{\beta_1},$$

$$Q_2^{\beta_1} = [0,0] = C_2\Pi^{\beta_1} = C_2\left(A^{-1}\Pi^{\beta_1} S^{\beta_1}\right) + C_2\left[(-A^{-1})B_{\text{d}} P^{\beta_1}\right] + g_{21}\Gamma_1^{\beta_1} + g_{22}\Gamma_2^{\beta_1}.$$

At this point we introduce an auxiliary variable $\widetilde{\Pi}^{\beta_1} = [\widetilde{\Pi}_1^{\beta_1},\ \widetilde{\Pi}_2^{\beta_1}]$ as the solution to the equation

$$A\widetilde{\Pi}^{\beta_1} = \Pi^{\beta_1} S^{\beta_1} - B_{\mathrm{d}} P^{\beta_1}.$$

Therefore we can replace the above equations by

$$\mathbf{0}_{1\times 2} = C_1 \left(\widetilde{\Pi}^{\beta_1}\right) + g_{11}\Gamma_1^{\beta_1} + g_{12}\Gamma_2^{\beta_1},$$

$$\mathbf{0}_{1\times 2} = C_2 \left(\widetilde{\Pi}^{\beta_1}\right) + g_{21}\Gamma_1^{\beta_1} + g_{22}\Gamma_2^{\beta_1},$$

which we can write as a single equation

$$G \begin{bmatrix} \Gamma_1^{\beta_1} \\ \Gamma_2^{\beta_2} \end{bmatrix} = -C\widetilde{\Pi}^{\beta_1},$$

that provides the solution

$$\begin{bmatrix} \Gamma_1^{\beta_1} \\ \Gamma_2^{\beta_2} \end{bmatrix} = -G^{-1}\big[C\widetilde{\Pi}^{\beta_1}\big].$$

Again, the algebro-differential system of equations

$$\Pi^{\beta_1} S^{\beta_1} = A\Pi^{\beta_1} + B_{\mathrm{in}}^1 \Gamma_1^{\beta_1} + B_{\mathrm{in}}^2 \Gamma_2^{\beta_1}, \tag{1.44}$$

$$A\widetilde{\Pi}^{\beta_1} = \Pi^{\beta_1} S^{\beta_1} - B_{\mathrm{d}} P^{\beta_1},$$

$$\begin{bmatrix} \Gamma_1^{\beta_1} \\ \Gamma_2^{\beta_2} \end{bmatrix} = -G^{-1}\big[C\widetilde{\Pi}^{\beta_1}\big], \tag{1.45}$$

is fully coupled.

We now present an example to demonstrate the geometric algorithm for a MIMO tracking and disturbance rejection problem for a one-dimensional heat equation on the spatial interval $(0, L)$.

Example 1.4 (MIMO Numerical Solution for the 1D Heat Equation)**.** We consider a 1D heat control problem on the rod depicted in Fig. 1.7.

$$z_t = z_{xx} + B_{\mathrm{d}}d + B_{\mathrm{in}}^1 u_1 + B_{\mathrm{in}}^2 u_2,$$
$$- z_x(0,t) + k_0 z(0,t) = 0, \quad k_0 > 0,$$
$$z_x(L,t) + k_1 z(L,t) = 0, \quad k_1 > 0,$$
$$z(x,0) = \varphi(x),$$
$$y_1 = C_1 z, \quad y_2 = C_2 z.$$

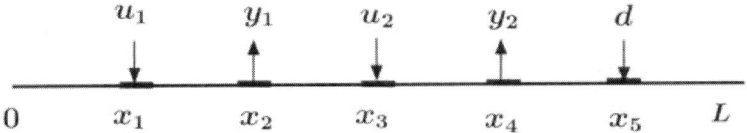

Fig. 1.7: One-Dimensional Rod.

Let $x_i \in (0, L)$ and $\nu_i > 0$ (sufficiently small) $i = 1, 2, \ldots, 5$. We denote the interval centered at x_i of width $2\nu_i$, inside $(0, L)$ by I_i, i.e.,

$$I_i = \{x \in (0, 1) : x_i - \nu_i < x < x_i + \nu_i\}.$$

With this we define the output operators by

$$y_1(t) = C_1 z = \frac{1}{|I_2|} \int_{I_2} z(x, t)\, dx \quad \text{and} \quad y_2(t) = C_2 z = \frac{1}{|I_4|} \int_{I_4} z(x, t)\, dx,$$

the control inputs by

$$B_{\text{in}}^1 \varphi = \frac{1}{|I_1|} \mathbf{1}_{I_1}(x)\varphi \quad \text{and} \quad B_{\text{in}}^2 \varphi = \frac{1}{|I_3|} \mathbf{1}_{I_3}(x)\varphi,$$

and the disturbance operator by

$$B_{\text{d}} = \frac{1}{|I_5|} \mathbf{1}_{I_5}(x).$$

Here $\mathbf{1}_{I_j}$ denotes the characteristic function (or indicator function) of the interval I_j defined in (1.10).

Our control objective is to find two controls u_1 and u_2 so that the outputs y_1 and y_2 track the reference signals

$$y_{rj}(t) = M_{rj} + A_{rj} \sin(\alpha_j t) \quad j = 1, 2,$$

while rejecting the disturbance

$$d(t) = M_d + A_d \sin(\beta t).$$

For this example we have chosen $L = 1.2$, $x_i = 0.2 \times i$, $\nu_i = 0.025$, $\alpha_1 = 1$, $\alpha_2 = 2$, $\beta = 0.5$, $M_{r1} = 0.5$, $M_{r2} = 0.25$, $M_d = 1$, $A_{r1} = 0.25$, $A_{r2} = 0.5$, $A_d = 0.75$, $k_0 = k_1 = 1$ and initial data $z(x, 0) = \cos\left(\frac{\pi x}{L}\right)$.

Fig. 1.8 depicts the reference signal $y_{r1}(t)$ and the controlled output $y_1(t)$, while Fig. 1.9 depicts the error $e_1(t) = y_1(t) - y_{r1}(t)$. Fig. 1.10 depicts the reference signal $y_{r2}(t)$ and the controlled output $y_2(t)$, while Fig. 1.11 depicts the error $e_2(t) = y_2(t) - y_{r2}(t)$. Finally, Fig. 1.12 contains the numerical solution for the temperature distribution $z(x, t)$ for $x \in [0, L]$ and $t \in [0, 20]$.

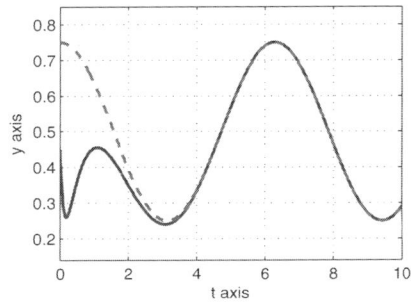

Fig. 1.8: $y_1(t)$ and $y_{r1}(t)$.

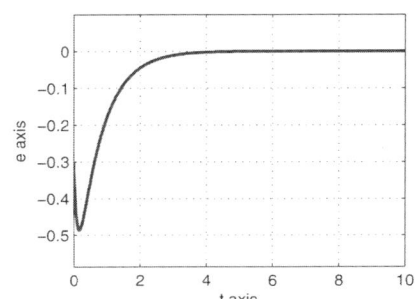

Fig. 1.9: Error $e_1(t)$.

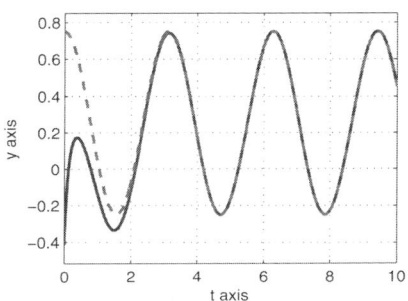

Fig. 1.10: $y_2(t)$ and $y_{r2}(t)$.

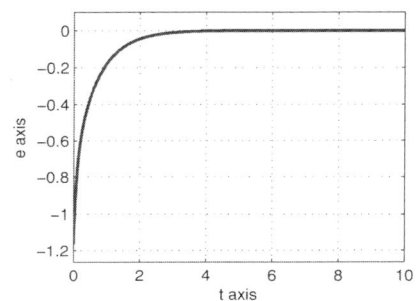

Fig. 1.11: Error $e_2(t)$.

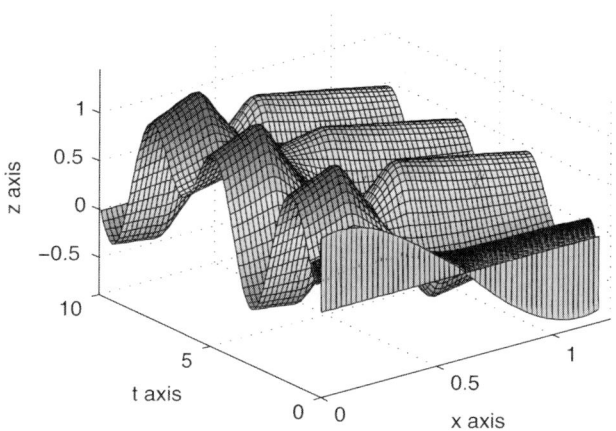

Fig. 1.12: Plot of Solution Surface.

1.5.3 Solution of the MIMO Regulator Equations Using the Transfer Function

We note that, just as in the SISO case considered in Example 1.2, it is possible to give explicit formulas for the solution of the above systems in terms of the transfer function of the system evaluated at the eigenvalues of the exosystem state matrix S. The advantage to the discussion given in Section 1.5.2 is that the algorithm presented there is easily adapted for numerical solution.

Nevertheless for esthetic reasons we show that even in the MIMO case it is still possible to obtain explicit formulas, analogous to those obtained in Example 1.2.

Let us now consider the details of solving explicitly either one of the problems (1.39)–(1.40) or (1.42)–(1.43). Actually the procedure we are to present can be easily modified to solve also the problem (1.44)–(1.45). Let us consider problem (1.39)–(1.40) written in a generic form with α_1 replaced by α. We also consider the original form of the regulator equations with generic superscripts α.

Thus we consider a system

$$\Pi^\alpha S^\alpha w^\alpha = A\Pi^\alpha w^\alpha + B^1_{\text{in}}\Gamma^\alpha_1 w^a + B^2_{\text{in}}\Gamma^\alpha_2 w^a,$$
$$C\Pi^\alpha w^a = Q^\alpha w^a,$$

where

$$w^\alpha = [w^\alpha_1, w^\alpha_2]^\mathsf{T} \in \mathbb{R}^2, \quad \Pi^\alpha = [\Pi^\alpha_1, \Pi^\alpha_2], \quad S^\alpha = \begin{bmatrix} 0 & \alpha \\ -\alpha & 0 \end{bmatrix}, \quad Q^\alpha = [1, 0].$$

To solve for the unknowns

$$\Gamma^\alpha_1 = [\Gamma^\alpha_{1,1}, \Gamma^\alpha_{1,2}] \quad \text{and} \quad \Gamma^\alpha_2 = [\Gamma^\alpha_{2,1}, \Gamma^\alpha_{2,2}]$$

we proceed as we did in Section 1.4.3 .

We first set $w^\alpha_1 = 1$ and $w^\alpha_2 = 0$, and then set $w^\alpha_1 = 0$ and $w^\alpha_2 = 1$ to obtain the two equations

$$-A\Pi^\alpha_1 - \alpha\Pi^\alpha_2 = B^1_{\text{in}}\Gamma^\alpha_{1,1} + B^2_{\text{in}}\Gamma^\alpha_{2,1}, \tag{1.46}$$
$$\alpha\Pi^\alpha_1 - A\Pi^\alpha_2 = B^1_{\text{in}}\Gamma^\alpha_{1,2} + B^2_{\text{in}}\Gamma^\alpha_{2,2}. \tag{1.47}$$

We multiply $i = \sqrt{-1}$ times (1.47) and add the result to (1.46) to obtain

$$(i\alpha - A)[\Pi^\alpha_1 + i\Pi^\alpha_2] = B^1_{\text{in}}\left(\Gamma^\alpha_{1,1} + i\Gamma^\alpha_{1,2}\right) + B^2_{\text{in}}\left(\Gamma^\alpha_{2,1} + i\Gamma^\alpha_{2,2}\right).$$

Next apply $(i\alpha - A^{-1})$ to both sides to get

$$[\Pi^\alpha_1 + i\Pi^\alpha_2] = (i\alpha - A^{-1})\{B^1_{\text{in}}\left(\Gamma^\alpha_{1,1} + i\Gamma^\alpha_{1,2}\right) + B^2_{\text{in}}\left(\Gamma^\alpha_{2,1} + i\Gamma^\alpha_{2,2}\right)\}.$$

Applying C_1 and C_2 to both sides of this equation and using the second regulator equation produces

$$C_1^\alpha \Pi_1^\alpha = 1, \quad C_1^\alpha \Pi_2^\alpha = 0, \quad C_2^\alpha \Pi_1^\alpha = 0, \quad C_2^\alpha \Pi_2^\alpha = 0,$$

and we have

$$1 = g_{11} \left(\Gamma_{1,1}^\alpha + i\Gamma_{1,2}^\alpha \right) + g_{1,2} \left(\Gamma_{2,1}^\alpha + i\Gamma_{2,2}^\alpha \right),$$
$$0 = g_{21} \left(\Gamma_{1,1}^\alpha + i\Gamma_{1,2}^\alpha \right) + g_{2,2} \left(\Gamma_{2,1}^\alpha + i\Gamma_{2,2}^\alpha \right),$$

or in matrix form

$$G(i\alpha) \begin{bmatrix} (\Gamma_{1,1}^\alpha + i\Gamma_{1,2}^\alpha) \\ (\Gamma_{2,1}^\alpha + i\Gamma_{2,2}^\alpha) \end{bmatrix} = \begin{bmatrix} 1 \\ 0 \end{bmatrix},$$

where

$$G(i\alpha) = \begin{bmatrix} g_{11}(i\alpha) & g_{12}(i\alpha) \\ g_{21}(i\alpha) & g_{22}(i\alpha) \end{bmatrix}, \quad g_{ij}(i\alpha) = C_i (i\alpha I - A)^{-1} B_{\text{in}}^j,$$

and we have

$$G(i\alpha)^{-1} = \frac{1}{\Delta} \begin{bmatrix} g_{22}(i\alpha) & -g_{12}(i\alpha) \\ -g_{21}(i\alpha) & g_{11}(i\alpha) \end{bmatrix}, \quad \text{with} \quad \Delta = \det(G(i\alpha)).$$

Applying $G(i\alpha)^{-1}$ to both sides and denoting

$$\begin{bmatrix} \widetilde{g}_1 \\ \widetilde{g}_2 \end{bmatrix} = G(i\alpha)^{-1} \begin{bmatrix} 1 \\ 0 \end{bmatrix} = \frac{1}{\Delta} \begin{bmatrix} g_{22}(i\alpha) \\ -g_{21}(i\alpha) \end{bmatrix},$$

we can write

$$\begin{bmatrix} \Gamma_{1,1}^\alpha + i\Gamma_{1,2}^\alpha \\ \Gamma_{2,1}^\alpha + i\Gamma_{2,2}^\alpha \end{bmatrix} = \begin{bmatrix} \operatorname{Re}(\widetilde{g}_1) + i \operatorname{Im}(\widetilde{g}_1) \\ \operatorname{Re}(\widetilde{g}_2) + i \operatorname{Im}(\widetilde{g}_2) \end{bmatrix}.$$

At this point we recall our assumption that the system is real which implies that the controls Γ_1^α and Γ_2^α are real 2×1 matrices. Therefore we see that

$$\Gamma_1^\alpha = \begin{bmatrix} \Gamma_{1,1}^\alpha & \Gamma_{1,2}^\alpha \end{bmatrix} = \left[\operatorname{Re}\left(\frac{g_{22}(i\alpha)}{\Delta} \right), \quad \operatorname{Im}\left(\frac{g_{22}(i\alpha)}{\Delta} \right) \right],$$

and

$$\Gamma_2^\alpha = \begin{bmatrix} \Gamma_{2,1}^\alpha & \Gamma_{2,2}^\alpha \end{bmatrix} = \left[\operatorname{Re}\left(\frac{-g_{21}(i\alpha)}{\Delta} \right), \quad \operatorname{Im}\left(\frac{-g_{21}(i\alpha)}{\Delta} \right) \right].$$

Once again it is clear from the above calculations that even in the MIMO case the control gains are computed in terms of the transfer function evaluated at the eigenvalues of the exosystem. And, in particular, we are forced to have $\det(G(s_0)) \neq 0$, for all eigenvalues s_0 of S. This is precisely the statement that no eigenvalue of the exosystem is an invariant zero of the control system which is a well known necessary condition for the solvability for a problem of output regulation.

Chapter 2

Linear Regulation with Unbounded Control and Sensing

2.1 Introduction

One of the main shortcomings of Chapter 1 is that the results are not applicable in modeling many important applications. In particular, in situations involving distributed parameter systems governed by partial differential equations it very often happens that the control inputs and the disturbances influence the system through the boundary, through point actuators inside the domain or through lower-dimensional surfaces inside the domain. Similarly it often happens that the measured output is taken as point evaluations or more general sensors with readings taken on lower-dimensional surfaces inside the domain or on the boundary of the domain.

Typical applications include control of fluid dynamic systems, systems with delays in the inputs or outputs, and control of tubular reactors often requires point actuators and sensors, or actuators and sensors supported on lower-dimensional submanifolds, or on the boundary of a spatial domain. These types of actuators and sensors produce unbounded input and output operators on the Hilbert state space and a rigorous mathematical analysis of such operators presents considerable technical difficulties even at the level of state space formulations and proofs of existence and regularity of the dynamics. Indeed, there has been, and continues to be, a significant research effort devoted to delineating classes of systems for which analogs of classical feedback control methodologies can be formulated. To date, the most general class of distributed parameter systems for which there is an established theory of inputs, outputs, transfer functions and feedback is the class of systems described in the series of papers [86, 88, 82, 87].

A rigorous development of this topic with complete definitions would require a lengthy discussion taking us considerably off course from the main intentions of this book. Our goal is to provide a roadmap to the application of the geometric theory in order to construct numerical solutions for problems of regulation for systems governed by partial differential equations. So while the most general abstract theory of unbounded control systems can involve rather pathological scenarios, in the case of systems governed by partial differential equations a fairly exhaustive theory of existence, uniqueness and regularity of solutions, as well as the long-time behavior of such systems is reasonably well known. Our main examples in this and the next chapter correspond to problems for which the theory is mostly well established.

2.2   Formulation of Control System and Interpolation Spaces

In order to keep the discussion as general as possible we will continue to describe the control system by an abstract distributed parameter control system,

$$z_t(t) = Az(t) + B_\mathrm{d}d(t) + B_\mathrm{in}u(t), \tag{2.1}$$

$$y(t) = Cz(t), \qquad\qquad\qquad \text{(measured output)} \tag{2.2}$$

$$z(0) = z_0, \tag{2.3}$$

where, just as in Chapter 1, z is the state of the system in a real or complex separable Hilbert space (state space) $\mathcal{Z}$ with inner product denoted by $\langle \cdot, \cdot \rangle$ and norm $\|\varphi\| = \langle \varphi, \varphi \rangle^{1/2}$. The term $d(t) \in \mathcal{D}$ is a disturbance, $u \in \mathcal{U}$ is a control input, and $y \in \mathcal{Y}$ is the measured output. $\mathcal{D}$, $\mathcal{U}$ and $\mathcal{Y}$ are real or complex separable (usually finite-dimensional) Hilbert disturbance, control and output spaces, respectively. A is an unbounded densely defined operator, with domain $\mathcal{D}(A)$ in $\mathcal{Z}$ which is assumed to be the infinitesimal generator of an exponentially stable, strongly continuous semigroup on $\mathcal{Z}$. In almost all of our application examples the state operator A is an unbounded self-adjoint operator with a compact inverse in $\mathcal{Z}$.

In our intended applications $\mathcal{Z} = L^2(\Omega)$ for Ω a bounded domain in $\mathbb{R}^n$ and the control input operators B_in and disturbance input operators B_d may be unbounded operators in the Hilbert state space. That is, they may correspond to boundary control operators corresponding to controls or disturbances that enter through boundary conditions on hypersurfaces on the boundary of the spatial domain or at points or on hypersurfaces inside the domain. This should be compared with the results described in Section 1.1.4. Therefore the main difference in Chapters 1 and 2 is that the operators B_in, B_d and C may be given in terms of unbounded analogs of the expressions in Section 1.1.4.

More precisely, in the systems (2.1)–(2.3) considered in this chapter there may be some terms B_d^j and B_in^j that correspond to unbounded operators. In order to distinguish such terms (1.11) let the sequences B_d^j and B_in^j be ordered so that the first n_d^b and n_in^b elements correspond to boundary operators $\mathcal{B}_\mathrm{d}^j$ and $\mathcal{B}_\mathrm{in}^j$ defined on hypersurfaces in Ω or on its boundary. Recall that in Chapter 1 the input operators B_d and B_in were given in (1.11) with all the functions B_in^j and B_d^j in $\mathcal{Z}$. In contrast, in Chapter 2 we have the same form for the input operators

$$B_\mathrm{d}d(t) = \sum_{j=1}^{n_\mathrm{d}} B_\mathrm{in}^j(x)\, d_j(t), \quad B_\mathrm{in}u = \sum_{j=1}^{n_\mathrm{in}} B_\mathrm{d}^j(x)\, u_j(t), \tag{2.4}$$

where some of the terms B_in^j and B_d^j are distributional expressions which are not in $\mathcal{Z}$.

In practical problems in which some of the control inputs and disturbances enter the system (2.1)–(2.3) via unbounded operators, the system is most often given in its the equivalent form

$$z_t(x,t) = A_0 z(x,t) + \sum_{j=n_d^b+1}^{n_d} (B_d^j d_j)(x,t) + \sum_{j=n_{in}^b+1}^{n_{in}} (B_{in}^j u_j)(x,t), \qquad (2.5)$$

$$(\mathcal{B}_d^j z)(x,t) = \mathcal{B}_{in}^j(x)\, d_j(t), \quad x \in \mathcal{S}_{dj}, \quad j = 1, \ldots, n_d^b, \qquad (2.6)$$

$$(\mathcal{B}_{in}^j z)(x,t) = \mathcal{B}_d^j(x)\, u_j(t), \quad x \in \mathcal{S}_{inj}, \quad j = 1, \ldots, n_{in}^b. \qquad (2.7)$$

where (2.6), (2.7) are the disturbance and control input terms acting on hypersurfaces inside the domain or on the boundaries of the domain. These replace the homogeneous boundary conditions which are contained in the definition of $\mathcal{D}(A)$ in (1.1) (see also below (2.9)).

We note that in this case the functions $\mathcal{B}_{in}^j(x)$ in (2.6) and $\mathcal{B}_d^j(x)$ in (2.7) are not the same as the distributional expressions B_{in}^j for $j = 1, \ldots, n_d^b$ and B_d^j for $j = 1, \ldots, n_{in}^b$ given in (2.4). Indeed, the functions $\mathcal{B}_{in}^j(x)$ and $\mathcal{B}_d^j(x)$ are typically smooth or piecewise smooth functions defined on a subregion of $\overline{\Omega}$. The reformulation of problem (1.1)–(1.3) into the form (2.5)–(2.7) is discussed in a variety of sources, see e.g., [6, 7, 78, 84]. In particular, it is well known that under suitable assumptions there exist operators B_d^j and B_{in}^j so that system (2.5)–(2.7) can be written in the form (1.1)–(1.3) (cf. [6, 7, 78, 84]).

Here we assume that the boundary of Ω, denoted by $\partial\Omega$, is piecewise C^2 and is represented by the union of $(n-1)$-dimensional connected hypersurfaces $\mathcal{S}_j$, which are subsets of $\partial\Omega$ and whose interiors are pairwise disjoint.

Let the surface

$$\mathcal{S}_0 = \bigcup_{j_0=1}^{n_0} \mathcal{S}_0^j$$

be the union of all the exterior surfaces where only homogeneous boundary conditions, of all types, are set. The operator A_0 is typically a linear elliptic partial differential operator (as given in (1.8)) engendered with a partial set of boundary conditions only on $\mathcal{S}_0$. Let $\mathcal{B}_0^j$ be the boundary operator (generally Neumann, Robin or Dirichlet) acting on the surface $\mathcal{S}_0^j$. We denote the dense domain of A_0 in $\mathcal{Z}$ by $\mathcal{D}(A_0)$, where

$$\mathcal{D}(A_0) = \{\varphi : \mathcal{B}_0^j \varphi = 0 \ (j_0 = 1, \cdots, n_0)\}. \qquad (2.8)$$

The operator A is the same linear elliptic partial differential operator A_0 with domain, denoted by $\mathcal{D}(A)$, given by

$$\mathcal{D}(A) = \{\varphi : \mathcal{B}_d^j \varphi = 0 \ (j = 1, \ldots, n_d^b), \qquad (2.9)$$
$$\mathcal{B}_{in}^j \varphi = 0 \ (j = 1, \ldots, n_{in}^b)\} \cap \mathcal{D}(A_0).$$

The structure of the boundary operators depends on the structure of the operator A and the other physical properties of the particular problem. Generally speaking the boundary operators $\mathcal{B}_d^j$ and $\mathcal{B}_{in}^j$ can represent any of the classical boundary conditions, including Dirichlet, Neumann and Robin, etc. The explicit examples in this book will provide a clear description of the general types of boundary conditions that can be handled with this methodology.

We note that in the applications considered in this chapter the output operators C are often given as point evaluation or as the weighted average of the state $z(x, t)$ on a hypersurface $\mathcal{S}$ inside or on the boundary of Ω, i.e.,

$$y_i(t) = Cz = \frac{1}{|\mathcal{S}|} \int_{\mathcal{S}} z(x, t) \, d\sigma_x,$$

where by $d\sigma_x$ we denote the natural hypersurface measure on $\mathcal{S}$ and $|\mathcal{S}|$ denotes the hypersurface measure of $\mathcal{S}$. For example it could be that $\mathcal{S}_i$ is one of the boundary patches $\mathcal{S}_j$. We note that the operators C are well defined in our setting since in a typical parabolic problem the state $z(\cdot, t)$ for $t > 0$ is contained in $C^\infty(\overline{\Omega})$ so that the trace on the boundary of Ω is a continuous function. But we also note that such operations do not produce bounded operators in the Hilbert state space $\mathcal{Z} = L^2(\Omega)$.

2.2.1 A Model Class of Systems

In this subsection we show how theoretical results from functional analysis can be used to describe the types of unboundedness allowed in a particular problem. As is often the case in practical applications we assume that the operator A, as described in (1.8), is an unbounded negative self-adjoint operator with compact inverse. Under this assumption the spectrum of $(-A)$ consists entirely of point spectrum (i.e., $\sigma(-A) = \sigma_p(-A)$) consisting of an infinite set of real eigenvalues, of finite multiplicity, which we denote by $\{\lambda_j\}_{j=1}^\infty$. These eigenvalues tend to plus infinity, and are assumed to be ordered so that

$$0 < \lambda_1 \le \lambda_2 \le \lambda_3 \le \cdots, \quad \lim_{j \to \infty} \lambda_j = \infty.$$

Corresponding to these eigenvalues there is a complete set of orthonormal eigenfunctions $\{\psi_j\}_{j=1}^\infty$, i.e.,

$$-A\psi_j = \lambda_j \psi, \quad \langle \psi_j, \psi_k \rangle = \delta_{jk} \text{ (Kronecker delta)}.$$

In addition we will assume that the eigenfunctions are uniformly bounded in $\mathcal{Z}$. Indeed, in most applications the eigenfunctions are contained in $C(\overline{\Omega})$ and are uniformly bounded, i.e., there exists $C_0 > 0$ so that

$$\|\psi_j\|_\infty = \operatorname{ess\,sup}_{x \in \overline{\Omega}} |\psi_j(x)| \le C_0 \ \forall \ j = 1, 2, \cdots.$$

In this case A generates an analytic semigroup $T(t)$ which can be represented in terms of its eigenvalues and orthonormal eigenfunctions as

$$T(t)f = e^{At}f = \sum_{j=1}^\infty e^{-\lambda_j t} \langle f, \psi_j \rangle \psi_j, \quad \text{with } \langle f, \psi_j \rangle = \int_\Omega f(x)\psi_j(x) \, dx.$$

A straightforward calculation based on this explicit formula for the semi-group shows it to be exponentially stable in $L^2(0,1)$, i.e.,

$$\|T(t)\| \le e^{-\lambda_1 t}.$$

Here and below we use the same notation $\|\cdot\|$ for the norm in $L^2(\Omega)$

$$\|\phi\| = \left(\int_\Omega |\phi(x)|^2\, dx\right)^{1/2},$$

and also for the operator norm for a bounded operator M on $L^2(\Omega)$,

$$\|M\| = \sup_{\|\varphi\|=1} \|M\varphi\|.$$

Thus the system (2.22)–(2.23) with $d=0$ and $u=0$ has solution $z(t) = \exp(At)z_0$ satisfying

$$\|z(t)\| = \|T(t)\varphi\| \le e^{-\lambda_1 t}\|z_0\|.$$

Another important consequence of the spectral theorem is that the operator $(-A)$ defines an infinite scale of Hilbert spaces $\mathcal{H}^\alpha$ ($\alpha \in \mathbb{R}$). For each $\alpha \ge 0$ the space $\mathcal{H}^\alpha$ consists of vectors $\phi \in \mathcal{H}^0 = L^2(\Omega)$ such that

$$\|\phi\|_\alpha = \left(\sum_{j=1}^\infty (\lambda_j)^\alpha \langle \phi, \psi_j\rangle^2\right)^{1/2} < \infty. \tag{2.10}$$

For $\alpha < 0$ the space $\mathcal{H}^\alpha$ is the dual to $\mathcal{H}^{-\alpha}$ with respect to the space $\mathcal{H}^0$. The norm in this case is given by the formula (2.10) but now the expression $\langle \phi, \psi_j\rangle$ is to be interpreted as the action of the linear functional ϕ applied to the vector ψ_j (note that $\psi_j \in \mathcal{H}^\alpha$ for all α).

The same spaces can also be described in a different way. Namely, the space $\mathcal{H}^\alpha$ is the domain of the operator $(-A)^{\alpha/2}$ with inner product given by

$$\langle \phi, \psi\rangle_\alpha = \left\langle (-A)^{\alpha/2}\phi, (-A)^{\alpha/2}\psi\right\rangle, \tag{2.11}$$

where the operator $(-A)^{\alpha/2}$ is defined on $\mathcal{H}^\alpha$ by the formula

$$(-A)^{\alpha/2}\phi = \sum_{j=1}^\infty (\lambda_j)^{\alpha/2}\langle \phi, \psi_j\rangle \psi_j.$$

The inner product (2.11) naturally defines the norm

$$\|\phi\|_\alpha = \sqrt{\langle \phi, \phi\rangle_\alpha},$$

which is the same as (2.10).

Lemma 2.1. *The spaces $\mathcal{H}^\alpha$ have the following properties:*

1. *If $\beta > \alpha$ then $\mathcal{H}^\beta \subset \mathcal{H}^\alpha$ and*

$$\|\phi\|_\alpha \le (\lambda_1)^{(\alpha-\beta)/2}\|\phi\|_\beta,$$

   for all $\phi \in \mathcal{H}^\beta$;

2. *$\mathcal{H}^\beta$ is dense in $\mathcal{H}^\alpha$;*

3. *The imbedding $\mathcal{H}^\beta \subset \mathcal{H}^\alpha$ is compact.*

The following results are an immediate consequence of a more general result in [19]:

Lemma 2.2. *The following estimates hold*

1. *If $g \in \mathcal{H}^\alpha$, then for all $t > 0$, we have*

$$\|T(t)g\|_\alpha \le e^{-\lambda_1 t}\|g\|_\alpha.$$

2. *For $\phi \in L^2(\Omega)$ and all $t > 0$*

$$\|T(t)\phi\|_\alpha \le M_\alpha \frac{1}{t^{\alpha/2}} e^{-\lambda_1 t/2}\|\phi\|,$$

   where

$$M_\alpha = \alpha^{\alpha/2} e^{-\alpha/2}.$$

   In the particular case $\alpha = 1$ we have

$$\|T(t)\phi\|_1 \le \frac{1}{\sqrt{e}} \frac{1}{t^{1/2}} e^{-\lambda_1 t/2}\|\phi\|. \tag{2.12}$$

Remark 2.1. We notice that $T(t) : L^2(\Omega) \to L^2(\Omega)$ is compact for all $t > 0$ from both (2.12) and the compactness of the embedding $\mathcal{H}^1 \subset L^2(\Omega)$ (Lemma 2.1).

With this notation we will assume that unbounded input and output operators are A-bounded in the sense of norms in the spaces $\mathcal{H}^\alpha$. Recall that our assumption here is that A^{-1} is bounded. Then we define the following standard Gelfand triple of spaces:

1. The space $\mathcal{Z}_1$ is defined by

$$\mathcal{Z}_1 = \mathcal{D}(A) \text{ with } \|z\|_1 = \|Az\|. \tag{2.13}$$

2. The space $\mathcal{Z}_{-1}$ is the completion of $\mathcal{Z}$ with respect to the norm

$$\|z\|_{-1} = \|A^{-1}z\|. \tag{2.14}$$

3. The space $\mathcal{Z}_{-1}$ can be identified with $\mathcal{D}(A^*)^*$. So, if $A = A^*$, then $\mathcal{Z}_{-1} = (\mathcal{Z}_1)^*$.

4. We have the dense embeddings

$$\mathcal{Z}_1 \hookrightarrow \mathcal{Z} \hookrightarrow \mathcal{Z}_{-1}.$$

5. The semigroup $T(t)$ extends to $\mathcal{Z}_{-1}$, and the generator of the extended semigroup is an extension of A to an operator in $\mathcal{L}(\mathcal{Z}, \mathcal{Z}_{-1})$. Recall that our notation $\mathcal{L}(H_1, H_2)$ describes the space of all bounded linear operators from a Hilbert space H_1 to a Hilbert space H_2.

We can also describe the relationship between the Hilbert scale $\mathcal{H}^\alpha$ and the spaces $\mathcal{Z}_1$ and $\mathcal{Z}_{-1}$ given as follows

$$\mathcal{Z}_1 = D(A) = \mathcal{H}^2,$$

and

$$\mathcal{Z}_{-1} = D(A^{-1}) = \mathcal{H}^{-2}.$$

The operator A has the following useful properties:

$$A \in \mathcal{L}(\mathcal{Z}, \mathcal{Z}_{-1}) = \mathcal{L}(\mathcal{H}^0, \mathcal{H}^{-2}), \quad A \in \mathcal{L}(\mathcal{Z}_1, \mathcal{Z}) = \mathcal{L}(\mathcal{H}^2, \mathcal{H}^0), \quad A \in \mathcal{L}(\mathcal{H}^1, \mathcal{H}^{-1}).$$

With the above we are in a position to describe the types of unboundedness of input and output operators that can be accommodated.

Assumption 2.1. *We assume that*

1. *$B_{in} \in \mathcal{L}(\mathcal{U}, \mathcal{Z}_{-1})$, $B_d \in \mathcal{L}(\mathcal{D}, \mathcal{Z}_{-1})$.*

2. *The operator C defined on some domain, denoted by $\mathcal{D}(C)$, satisfies $C \in \mathcal{L}(Z_1, \mathcal{Y})$.*

3. *The operators B_{in} and C have the property that for all u, $\big((sI - A)^{-1}B_{in}\big)u \in \mathcal{D}(C)$ and $C(sI - A)^{-1}B_{in} \in \mathcal{L}(\mathcal{U}, \mathcal{Y})$ for all $s \in \rho(A)$. In this case we define the transfer function by*

$$G(s) = C(sI - A)^{-1}B_{in}, \quad \forall\, s \in \rho(A).$$

4. *We also assume $(sI - A)^{-1}B_d \in \mathcal{D}(C)$ and $C(sI - A)^{-1}B_d \in \mathcal{L}(\mathcal{D}, \mathcal{Y})$ for all $s \in \rho(A)$.*

5. *In addition we make the following assumptions concerning the boundedness of the transfer function in a closed right half plane $\overline{\mathbb{C}}_\delta = \{\sigma \in \mathbb{C} : \mathrm{Re}\,(s) \geq \delta\}$. We assume*

   (a) *There exists a $\delta > 0$ so that*

   $$\sup_{s \in \overline{\mathbb{C}}_\delta} \|G(s)\|_{\mathcal{L}(\mathcal{U}, \mathcal{Y})} < \infty.$$

(b) The limit of $G(s)$ for s on the positive real axis exists and

$$\lim_{s \to +\infty} \|G(s)\|_{\mathcal{L}(\mathcal{U}, \mathcal{Y})} = 0.$$

While the theoretical discussion involving the Hilbert scales $\mathcal{H}^\alpha$ is useful to guarantee existence and provide certain theoretical estimates a major drawback from a practical point of view is that it requires the availability of the eigenvalues and eigenvectors of the system which for multi-dimensional partial differential operator A can be extremely difficult to obtain. An alternative approach leading to essentially the same types of results is straightforward elliptic estimates in standard Sobolev spaces $H^s(\Omega)$. To see this type of results we refer the reader to the work of A. Cheng and K. Morris [32]. In [32] the authors show precisely what is required for well-posedness and boundedness of the input/output map in the case of parabolic boundary control systems in higher-dimensional spatial domains. They consider boundedness of the input/output map for several large classes of problems with Dirichlet, Neumann or Robin boundary control when the state operator is given by a uniformly elliptic linear differential operator which includes most of the standard examples contained in this book. For more details we refer the reader to the paper [32].

In summary, a rigorous mathematical justification of the regulator equations and their role in solving tracking and disturbance rejection problems can be given, either in terms of Sobolev spaces or the infinite Hilbert scale of spaces generated by the positive self-adjoint operator $(-A)$ in $L^2(\Omega)$.

2.2.2 Well-Posedness of the Boundary Control System

For the systems considered in Chapter 1 with bounded control and disturbance operators the solution to (1.1)–(1.3) can be expressed using the variation of parameters formula in $\mathcal{Z}$ as

$$z(t) = e^{At} z_0 + \int_0^t e^{A(t-\tau)} \left(B_d d(\tau) + B_{in} u(\tau) \right) d\tau,$$

and there is no concern over the existence of any of the terms, or the convergence of the integral. Furthermore, one can apply the bounded output operator C to obtain

$$y(t) = Cz(t) = Ce^{At} z_0 + \int_0^t Ce^{A(t-\tau)} \left(B_d d(\tau) + B_{in} u(\tau) \right) d\tau,$$

where once again all the terms exist. Notice that we are freely allowed to move the bounded operator C inside the integral. In the case $z_0 = 0$ and with no disturbance we can apply the Laplace transform for Re $(s) > 0$ to obtain

$$\widehat{y}(s) = C(sI - A)^{-1} B_{in} \widehat{u}(s).$$

The operator (which we have already defined as the transfer function of the control system)

$$G(s) = C(sI - A)^{-1}B_{in},$$

is a well defined bounded operator in $\mathcal{L}(\mathcal{U}, \mathcal{Y})$. Similar statements hold for the operator $C(sI - A)^{-1}B_d$ in $\mathcal{L}(\mathcal{D}, \mathcal{Y})$.

In an effort to understand general conditions on a system with unbounded inputs and outputs that would allow the above statements to hold in more general spaces (like $\mathcal{Z}_{-1}$), G. Weiss (and several others authors) (cf. [33], [85]–[87]) have defined the class of regular linear systems. The definition of a regular system hinges on conditions on (A, B, C) for which the variation of parameters formula and the other statements above still hold. Basically, a regular system is one for which the following hold:

1. B is an *admissible control* [85] for the semigroup T, i.e., the system (2.1)–(2.3) is well posed for any $u \in L^2((0, \infty), \mathcal{U})$ in the sense that for any $z_0 \in \mathcal{Z}$ it has a strong solution $z(t)$ and the equation (2.1) holds in the sense of $\mathcal{Z}_{-1}$.

2. C is an *admissible input* operator for T, i.e., the output $y(t) = CT(t)\varphi$ is defined for almost every $t \in [0, \infty)$ for any $\varphi \in \mathcal{Z}$.

3. The transfer function $G(s) = C(sI - A)^{-1}B_{in}$, $s \in \rho(A)$ is well defined as an operator on $\mathcal{U}$ and is bounded in some right half plane $\mathbb{C}_a^+ = \{\zeta \in \mathbb{C} : \text{Re}(\zeta) > a$ for some $a \in \mathbb{R}\}$.

We note that the above definitions certainly hold in the case of the systems considered in Section 2.2.1. This class will cover many of the examples considered in this book.

In any case all these technical issues can be addressed, in a one-by-one manner, for explicit examples. The good news is that in the actual solution of the regulator problems we consider this is not necessary and only arises in an attempt to formulate a boundary control problem in the standard state space form (1.1)–(1.3) and then prove well-posedness, etc. In practice, and especially for linear problems, this is not usually necessary due to already existing theory of existence, uniqueness, regularity and stability.

The following example for a one-dimensional heat equation is presented to show how one can carry out the transformation between the boundary control form of a system as given in (2.5)–(2.7) and (2.8) to the standard state space form (1.1)–(1.3).

Example 2.1. We recall that in Example 1.1 the operators B_{in}, B_d and C were bounded. In this example we consider a very similar control system modeled by a one-dimensional heat equation on a unit interval, $0 < x < 1$. We assume that the left end of the rod is held at a constant temperature $d = M_d$ (a constant disturbance) and we are able to control the flow of heat at the right end of the rod (a flux boundary control). This system, with initial temperature distribution $z_0(x)$ would normally be written as an initial boundary value problem in the Hilbert state space $\mathcal{Z} = L^2(0, 1)$ as

$$z_t(x, t) = z_{xx}(x, t), \tag{2.15}$$
$$z(0, t) = d,$$
$$z_x(1, t) = u,$$
$$z(x, 0) = z_0(x). \tag{2.16}$$

In this example we consider essentially the same tracking disturbance rejection problem as in Examples 1.1 and 1.2. In particular we seek a control u in order to force the temperature at a given point, $0 < x_1 < 1$, to track a sinusoidal reference signal $y_r = A_r \sin(\alpha t)$, while rejecting a constant disturbance $d = M_d$. In this case our measured output is point evaluation $y = Cz(t) = z(x_1, t)$, an unbounded operation in $\mathcal{Z}$. The "boundary control" system (2.15)–(2.16) can be transformed in an equivalent form as the system

$$z_t(x, t) = Az(x, t) + \frac{d\delta_0}{dx}d + \delta_1 u, \tag{2.17}$$
$$z(x, 0) = z_0(x), \tag{2.18}$$

where δ_j for $j = 0, 1$ denote the Dirac delta function supported at $x = 0$ and $x = 1$, respectively. In this form the system now looks like a system written in standard system form as (1.1), where

$$B_{\mathrm{d}} = \frac{d\delta_0}{dx} \quad \text{and} \quad B_{\mathrm{in}} = \delta_1,$$

and

$$A = \frac{d^2}{dx^2} \quad \text{with domain } \mathcal{D} = \{\varphi \in H^2(0, 1) : \varphi(0) = 0, \quad \varphi'(1) = 0\}.$$

However the mathematical problem (2.17), (2.18), must be studied in a space of distributions rather than the more desirable space $L^2(0, 1)$.

It is easy to show that the operator $(-A)$ has all the properties needed to apply the general discussion in Section 2.2. In particular it is negative self-adjoint and has a compact inverse. Therefore it has a countable set of eigenvalues and orthonormal eigenfunctions obtained by analyzing the regular Sturm-Liouville problem

$$\frac{d^2\varphi}{dx^2} + \mu^2\varphi = 0, \quad \varphi(0) = 0, \quad \frac{d\varphi}{dx}(1) = 0.$$

A simple analysis shows that the eigenvalues come from the zeros of the characteristic equation

$$\cos(\mu) = 0,$$

which has infinitely many zeros $\{\mu_j\}_{j=1}^{\infty}$ satisfying

$$\mu_j = \frac{(2j - 1)\pi}{2}, \quad j = 1, 2, \cdots,$$

providing the eigenvalues of $(-A)$

$$\lambda_j = \mu_j^2,$$

and a complete set (in $L^2(0,1)$) of orthonormal eigenfunctions

$$\psi_j(x) = \sqrt{2}\sin(\mu_j x).$$

These eigenfunctions satisfy three important properties: (1) orthogonality, i.e., $\langle \psi_i, \psi_j \rangle = \delta_{i,j}$; (2) $-A\psi_j = \lambda_j \psi_j$ and (3) the entire family of functions are uniformly bounded on $[0,1]$, i.e., there exists a positive constant $C_0 = \sqrt{2}$ so that

$$\|\psi_j\|_\infty = \sup_{x \in [0,1]} |\psi_j(x)| \le \sqrt{2} \quad \forall \; j = 1, 2, \cdots .$$

We can also see that

$$\|\psi_j'\|_\infty = \sup_{x \in [0,1]} |\psi_j'(x)| \le \sqrt{2}\mu_j \quad \forall \; j = 1, 2, \cdots .$$

A generates an exponentially stable analytic semigroup $T(t)$ which can be represented in terms of its eigenvalues and orthonormal eigenfunctions as

$$T(t)f = e^{At}f = \sum_{j=1}^{\infty} e^{-\lambda_j t} \langle f, \psi_j \rangle \psi_j, \quad \text{with} \quad \langle f, \psi_j \rangle = \int_0^1 f(x)\psi_j(x)\,dx,$$

and

$$\|z(t)\| = \|T(t)z_0\| \le e^{-\lambda_1 t}\|z_0\|.$$

The operator $(-A)$ defines an infinite scale of Hilbert spaces $\mathcal{H}^\alpha$ $(\alpha \in \mathbb{R})$ as discussed in Section 2.2.

Using the fact that $\mathcal{D}(A)$ is a core* in $\mathcal{D}((-A)^{1/2}) = \mathcal{H}^1$ (cf, Theorem 3.35, [57]), it is easy to show that the norm in $\mathcal{H}^1$ can be written as

$$\|\varphi\|_1 = \|\varphi_x\|.$$

It can be shown that the space $\mathcal{H}^1$ is the same as the space

$$H_{0,*}^1(0,1) = \{\varphi \in H^1(0,1) \; : \; \varphi(0) = 0\},$$

where the norm in $H^1(0,1)$ is given by

$$\|\varphi\|_{H^1}^2 = \|\varphi\|^2 + \|\varphi_x\|^2.$$

Namely for each $\varphi \in H_{0,*}^1(0,1)$ we have

$$\varphi(x) = \int_0^x \varphi_\xi(\xi)\,d\xi,$$

---

*For a closed operator A, a linear subspace $S \subset \mathcal{D}(A)$ is called a core of M if the set $\{\varphi, A\varphi\}$ for $\varphi \in S$ is dense in the graph of A, $G(A)$.

which implies

$$\sup_{x \in [0,1]} |\varphi(x)| \leq \|\varphi_x\|,$$

and consequently

$$\|\varphi\| \leq \|\varphi_x\|.$$

Thus

$$\|\varphi\|_1 \leq \|\varphi\|_{H^1} \leq \sqrt{2}\|\varphi\|_1,$$

which gives the equivalency between the norms of $\mathcal{H}^1$ and $H_{0,*}^1(0,1)$.

The output and input operators C, B_{in} are in $\mathcal{H}^{-1} \subset \mathcal{H}^{-2} = \mathcal{Z}_{-1}$. Indeed, we can show that for any $x_0 \in [0,1]$ we have $\delta_{x_0} \in \mathcal{H}^{-1}$. In fact we have

$$\|\delta_{x_0}\|_{-1}^2 = \sum_{j=1}^{\infty} \lambda_j^{-1} |\langle \delta_{x_0}, \psi_j \rangle|^2 \leq \frac{\sqrt{2}}{2}.$$

We can also show that B_{d} is in $\mathcal{Z}_{-1}$, namely

$$\|\delta_0\|_{-2}^2 = \sum_{j=1}^{\infty} \lambda_j^{-2} \left|\langle \frac{d}{dx}\delta_0, \psi_j \rangle\right|^2 = \sum_{j=1}^{\infty} \lambda_j^{-2} \left|\langle \delta_0, \frac{d}{dx}\psi_j \rangle\right|^2 = 1.$$

A straightforward calculation using elementary spectral theory gives

$$G(s) = \sum_{j=1}^{\infty} \frac{\langle \delta_{x_1}, \psi_j \rangle \langle \psi_j, \delta_1 \rangle}{(s + \lambda_j)},$$

and therefore we have

$$\|G(s)\|_{\mathcal{L}(\mathcal{U},\mathcal{Y})} = \left| \sum_{j=1}^{\infty} \frac{\langle \delta_{x_1}, \psi_j \rangle \langle \psi_j, \delta_{x_1} \rangle}{(s + \lambda_j)} \right| \leq A_r^2 \sum_{j=1}^{\infty} \frac{1}{(s + \lambda_j)}$$

$$= \frac{A_r^2}{2} \frac{\tanh(\sqrt{s})}{\sqrt{s}} \leq \frac{A_r^2}{2\sqrt{s}}.$$

From this we see that all the conditions of Assumption 2.1 are satisfied and, in particular, we have

$$\sup_{s \in \overline{\mathbb{C}_\delta}} \|G(s)\|_{\mathcal{L}(\mathcal{U},\mathcal{Y})} < \infty \quad \text{and} \quad \lim_{s \to +\infty} \|G(s)\|_{\mathcal{L}(\mathcal{U},\mathcal{Y})} = 0.$$

We also note that a straightforward calculation using the Laplace transform gives a simpler alternate formula for $G(s)$

$$G(s) = \frac{\sinh(\sqrt{s}x_1)}{\sqrt{s}\cosh(\sqrt{s})}.$$

We now turn to the main topic in this chapter.

2.2.3   Regulation Problem for the Unbounded Case

In this book we are primarily interested in providing the necessary tools for the reader to be able to solve problems of regulation numerically. But we also want to provide a certain amount of theoretical development without going overboard with mathematical formalisms. In particular it would be nice to present an analog to Theorem 1.1 providing necessary and sufficient conditions for solvability of the regulator problem in terms of solvability of some analog of the regulator equations in (1.19)–(1.20). Unfortunately, in order to state such a result we would need to introduce a significant amount of theoretical machinery which would not in any way enhance the most important point of this book.

The main theorem in the work [66] provides an analog of Theorem 1.1 for a class of boundary control systems governed by the class of regular systems. We will not describe all the necessary technical issues needed to present the most general result found in [66]. Rather we will state (without proof) a version of the result appropriate for the class of systems satisfying the conditions in Assumption 2.1 from Section 2.2.1. First we restate the linear state feedback regulator problem for Chapter 2.

Problem 2.1 (The Linear State Feedback Regulator Problem). *Find a feedback control law in the form* $u(t) = \Gamma w(t)$, $\Gamma \in \mathcal{L}(\mathcal{W}, \mathcal{U})$, *such that for the closed-loop system*

$$\frac{d}{dt}\begin{bmatrix} z \\ w \end{bmatrix} = \begin{bmatrix} A & (B_d P + B_{in}\Gamma) \\ 0 & S \end{bmatrix}\begin{bmatrix} z \\ w \end{bmatrix}, \tag{2.19}$$

the norm of the error

$$\|e(t)\| = \|y(t) - r(t)\| = \|Cz(t) - Qw(t)\| \to 0, \quad as \ \ t \to \infty,$$

for any initial conditions $z(0) = z_0 \in Z$ *and* $w(0) = w_0 \in \mathcal{W}$.

We now state a simplified version of the main result in [66] based on appropriate assumptions on the control system operators (A, B_{in}, B_d, C) and the exosystem.

Theorem 2.1. *Under appropriate conditions (see Assumption 2.1) guaranteeing the well-posedness of the closed loop system* (2.19), *existence and boundedness of the transfer function, the state feedback regulator problem (Problem 2.1) is solvable if and only if there exist mappings* $\Pi \in \mathcal{L}(\mathcal{W}, Z)$ *and* $\Gamma \in \mathcal{L}(\mathcal{W}, \mathcal{U})$ *satisfying the "regulator equations"*

$$\Pi S = A\Pi + (B_{in}\Gamma + B_d P), \tag{2.20}$$

$$C\Pi = Q. \tag{2.21}$$

The first regulator equation holds in $\mathcal{L}(\mathcal{W}, \mathcal{Z})$ *and the second holds in* $\mathcal{L}(\mathcal{W}, \mathcal{Y})$.

If the regulator equations are solvable a feedback law solving the state feedback regulator problem is given by

$$u(t) = \Gamma w(t).$$

Remark 2.2. As we have already mentioned, a detailed proof of an even more general version of Theorem 2.1 can be found in [66].

2.2.4 Computation of the Transfer Function for a BCS

From a practical point of view the prospects of solving the regulator equations (2.20)–(2.21) appear to be rather difficult, since we usually start with a boundary control system (2.5)–(2.7). It would seem that we first need to find the distributional terms B_{d}^j for $j = 1, \ldots, n_{\mathrm{d}}^b$ and B_{in}^j for $j = 1, \ldots, n_{\mathrm{in}}^b$ in order to transform the problem (2.5)–(2.7) into the problem (2.1)–(2.3), so that we can apply Theorem 2.1. Fortunately, it turns out this is not necessary at all as we will see below.

In the case of bounded or unbounded input and output operators, the most important construct needed to solve the regulator equations and thereby obtain $u = \Gamma w$ is the transfer function $G(s) = C(sI - A)^{-1} B_{\mathrm{in}}$. For a boundary control system it is not necessary to find the complicated expressions for the distributions needed to represent the boundary control operator.

In order to simplify the exposition let us assume in this subsection that all the control inputs and outputs are unbounded, i.e., $n_{\mathrm{d}}^b = n_{\mathrm{in}}$ in (2.6) and (2.7). In this case the transfer function with n_{in} inputs and n_{out} outputs is given by an $n_{\mathrm{out}} \times n_{\mathrm{in}}$ matrix in the form

$$G(s) = [g_{ij}(s)]_{i=1, j=1}^{n_{\mathrm{out}}, n_{\mathrm{in}}},$$

where

$$g_{ij}(s) = C_i(sI - A)^{-1} B_{\mathrm{in}}^j, \quad i = 1, \ldots, n_{\mathrm{out}}, \; j = 1, \ldots, n_{\mathrm{in}}.$$

For a fixed term j_0, corresponding to a boundary control input B_{in}^j, rather than using the formula above, which requires knowing B_{in}^j explicitly, we will solve

$$(sI - A_{j_0})X_{j_0} = 0, \quad \mathcal{B}_{\mathrm{in}_{j_0}} X_{j_0} = 1,$$

where A_{j_0} is the operator obtained from A_0 with domain

$$\mathcal{D}(A_{j_0}) = \{\varphi : \mathcal{B}_{\mathrm{d}}^j \varphi = 0 \; (j = 1, \ldots, n_{\mathrm{d}}^b),$$
$$\mathcal{B}_{\mathrm{in}}^j \varphi = 0 \; (j = 1, \ldots, n_{\mathrm{in}}^b, \; j \neq j_0)\} \cap \mathcal{D}(A_0).$$

We then apply the output operator C_i to obtain the entry

$$g_{ij_0}(s) = C_i(sI - A)^{-1} B_{j_0} = C_i X_{j_0}.$$

Letting j_0 run through all possible indices and denoting the entries by $g_{ij}(s)$ without the subscript 0 we obtain the transfer function, i.e.,

$$G(s) = [g_{ij}(s)]_{i=1, j=1}^{n_c, n_c}.$$

This provides an extremely simple procedure that allows us to solve the regulator equations using the exact same methodology as employed in Chapter 1 but in the calculation of the transfer function we proceed using the above algorithm.

2.3   Examples with Unbounded Sensing and Control

In this section we consider several examples. The first example is relatively simple and we will provide all the details involved in converting the problem from the boundary control form (2.5)–(2.7) to the distributional form (2.1)–(2.3). The second example is similar to the first one, so we will skip the description of the distributional spaces and transfer function. The most interesting point of this example is that we are going to track an approximation of a periodic triangle function. The triangle function under consideration is continuous but has a Fourier series representation containing infinitely many harmonics. To capture this function exactly we would need an infinite-dimensional exosystem. Nevertheless by simply truncating the Fourier series, we obtain a very accurate approximation of the signal to be tracked, which in turn allows us to use a finite-dimensional exosystem. In the last two examples we revisit some of the examples described in Chapter 1, where now the input/output operators will be unbounded. The initial setup for the first example is similar to Example 2.1 except that here we will actually solve the regulation problem instead of simply formulating the various spaces and operators. We refer to Example 2.1 for details concerning the definitions of the unbounded operators B_{in}, Bd and C.

Example 2.2. Consider a one-dimensional heat equation on a finite interval $[0,1]$ with a scalar Neumann boundary control $u(t)$ at the right end point $(x = 1)$ and a boundary disturbance $d(t)$ entering through a Robin boundary condition at the left end point $(x = 0)$. Notice that this control actuation and disturbance injection correspond to unbounded operators in the Hilbert state space $\mathcal{Z} = L^2(0,1)$.

$$z_t(x,t) = z_{xx}(x,t), \quad x \in (0,1), t > 0, \tag{2.22}$$
$$z(x,0) = z_0(x),$$
$$\mathcal{B}_0(z)(t) = -z_x(0,t) + kz(0,t) = d(t), \quad k > 0,$$
$$\mathcal{B}_1(z)(t) = z_x(1,t) = u(t). \tag{2.23}$$

Assume that the output $y(t)$ is defined as a point evaluation of the state $z(x,t)$ at a prescribed point $x_1 \in [0,1]$

$$y(t) = C(z)(t) = z(x_1,t), \quad x_1 \in [0,1].$$

Once again the operator given by point evaluation is not only an unbounded operator in $\mathcal{Z}$ but it is not even closable.

Our objective is to design a feedback law for which the output of the closed loop system tracks a periodic reference signal $y_r = A_r \sin(\alpha t)$, while rejecting a constant disturbance $d(t) = M_d$ entering through the left endpoint of the rod.

For the present problem we consider an exosystem

$$\frac{dw}{dt} = Sw, \ \ S = \begin{bmatrix} S^1 & 0 \\ 0 & S_2 \end{bmatrix}, \tag{2.24}$$

$$S^1 = \begin{bmatrix} 0 & \alpha \\ -\alpha & 0 \end{bmatrix}, \ \ S_2 = 0, \tag{2.25}$$

$$w = \begin{bmatrix} w^1 \\ w_3 \end{bmatrix} \in \mathbb{R}^3, \ w^1 = \begin{bmatrix} w_1 \\ w_2 \end{bmatrix} = \begin{bmatrix} A_r \sin(\alpha t) \\ A_r \cos(\alpha t) \end{bmatrix}, \ w_3 = M_d,$$

$$w(0) = w_0 = \begin{bmatrix} w^1 \\ w_3 \end{bmatrix}(0), \ \ w^1(0) = \begin{bmatrix} 0 \\ A_r \end{bmatrix}, \ \ w_3(0) = \begin{bmatrix} M_d \end{bmatrix}, \tag{2.26}$$

$$y_r(t) = Qw(t) \equiv \begin{bmatrix} 1 & 0 & 0 \end{bmatrix} w(t) = w_1(t) = A_r \sin(\alpha t),$$

$$d(t) = Pw(t) \equiv \begin{bmatrix} 0 & 0 & 1 \end{bmatrix} w(t) = w_3(t) = M_d. \tag{2.27}$$

For this example the input space $\mathcal{U}$, the disturbance space $\mathcal{D}$ and the output space $\mathcal{Y}$ are all given by $\mathbb{R}$.

Referring to our development of the Hilbert scales discussed in Section 2.2 we note that the operator A in this case is $A = d^2/dx^2$ with domain $\mathcal{D}(A)$ in $\mathcal{Z} = L^2(0,1)$ given by

$$\mathcal{D}(A) = \{\varphi \in H^2(0,1) \ : \ -\varphi_x(0) + k\varphi(0) = 0, \ \ \varphi_x(1) = 0\}.$$

This example is similar to Example 2.1 in many aspects, however, there are a few subtle differences. We will present all the important parts that differ.

Once again in this case the operator $(-A)$ is a positive, self-adjoint operator in $\mathcal{Z} = L^2(0,1)$ with compact inverse whose eigenvalues and eigenfunctions are generated by analyzing a regular Sturm-Liouville problem.

$$\frac{d^2\varphi}{dx^2} + \mu^2\varphi = 0, \ \ \frac{d\varphi}{dx}(0) - k\varphi(0) = 0, \ \ \frac{d\varphi}{dx}(1) = 0.$$

A simple analysis shows that the eigenvalues of $-A$ can be written as $\lambda_j = \mu_j^2$ where the μ_j are the roots of

$$\cos(\mu) - \frac{\mu}{k}\sin(\mu) = 0.$$

This equation has infinitely many roots $\{\mu_j\}_{j=1}^{\infty}$ satisfying

$$(j-1)\pi < \mu_j < \left(j - \frac{1}{2}\right)\pi, \ j = 1, 2, \cdots, \ \ \text{and} \ \mu_j - (j-1)\pi \xrightarrow{j\to\infty} 0. \tag{2.28}$$

These provide the eigenvalues of $(-A)$ as

$$\lambda_j = \mu_j^2, \quad j = 1, 2, \cdots .$$

There exists a complete orthonormal system of eigenfunctions $\{\psi_j(x)\}_{j=1}^\infty$ in $L^2(0,1)$, with the eigenvalues λ_j corresponding to $\psi_j(x)$. These eigenfunctions can be easily computed explicitly, but mainly we are interested in three important properties:

1. orthogonality, i.e., $\langle \psi_i, \psi_j \rangle = \delta_{i,j}$,

2. $-A\psi_j = \lambda_j \psi_j$,

3. the entire family of functions are uniformly bounded on $[0, 1]$, i.e., there exists $C_0 > 0$ such that

$$\|\psi_j\|_\infty = \operatorname*{ess\,sup}_{x \in [0,1]} |\psi_j(x)| \le C_0 \ \forall \ j = 1, 2, \cdots . \tag{2.29}$$

As in Section 2.2 the operator A generates an analytic semigroup $T(t)$ represented in terms of its eigenvalues and orthonormal eigenfunctions as

$$T(t)f = e^{At}f = \sum_{j=1}^\infty e^{-\lambda_j t} \langle f, \psi_j \rangle \psi_j, \ \text{ with } \langle f, \psi_j \rangle = \int_0^1 f(x)\psi_j(x)\,dx. \tag{2.30}$$

The semigroup in (2.30) is exponentially stable in $L^2(0,1)$, i.e.,

$$\|T(t)\| \le e^{-\lambda_1 t}.$$

Thus the system (2.22)–(2.23) with $d = 0$ and $u = 0$ has solution $z(t) = \exp(At)z_0$ satisfying

$$\|z(t)\| = \|T(t)\varphi\| \le e^{-\lambda_1 t} \|z_0\|.$$

Further, the operator $(-A)$ defines an infinite scale of Hilbert spaces as in Section 2.2. The first main difference between this example and Example 2.1 is that the norm in $\mathcal{H}^1$ can be written as

$$\|\varphi\|_1^2 = \|\varphi_x\|^2 + k|\varphi(0)|^2. \tag{2.31}$$

We will now show that in this case we have $\mathcal{H}^1 = H^1(0,1)$.

Lemma 2.3. *For any $k > 0$ and $\varphi \in \mathcal{H}^1$ we have*

$$\|\varphi\|^2 \le \lambda_1^{-1} \|\varphi\|_1^2,$$

where $0 < \lambda_1 < \pi^2$ is the first eigenvalue of $(-A)$, $\|\cdot\|$ is the norm in $L^2(0,1)$ and $\|\cdot\|_1$ is defined in (2.31).

It follows that for all $\varphi \in \mathcal{H}^1$,

$$\|\varphi\|_{H^1(0,1)}^2 \le \left(1 + \lambda_1^{-1}\right) \|\varphi\|_1^2$$

and therefore $\mathcal{H}^1 \subset H^1(0,1)$.

In order to show the reverse containment $H^1(0,1) \subset \mathcal{H}^1$, we recall the following well known relationship between the $L^\infty(0,1)$ norm

$$\|z\|_{L^\infty(0,1)} := \operatorname*{ess\,sup}_{x \in [0,1]} |z(x)|,$$

and the norm in $H^1(0,1)$.

Lemma 2.4. *For $z \in H^1(0,1)$ we have the estimate*

$$\|z\|_{L^\infty(0,1)} \leq \sqrt{2}\|z\|_{H^1(0,1)},$$

and, hence, for $z \in \mathcal{H}^1$,

$$\|z\|_{L^\infty(0,1)} \leq c\|z\|_1,$$

where

$$c = \sqrt{2}(1 + \lambda_1^{-1})^{1/2}. \tag{2.32}$$

In general such results follow from the classical Sobolev embedding theorem but this special case can easily be established using elementary calculus, the Cauchy-Schwartz inequality and Lemma 2.3.

We readily conclude that $H^1(0,1) \subset \mathcal{H}^1$, since for $\varphi \in H^1(0,1)$ we can write

$$\|\varphi\|_1^2 = \|\varphi_x\|^2 + k\|\varphi(0)\|^2 \leq \|\varphi_x\|^2 + k\|\varphi\|_\infty^2 \leq (1 + 2k)\|\varphi\|_{H^1(0,1)}^2.$$

From Lemma 2.3 and Lemma 2.4 it follows that the respective norms in $H^1(0,1)$ and $\mathcal{H}^1$ are equivalent, thus $H^1(0,1) = \mathcal{H}^1$.

Lemma 2.4 also allows us to show that $C \in \mathcal{L}(\mathcal{H}^1, \mathbb{R})$. In particular, let $\varphi \in \mathcal{H}^1 = H^1(0,1)$ (so that $\varphi \in C[0,1]$), then we have

$$|C(\varphi)| = |\varphi(x_1)| \leq \|\varphi\|_\infty \leq c\|\varphi\|_1,$$

which implies $\|C\| \leq c$ (with c in (2.32)) and therefore $C \in \mathcal{L}(\mathcal{H}^1, \mathbb{R})$.

We can also obtain the relationship between the Hilbert scale $\mathcal{H}^\alpha$ and the spaces $\mathcal{Z}_1$ and $\mathcal{Z}_{-1}$ given by

$$\mathcal{Z}_1 = D(A) = \mathcal{H}^2,$$

and

$$\mathcal{Z}_{-1} = D(A^{-1}) = \mathcal{H}^{-2}.$$

Lemma 2.5. *The output and input operators C, B_{in} and B_d are in $\mathcal{H}^{-1} \subset \mathcal{H}^{-2} = \mathcal{Z}_{-1}$.*

We now show that for any $x_0 \in [0,1]$, $\delta_{x_0} \in \mathcal{H}^{-1}$. We have

$$\|\delta_{x_0}\|_{-1}^2 = \sum_{j=1}^\infty \lambda_j^{-1} |\langle \delta_{x_0}, \psi_j \rangle|^2 = \sum_{j=1}^\infty \lambda_j^{-1} |\psi_j(x_0)|^2 \leq cC_0^2 < \infty,$$

where we have used

$$\sum_{j=1}^{\infty} \lambda_j^{-1} < \lambda_1^{-1} + \sum_{j=2}^{\infty} \frac{1}{\pi^2(j-1)} = \lambda_1^{-1} + \frac{1}{6} = c$$

and

$$\|\psi_j\|_\infty \leq C_0 \quad \forall \ j = 1, 2, \cdots .$$

(see (2.28) and (2.29)).

We are now in a position to explain the equivalence between the system

$$z_t = Az + \delta_0 Pw + \delta_1 u, \tag{2.33}$$

$$z(x,0) = z_0(x), \tag{2.34}$$

and the system (2.22)–(2.23).

We note that $\mathcal{D}(A)$ is a core in $\mathcal{D}(A^{1/2}) = H^1(0,1)$ (cf, Theorem 3.35, [57]) so we establish the result for test functions $\eta \in \mathcal{D}(A)$. The weak solution $z(x,t)$ of (2.22)–(2.23) is a function $z(\cdot,t) \in H^1(0,1)$, $z_t(\cdot,t) \in L^2(0,1)$ for almost every $t \in [0,\infty)$, which for all functions $\eta \in \mathcal{D}(A)$ satisfies the integral relations obtained by integration by parts

$$\int_0^1 z_t(x,t)\eta(x)\,dx = -\int_0^1 z_x(x,t)\eta_x(x)\,dx + z_x(x,t)\eta(x)\big|_0^1$$

$$= \int_0^1 z(x,t)\eta_{xx}(x)\,dx - z(x,t)\eta_x(x)\big|_0^1 + z_x(x,t)\eta(x)\big|_0^1 .$$

The boundary terms can be rewritten as

$$z_x(x,t)\eta(x)\big|_0^1 = u(t)\eta(1) + (-kz(0,t) + d(t))\eta(0),$$

and

$$-z(x,t)\eta_x(x)\big|_0^1 = z(0,t)\eta_x(0) = z(0,t)k\eta(0),$$

where we have used $\eta_x(1) = 0$ and $\eta_x(0) - k\eta(0) = 0$. Then adding the boundary terms together we have

$$z_x(x,t)\eta(x)\big|_0^1 - z(x,t)\eta_x(x)\big|_0^1 = u(t)\eta(1) + (-kz(0,t) + d(t))\eta(0)$$
$$+ z(0,t)k\eta(0) = u(t)\eta(1) + d(t)\eta(0).$$

So we have

$$\int_0^1 z_t(x,t)\eta(x)\,dx = \langle z, A\eta \rangle + u(t)\eta(1) + d(t)\eta(0)$$
$$= \langle z, A\eta \rangle + u(t)\langle \delta_1, \eta \rangle + d(t)\langle \delta_0, \eta \rangle$$
$$= \langle z, A\eta \rangle + \langle B_d d(t), \eta \rangle + \langle B_{in} u(t), \eta \rangle,$$

for any test function $\eta \in \mathcal{D}(A)$.

Now let us suppose that z is a classical solution to the problem

$$z_t(x,t) = z_{xx}(x,t) + B_d d(t) + B_{in} u(t), \quad x \in (0,1), \tag{2.35}$$

$$z(x,0) = z_0(x), \tag{2.36}$$

$$-z_x(0,t) + kz(0,t) = 0, \quad k > 0, \tag{2.37}$$

$$z_x(1,t) = 0. \tag{2.38}$$

Then we have $z(\cdot,t) \in \mathcal{D}(A)$ and for every η, as above, we have

$$\langle z, A\eta \rangle = \langle Az, \eta \rangle$$

and by the uniqueness of solutions, the two problems (2.22)–(2.23) and (2.35)–(2.38) are equivalent.

A straightforward calculation shows that for all $s \in \overline{\mathbb{C}_0^+}$ (the closed right half complex plane) and $x_1 \in [0,1]$, the closed loop transfer function for the system (2.33) is given by (see [23, 24])

$$G_k(s) = \frac{\cosh(x_1\sqrt{s}) + k\sinh(x_1\sqrt{s})/\sqrt{s}}{\sqrt{s}\sinh(\sqrt{s}) + k\cosh(\sqrt{s})}. \tag{2.39}$$

Since the transfer function (2.39) is a meromorphic function in $\mathbb{C}$ with simple poles on the negative real axis, it is easy to see from elementary complex analysis, that for every $k > 0$ there exists a constant C_k so that

$$\sup_{s \in \overline{\mathbb{C}^+}} |G(s)| \le C_k.$$

Furthermore it is also straightforward to see that

$$\lim_{s \to \infty} G(s) = 0, \quad \text{for all } k \ge 0.$$

For the system (2.33), (2.34) with exosystem defined in (2.24), disturbance (2.27), reference signal (2.26), and output (2.34), we seek a control in the form $u = \Gamma w$ to solve the regulator problem. Thus we seek mappings Π and Γ,

$$\Pi = \begin{bmatrix} \Pi^1 & \Pi_3 \end{bmatrix}, \quad \Pi^1 = \begin{bmatrix} \Pi_1 & \Pi_2 \end{bmatrix},$$

with $\Pi_j \in \mathcal{Z}$ and

$$\Gamma = \begin{bmatrix} \Gamma^1 & \Gamma_3 \end{bmatrix}, \quad \Gamma^1 = \begin{bmatrix} \Gamma_1 & \Gamma_2 \end{bmatrix},$$

with $\Gamma_j \in \mathbb{R}$. These mappings are to satisfy the regulator equations given by

$$\Pi S w = A\Pi w + B_{in}\Gamma w + B_d P w, \tag{2.40}$$

$$0 = C\Pi w - Q w, \quad \text{for all } w \in \mathbb{R}^3. \tag{2.41}$$

The first regulator equation (2.40) naturally decomposes into two systems of equations. Namely, we have

$$\Pi^1 S^1 w^1 = A\Pi^1 w^1 + \delta_1 \Gamma^1 w^1, \tag{2.42}$$

$$\Pi_3 S_2 w_3 = A\Pi_3 w_3 + \delta_1 \Gamma_3 w_3 + \delta_0 w_3. \tag{2.43}$$

The second regulator equation (2.41) immediately gives

$$1 = C\Pi_1 = \Pi_1(x_1), \quad 0 = C\Pi_j = \Pi_j(x_1), \quad j = 2, 3. \tag{2.44}$$

We turn to a more detailed examination of equations (2.42) and (2.43). First consider (2.43). Writing out the terms we arrive at

$$0 = A\Pi_3 + \delta_1 \Gamma_3 + \delta_0. \tag{2.45}$$

From (2.45) we have

$$\Pi_3 = -A^{-1} (\delta_1 \Gamma_3 + \delta_0). \tag{2.46}$$

Applying C to (2.46) and using (2.44) we have

$$0 = C\Pi_3 = C\left(-A^{-1}\delta_1\right) \Gamma_3 - C\left(A^{-1}\delta_0\right),$$

or

$$\Gamma_3 = \frac{C\left(A^{-1}\delta_0\right)}{C\left(-A^{-1}\delta_1\right)}. \tag{2.47}$$

Note that the denominator is actually $G(0)$ where

$$G(s) = C\left(sI - A\right)^{-1} B_{\mathrm{in}}, \quad s \in \rho(A),$$

is the transfer function for the system (A, B_{in}, C). We note, in this calculation, that 0 (an eigenvalue of S) is not a transmission zero of the closed loop system. Note that the evaluation of Γ_3 in Equation (2.47) is exactly the same as its corresponding bounded input/output counterpart previously obtained in Equation (1.25), where $\delta_1 = B_{\mathrm{in}}$ and $\delta_0 = B_{\mathrm{d}}$.

It is easy to compute directly (or as a simple calculation using (2.39) that

$$G(0) = \frac{1 + kx_1}{k}. \tag{2.48}$$

A straightforward calculation shows that for every $x_0 \in [0, 1]$,

$$\left(A^{-1}\delta_{x_0}\right)(x) = \begin{cases} -\dfrac{(1 + kx_0)}{k}, & x > x_0, \\[2ex] -\dfrac{(1 + kx)}{k}, & x < x_0. \end{cases} \tag{2.49}$$

Thus we see that

$$\left(A^{-1}\delta_0\right)(x) = -1/k.$$

From this and from Equation (2.49) we find that

$$\Gamma_3 = \frac{C\left(A^{-1}\delta_0\right)}{C\left(-A^{-1}\delta_1\right)} = -\frac{1}{1 + kx_1}.$$

Next we turn to the calculation of Γ_1 and Γ_2 from (2.42). It is easy to see that this equation can be written as a coupled system of two-point boundary value problems with a pair of side constraints given by the conditions imposed by the second regulator equation (2.44). Namely we have

$$A\Pi_1 + \alpha\Pi_2 = -\delta_1\Gamma_1,$$
$$A\Pi_2 - \alpha\Pi_1 = -\delta_1\Gamma_2.$$

This equation can be solved, by completely elementary means, exactly as in the corresponding bounded input/output counterpart discussed in detail in Section 1.4.3. We obtain

$$\Gamma_1 = \text{Re } (G_k^{-1}(i\alpha)), \qquad (2.50)$$

$$\Gamma_2 = \text{Im } (G_k^{-1}(i\alpha)), \qquad (2.51)$$

$$\Gamma_3 = \frac{1}{1 + kx_1}, \qquad (2.52)$$

where, from (2.39), we have

$$G_k(i\alpha)) = \frac{\cosh(\sqrt{i\alpha}x_1) + k\sinh(\sqrt{i\alpha}x_1)/\sqrt{i\alpha}}{(\sqrt{i\alpha}\sinh(\sqrt{i\alpha}) + k\cosh(\sqrt{i\alpha}))}.$$

For the numerical simulation we have taken as initial condition $z_0(x) = \cos(\pi x)$, $k = 1$, $d(t) = 1$ and signal to be tracked $y_r(t) = \sin(2t)$.

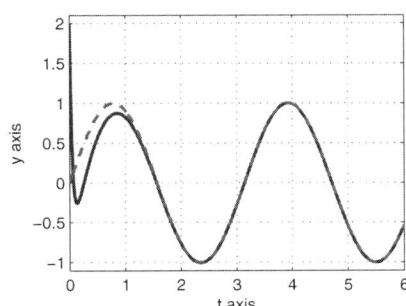

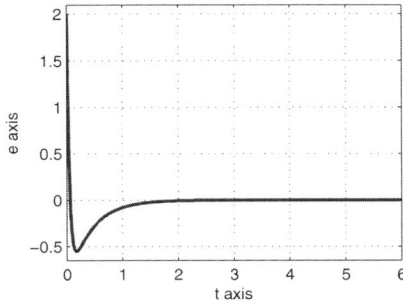

Fig. 2.1: $y(t)$ and $y_r(t)$.   **Fig. 2.2**: Error $e(t)$.

Fig. 2.1 depicts the reference signal $y_r(t)$ and the controlled output $y(t) = Cz$. Fig. 2.2 depicts the error $e(t) = y(t) - y_r(t)$, and finally Fig. 2.3 contains the numerical solution for the displacement z for $x \in [0,1]$ and $t \in [0,6]$.

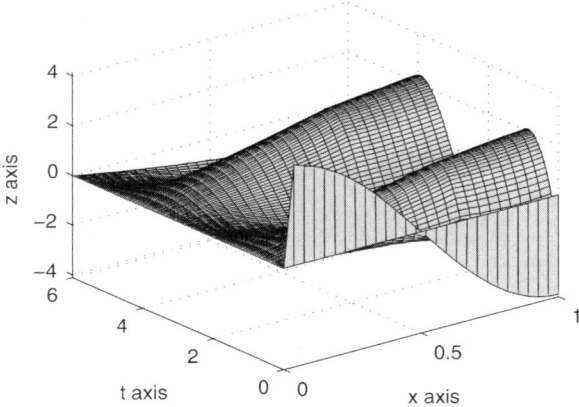

Fig. 2.3: Plot of Solution Surface.

Example 2.3. This example is related to the previous one, however, in this case we want to track a continuous function that can be approximated by a Fourier series. In order to apply the geometric theory with a finite-dimensional exosystem we truncate the Fourier series expansion to obtain an approximate reference signal. In this example we consider the periodic triangle function defined to be $f(t) = |t|$ on the interval $-1 \leq t < 1$ and then extended periodically to the whole real line. The Fourier series expansion for $f(t)$ is given by

$$f(t) = \frac{1}{2} - \frac{4}{\pi^2} \sum_{j=1}^{\infty} \frac{\cos((2j-1)\pi t)}{(2j-1)^2}.$$

We take $y_r(t)$ to be a finite truncation of this infinite sum

$$y_r(t) = f_N(t) = \frac{1}{2} - \frac{4}{\pi^2} \sum_{j=1}^{N} \frac{\cos((2j-1)\pi t)}{(2j-1)^2}. \tag{2.53}$$

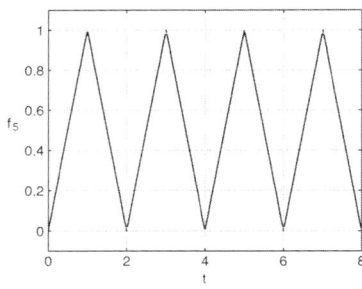

Fig. 2.4: Plot of $f(t)$ and $f_5(t)$.

In Fig. 2.4 we have plotted both $f(t)$ and its approximation $f_5(t)$. Note that, already for $N = 5$, the two functions are visually indistinguishable.

For this example we do not consider any disturbance, i.e., $d(t) = 0$. From Equation (2.53) we seek a solution in the form $z = \Pi w$, where $w \in \mathbb{R}^{(1+2N)}$ is given by

$$w = [w_0, \ w^1, \ w^2, \cdots, \ w^N]^\mathsf{T}.$$

Here $w_0 \in \mathbb{R}$ is responsible for the set-point tracking, i.e.,

$$w_0 = \frac{1}{2},$$

while $w^j \in \mathbb{R}^2$ is responsible for the harmonic tracking of the j−component in the sum, i.e.,

$$\begin{cases} w^j(t) = [w_{(2j-1)}(t), \ w_{(2j)}(t)]^\mathsf{T} = \left[-\dfrac{4}{\alpha_j^2}\sin(\alpha_j t), \ -\dfrac{4}{\alpha_j^2}\cos(\alpha_j t)\right]^\mathsf{T}, \\ \alpha_j = (2j-1)\pi, \quad j = 1, \cdots, N. \end{cases}$$

The exosystem satisfies

$$\frac{dw}{dt} = Sw,$$

with initial conditions

$$w_0(0) = \frac{1}{2}, \quad \text{and} \quad w^j(0) = \left[0, \ -\frac{4}{\alpha_j^2}\right]^\mathsf{T}, \quad j = 1, \cdots, N.$$

Here S is the $(1 + 2N) \times (1 + 2N)$ block diagonal matrix defined by

$$S = \operatorname{diag}[S_0, S^1, S^2, \cdots, S^N],$$

where $S_0 = 0$ and each S^j is a 2×2 harmonic oscillator matrix given in Equation (2.25).

With this the signal to be tracked can be written as

$$y_r = Qw = [1, Q^1, Q^2, \cdots, Q^N]\, w, \quad \text{with } Q^j = [0, 1]\,, j = 1, \cdots, N.$$

We have to find a control in the form $u = \Gamma w$, where

$$\Gamma = [\Gamma_0, \Gamma^1, \Gamma^2, \cdots \Gamma^N], \quad \text{with } \Gamma_0 \in \mathbb{R} \text{ and } \Gamma^j \in \mathbb{R}^2, \ j = 1, \cdots N.$$

For the control system we consider a one-dimensional heat equation on a finite interval $[0, 1]$ with a Neumann boundary control $u(t)$ at the right end point $(x = 1)$ and homogeneous Robin boundary condition at the left end point $(x = 0)$

$$z_t(x, t) = z_{xx}x(x, t), \quad x \in (0, 1),$$
$$z(x, 0) = z_0(x),$$

$$\mathcal{B}_0(z)(t) = -z_x(0,t) + kz(0,t) = 0, \quad k > 0,$$
$$\mathcal{B}_1(z)(t) = z_x(1,t) = u(t).$$

We also assume that the output $y(t)$ is given as a point evaluation of the state $z(x,t)$ at a prescribed point $x_1 \in [0,1]$ in the domain

$$y(t) = C(z)(t) = z(x_1,t), \quad x_1 \in [0,1].$$

To evaluate the set point tracking control Γ_0 and harmonic tracking controls Γ^j we use Equation (2.48) and Equations (2.50)–(2.52), respectively, derived in the previous example. This yields

$$\Gamma_0 = \frac{1}{G_k(0)} = \frac{k}{1 + kx_1}$$

and

$$\Gamma_1^j = -\text{Im}\,(G_k(i\alpha)^{-1}), \tag{2.54}$$

$$\Gamma_2^j = \text{Re}\,(G_k(i\alpha)^{-1}), \tag{2.55}$$

for $j = 1, \cdots, N$, where, from (2.39), we have

$$G(i\alpha)) = \frac{\cosh(\sqrt{i\alpha}x_1) + k\sinh(\sqrt{i\alpha}x_1)/\sqrt{i\alpha}}{(\sqrt{i\alpha}\sinh(\sqrt{i\alpha}) + k\cosh(\sqrt{i\alpha}))}.$$

Notice that in this example we track cos-functions rather then sin-functions; this explains the different form of Equations (2.54)–(2.55) with respect to Equations (2.50)–(2.51).

With $k = 1$, $N = 5$ and $x_1 = .25$ we obtain

$$\Gamma = \big[0.8,\ [-3.5789,\ -0.5238],\ [-7.8138,\ -10.7485],$$
$$[-3.5284,\ -29.2650],\ [14.1409,\ -52.2916],\ [48.8700,\ -74.3707]\big].$$

We chose as initial date $\varphi(x) = \cos(\pi x)$. Fig. 2.5 depicts the reference signal $y_r(t)$ and the controlled output $y(t) = Cz$. Fig. 2.6 depicts the error $e(t) = y(t) - y_r(t)$, Fig. 2.7 depicts the control $u = \Gamma w$, and finally Fig. 2.8 contains the numerical solution for the displacement z for $x \in [0,1]$ and $t \in [0,8]$.

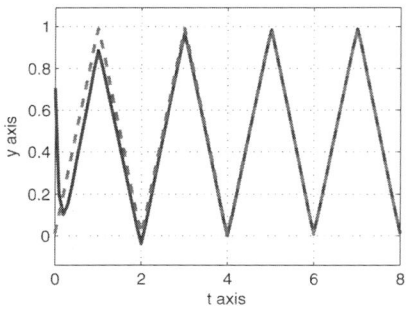

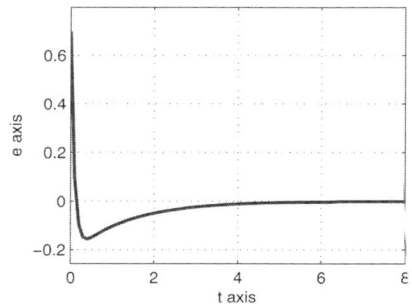

Fig. 2.5: $y(t)$ and $y_r(t)$.   **Fig. 2.6**: Plot of Error $e(t)$.

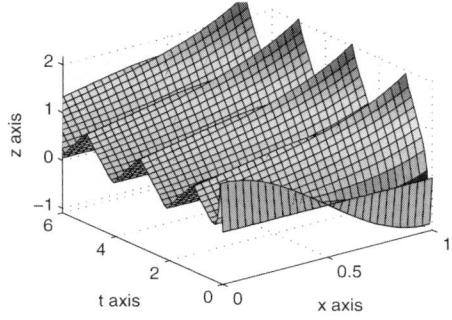

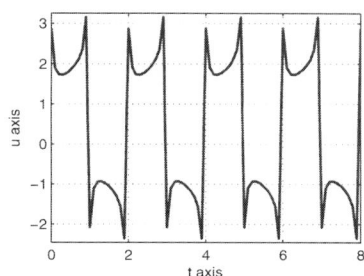

Fig. 2.7: Plot of Solution Surface.   **Fig. 2.8**: Plot of $u = \Gamma w$.

2.3.1 Limits of Bounded Input and Output Operators

Next we want to draw attention to the relation between the examples in Chapter 1 and the analog of those examples within the context of unbounded actuation and observation. In particular in Chapter 1 we gave examples of bounded output and input operators in (1.9) and (1.11). The SISO examples given there are not most general. But they do represent simple examples of functions referred to as "Delta Sequences" [83]. These are also not the most general such sequences but they serve the purpose that we intend. We want to consider control inputs and measured outputs defined in terms of a limit of these Delta Sequences.

Consider the sequence of functions centered at a point $x_1 \in (a, b)$ for some interval (a, b)

$$\Psi_n(x) = \left(\frac{n}{2}\right) \mathbf{1}_{(x_1 - 1/n, \, x_1 + 1/n)}(x),$$

$$y(t) = (Cz)(t) = \lim_{n \to \infty} \int_a^b z(x, t) \Psi_n(x) \, dx. \qquad (2.56)$$

We note that (see [83], page 110)

$$\Psi_n(x) \geq 0 \text{ for all } x \in (a, b), \quad \text{and} \quad \int_a^b \Psi_n(x) \, dx = 1 \text{ for sufficiently large } n.$$

From this we can conclude that the sequence $\Psi_n(x)$ is a delta sequence and for any continuous function φ we have

$$\lim_{n \to \infty} \int_a^b \varphi(x) \Psi_n(x) \, dx = \varphi(x_1).$$

In particular for a smooth solution $z(x, t)$ the limit (2.56) gives

$$y(t) = (Cz)(t) = z(x_1, t)$$

which is an unbounded functional in $\mathcal{Z}$, i.e., it is not continuous. In particular this represents the action of the delta distributional function supported at x_1. Fortunately, the solutions z of parabolic initial boundary value problems such as we have here are continuous so that $z(\cdot, t) \in \mathcal{D}(C)$ for all $t > 0$.

For the input operators given in terms of the expressions in (1.11) in Chapter 1, if we pass to the limit we obtain an initial boundary value problem in the state space $H^{-1}(0, 1)$.

Namely, if the sequences of functions $(B_{\text{in}})_n$ and $(B_{\text{d}})_n$ in $\mathcal{Z}$ are given by expressions that define a delta sequence then in the limit we obtain a delta function. For example if Φ_n has the form

$$(B_{\text{in}})_n(x) = \left(\frac{n}{2}\right) \mathbf{1}_{(x_0 - 1/n, x_0 + 1/n)}(x)$$

then for any continuous function (and therefore for any test function) we have (as above)

$$\lim_{n \to \infty} \langle (B_{\text{in}})_n, \varphi \rangle = \varphi(x_0)$$

which implies that the heat problem with any standard BCs

$$z_t = z_{xx} + (B_{\text{in}})_n u,$$

for any test function η takes the form

$$\langle z_t, \eta \rangle = -\langle z_x, \eta_x \rangle + \langle (B_{\text{in}})_n u, \eta \rangle.$$

Passage to the limit as n tends to infinity gives

$$\langle z_t, \eta \rangle = -\langle z_x, \eta_x \rangle + \eta(x_0) u,$$

or

$$\langle z_t, \eta \rangle = -\langle z_x, \eta_x \rangle + \langle \delta_{x_0}, \eta \rangle u.$$

In other words, for any standard boundary operators $\mathcal{B}_{\text{in}}^0$, $\mathcal{B}_{\text{in}}^1$ (for example, Dirichlet, Neumann or Robin) we obtain a boundary control system

$$z_t = z_{xx} + \delta_{x_0} u,$$
$$\mathcal{B}_{\text{in}}^0(z) = 0, \quad \mathcal{B}_{\text{in}}^1(z) = 0,$$
$$z(x, 0) = z_0(x),$$

which can be handled using the techniques of this chapter.

On the other hand, for computational purposes it is desirable to formulate the problem in the standard L^2 state space rather than in a space of distributions. This can be accomplished as follows. For a function φ defined on $(0, 1)$ and a point $x_0 \in (0, 1)$ we define

$$[\varphi]_{x_0} = \varphi(x_0^+) - \varphi(x_0^-),$$

where the notation $x^{\pm}$ corresponds to the usual meaning of left and right hand limits. With this notation the problem can be recast as the following:

$$z_t = z_{xx},$$
$$\mathcal{B}_0(z) = 0, \quad \mathcal{B}_1(z) = 0,$$
$$\mathcal{B}_{x_0}(z) = \left\{ \begin{array}{l} [z]_{x_0} = 0, \\ [z_x]_{x_0} = u(t), \end{array} \right.$$
$$z(x,0) = z_0(x),$$
$$y(t) = C(z)(t) = z(x_1, t).$$

Here the control enters as a jump in the flux at a point, which is equivalent to a delta source but this formulation allows for a simpler numerical solution.

In the next subsection we consider the analog of the heat problems considered in Chapter 1 when the bounded input and output operators are converged to the appropriate delta distributional limits.

2.3.2 Revisited Examples from Chapter 1

In light of the discussion in Section 2.3.1 we consider some unbounded analogs of the examples studied in Chapter 1, Sections 1.4 and 1.5.

Example 2.4 (Damped Wave Equation). Consider a controlled one-dimensional damped wave equation governing the small vibrations of a finite string

$$z_{tt} = z_{xx} - 2\gamma z_t, \tag{2.57}$$
$$z(0,t) = 0,$$
$$\mathcal{B}_d = z_x(1,t) + k_1 z(x,t) = d(t),$$
$$\mathcal{B}_{x_{in}}(z) = \left\{ \begin{array}{l} [z]_{x_{in}} = 0, \\ [z_x]_{x_{in}} = u(t), \end{array} \right.$$
$$y(t) = Cz(t) = z(x_{out}, t), \tag{2.58}$$
$$z(0) = \phi \in H^2(0,1),$$
$$z_t(0) = \psi \in L^2(0,1).$$

In the above system the operator $\mathcal{B}_{in}(z)$ is the condition of jump in the derivative for the solution z at the point $x_{in} \in (0,1)$, the disturbance operator $\mathcal{B}_d$ is given by a Robin boundary condition on the right extrema of the domain and the output operator C corresponds to a point evaluation inside the domain at $x_{out} \in (0,1)$.

As pointed out before the control input operator $\mathcal{B}_{x_{in}}(z)$ is equivalent to applying a point source force at x_{in}

$$B_{in} u(t) = \delta_{x_{in}} u(t).$$

Note that in this example the control, the disturbance and the output operators are all unbounded.

Our design objective is to track a prescribed constant signal $y_r(t) = M_r$ while rejecting a sinusoidal disturbance $d(t) = A_d \sin(\beta t)$.

At this point we rewrite the system (2.57)–(2.58) in its equivalent operator form as

$$z_{tt} = Az - 2\gamma z_t + B_{\mathrm d}d(t) + B_{\mathrm{in}}u(t), \qquad (2.59)$$

$$y(t) = Cz(t), \qquad (2.60)$$

where to the operator $A = d^2/dx^2$ we associate homogeneous Dirichlet boundary condition at $x = 0$ and homogeneous Robin boundary condition at $x = 1$.

The system (2.59)–(2.60) is now formally identical to the one described in Chapter 1. Thus to find the desired control $u(t) = \Gamma w$ we follow the description given in Chapter 1 to find

$$w = \begin{bmatrix} w_1 \\ w^\beta \end{bmatrix} = \begin{bmatrix} w_1 \\ w_2 \\ w_3 \end{bmatrix} = \begin{bmatrix} M_r \\ A_d \sin(\beta t) \\ A_d \cos(\beta t) \end{bmatrix}.$$

Setting

$$R_\beta = \mathrm{Re}\ (G(i\beta)), \quad I_\beta = \mathrm{Im}\ (G(i\beta)),$$

$$R_{\mathrm d,\beta} = \mathrm{Re}\ (G_{B_{\mathrm d}}(i\beta)), \quad I_{\mathrm d,\beta} = \mathrm{Im}\ (G_{B_{\mathrm d}}(i\beta)),$$

we can write

$$\Gamma = [\Gamma_1, \Gamma^\beta] = [\Gamma_1, [\Gamma_2, \Gamma_3]]$$

$$= \left[G(0)^{-1}, \left[\left(-R_\beta^{-1}\ R_{\mathrm d,\beta} + I_\beta^{-1}\ I_{\mathrm d,\beta} \right), \left(-R_\beta^{-1}\ I_{\mathrm d,\beta} - I_\beta^{-1}\ R_{\mathrm d,\beta} \right) \right] \right].$$

In our numerical example we have set $x_{\mathrm{out}} = 3/4$, $x_{\mathrm{in}} = 0.5$, $M_r = 1$, $A_d = 0.25$, $\beta = 0.5$, $\gamma = .5$, $k_1 = 1$ and chose initial conditions $\phi(x) = 16x^2(1-x)^2$ and $\psi = 0.5\sin^2(\pi x)$. For these values the numerical solution of the regulator equations produces the approximate control gains

$$\Gamma = [-3.2,\ -1.1927,\ -0.01526].$$

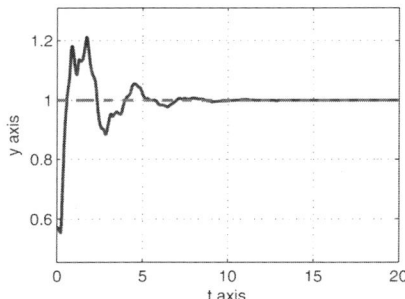

Fig. 2.9: $y(t)$ and $y_r(t)$.

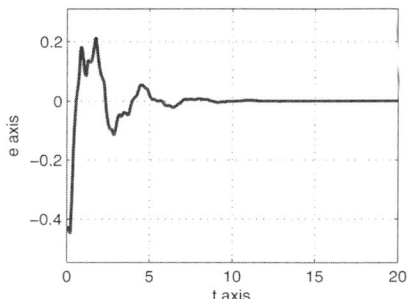

Fig. 2.10: Error $e(t)$.

Fig. 2.9 depicts the reference signal $y_r(t) = 1$ and the controlled output $y(t) = Cz$ for the closed loop system (1.35). Fig. 2.10 depicts the error $e(t) = y(t) - y_r(t)$, and finally Fig. 2.11 contains the numerical solution for the displacement z for $x \in [0, 1]$ and $t \in [0, 20]$.

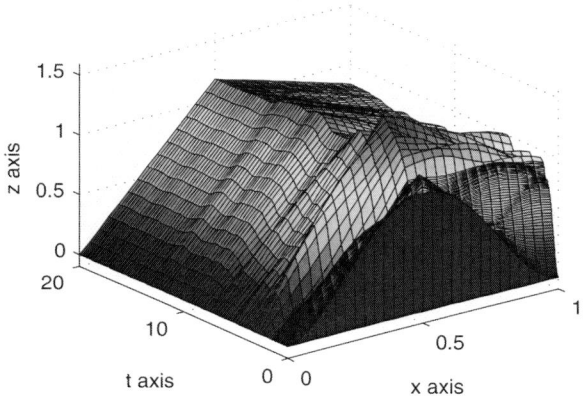

Fig. 2.11: Solution Surface $z(x, t)$.

Example 2.5 (MIMO Numerical Solution for the 1D Heat Equation). Consider the one-dimensional MIMO heat equation governed by

$$z_t(x, t) = z_{xx}(x, t), \tag{2.61}$$

$$- z_x(0, t) + k_0 z(0, t) = 0, \quad k_0 > 0, \tag{2.62}$$

$$z_x(L, t) + k_1 z(L, t) = 0, \quad k_1 > 0, \tag{2.63}$$

$$\mathcal{B}_{in}^1(z) = \begin{cases} [z]_{x_1} = 0, \\ [z_x]_{x_1} = u_1(t), \end{cases} \tag{2.64}$$

$$\mathcal{B}_{in}^2(z) = \begin{cases} [z]_{x_3} = 0, \\ [z_x]_{x_3} = u_2(t), \end{cases} \tag{2.65}$$

$$\mathcal{B}_d(z) = \begin{cases} [z]_{x_5} = 0, \\ [z_x]_{x_5} = d(t), \end{cases} \tag{2.66}$$

$$y_1 = C_1 z = z(x_2, t), \quad y_2 = C_2 z = z(x_4, t), \tag{2.67}$$

$$z(x, 0) = \varphi(x). \tag{2.68}$$

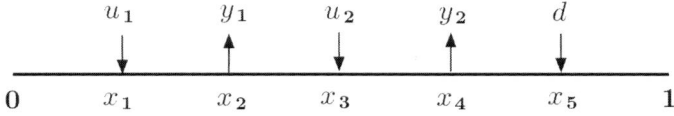

Fig. 2.12: One-Dimensional Rod.

In the above system the operators $\mathcal{B}_{\text{in}}^1(\cdot)$, $\mathcal{B}_{\text{in}}^2(\cdot)$, $\mathcal{B}_d(\cdot)$ correspond to the conditions of jump in the derivative of the solution at the points x_1, x_3, x_5, respectively, and are equivalent to the point sources $B_{\text{in}}^1 = \delta_{x_1}$, $B_{\text{in}}^2 = \delta_{x_2}$ and $B_d = \delta_{x_5}$ at the same points. The output operators C_1 and C_2 correspond to point evaluations respectively at x_2 and x_4.

With this the control system (2.61)–(2.68) can be rewritten in the standard state space form as

$$z_t = Az + B_d d(t) + B_{\text{in}}^1 u_1(t) + B_{\text{in}}^2 u_2(t), \tag{2.69}$$

$$y_1 = C_1 z, \quad y_2 = C_2 z, \tag{2.70}$$

where we associated to the operator $A = d^2/dx^2$ homogeneous Robin boundary conditions at both ends.

Our control objective is to find two controls u_1 and u_2 so that the outputs y_1 and y_2 track the reference signals

$$y_{rj}(t) = M_{rj} + A_{rj}\sin(\alpha_j t) \quad j = 1, 2,$$

while rejecting the disturbance

$$d(t) = M_d + A_d\sin(\beta t).$$

The system (2.69)–(2.70) is now formally identical to the one described in Example 2.5.1 for bounded operators, and to find the desired control $u(t) = \Gamma w$ we follow step by step the solution strategy described in detail in Section 1.5.2.

For this example we have chosen $L = 1.2$, $x_i = 0.2 \times i$, $\alpha_1 = 1$, $\alpha_2 = 2$, $\beta = 0.5$, $M_{r1} = 0.5$, $M_{r2} = 0.25$, $M_d = 1$, $A_{r1} = 0.25$, $A_{r2} = 0.5$, $A_d = .75$, $k_0 = k_1 = 1$ and initial data $z(x,0) = \cos\left(\dfrac{\pi x}{L}\right)$. For these values the computed control gains are

$$\Gamma = [[-6.66667, \ 5], \ [6.66667, \ -6.42857], \ [1, \ -1.71429],$$
$$[-6.66075, \ 4.99984], \ [0.73703, \ 0.03333], \ [6.64299, \ -6.46022],$$
$$[-1.47404, \ 1.97131], \ [1, 1.71368], \ [4.93217e - 14, \ -0.02311]].$$

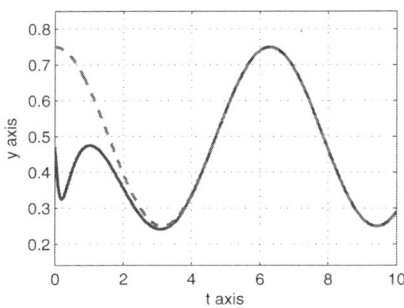

Fig. 2.13: $y_1(t)$ and $y_{r1}(t)$.

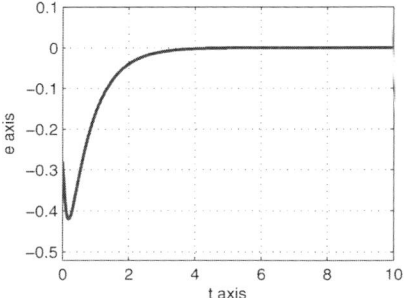

Fig. 2.14: Error $e_1(t)$.

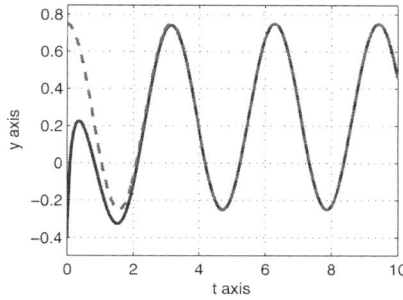

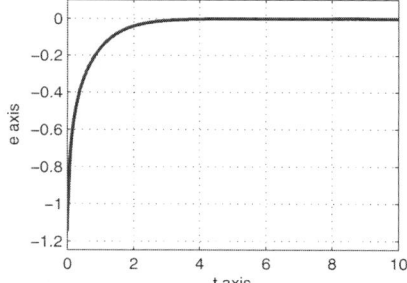

Fig. 2.15: $y_2(t)$ and $y_{r2}(t)$.   **Fig. 2.16**: Error $e_2(t)$.

Fig. 2.13 depicts the reference signal $y_{r1}(t)$ and the controlled output $y_1(t)$, while Fig. 2.14 depicts the error $e_1(t) = y_1(t) - y_{r1}(t)$. Fig. 2.15 depicts the reference signal $y_{r2}(t)$ and the controlled output $y_2(t)$, while Fig. 2.16 depicts the error $e_2(t) = y_2(t) - y_{r2}(t)$. Finally, Fig. 2.17 contains the numerical solution for the displacement z for $x \in [0, L]$ and $t \in [0, 20]$.

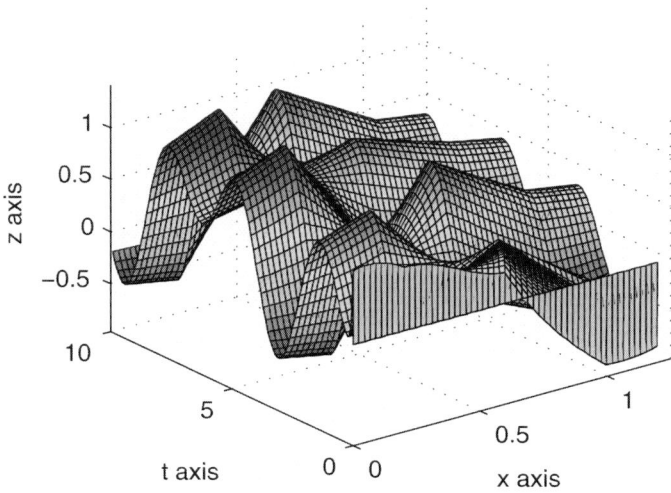

Fig. 2.17: Plot of Solution Surface.

Chapter 3

Examples Linear Regulation

3.1   Introduction

In this chapter we present a variety of examples that are more interesting and somewhat more challenging than the motivating examples given in Chapters 1 and 2. All the examples have been solved using the finite element package COMSOL Multiphysics [65], although we note that the numerical solution could have been found by using any alternative PDE package solver. Our choice of using COMSOL is motivated by its flexibility for solving coupled multiphysics problems.

In order for the reader to become familiar with the strengths and limitations of the algorithms, described in Chapters 1 and 2, to solve the regulator equations we present several examples. Each of these examples introduces new features and a detailed description that will hopefully allow readers to incorporate the ideas into their own research.

The first three examples deal with control problems in structural mechanics. In Section 3.2 we describe an example of "Harmonic Tracking for a Coupled Wave Equation" in which the control force is acting at the free end of a one-dimensional string. The desired output not only has to track a reference signal but also has to satisfy an extra ODE constraint. In the next example in Section 3.3 we present a regulation problem for "Control of a Damped Rayleigh Beam". This example features higher-order derivatives, not only in the fourth-order reaction term, but also in the mixed inertial and damping terms. For this example, even the output operator, given by the angular velocity at one point, is represented by a mixed second-order derivative. In the example in Section 3.4 we describe the "Vibration Regulation of a 2D Plate". This is the first example in the book where a two-dimensional problem is solved, and features a fourth-order reaction term and MIMO regulation problem.

The next three examples deal with control problems in computational fluid dynamics (CFD). In Section 3.5 we describe the "Control of a Linearized Stokes Flow in 2 Dimensions". In this example we show that the theory developed in the previous chapters also works for the linearized Stokes' operator, where the convective nonlinear term is neglected but the continuity incompressible constraint has to be verified. In Sections 3.6 and 3.7 we describe problems of "Thermal Control of a 2D Fluid Flow" and "Thermal Regulation in a 3D Room", respectively. These are sophisticated examples of thermodynamic regulation, where the fluid velocity is first found by solving the nonlinear Navier-Stokes equations and then this fluid velocity is used in the temperature equation. The example in Section 3.6 features a MIMO regulation problem, while the example in Section 3.7 features a regulation problem in a three-dimensional domain.

In Section 3.8 we describe a general control strategy for the design of control laws "Using Fourier Series for Tracking Periodic Signals". In this section we show that, even if the signals to be tracked or rejected are generated by an infinite-dimensional exosystem consisting of an infinite set of harmonic oscillators producing signals with infinite Fourier expansion, by truncating the Fourier series representation of the signals it is possible to find an accurate approximation of the desired controls. We also show that it is possible to build an algorithm to automate the whole procedure.

In Section 3.9 we introduce the "Zero Dynamics Inverse Design", where it is assumed that the output and input operators are co-located but can now be allowed to be infinite-dimensional in both time and space. Two examples are given to demonstrate the features of the method.

3.2 Harmonic Tracking for a Coupled Wave Equation

We consider a vertical string whose displacement is described by the one-dimensional wave equation on the domain $(0, \pi)$. The upper end (corresponding to $x = \pi$) is kept fixed and an object of mass M is attached at the lower end (corresponding to $x = 0$). The external input is a vertical force v acting on the object at $x = 0$. We also consider damping mechanisms for both the spring and the boundary control. The resulting system is described by the following equations for all $x \in (0, \pi)$ and for all $t \in (0, \infty)$:

$$q_{tt}(x,t) - (a(x)q_x)_x(x,t) + \beta q_t(x,t) = 0, \tag{3.1}$$

$$q(\pi, t) = 0, \tag{3.2}$$

$$- a(0)q_x(0,t) = -Mq_{tt}(0,t) - \beta q_t(0,t) + u(t), \tag{3.3}$$

$$q(x,0) = 0, \quad q_t(x,0) = 0, \tag{3.4}$$

$$y(t) = q_x(0,t). \tag{3.5}$$

Here $y(t)$ represents the desired measured output, $a(x)$ the spring stiffness and β the damping coefficient. An analogous problem without damping has been analyzed in detail in [84], where it was shown to be a regular linear system. Our control problem consists in finding $u(t)$ so that

$$y(t) \to y_r = A_r \sin(\alpha t) \quad \text{as } t \to \infty.$$

We proceed exactly as described in Chapters 1 and 2 for the harmonic oscillator case. Let $w = [w_1, w_2]^\mathsf{T} = [M_r \sin t, M_r \cos t]^\mathsf{T}$ be the solution of the exosystem

$$w_t = Sw = \begin{bmatrix} 0 & \alpha \\ -\alpha & 0 \end{bmatrix} w, \quad w(0) = \begin{bmatrix} 0 \\ M_r \end{bmatrix}.$$

Clearly, $w_{tt} = S^2 w = -\alpha^2 w$. Let $\Pi = [\Pi_1, \Pi_2]$, $\Gamma = [\Gamma_1, \Gamma_2]$.

In Equations (3.1)–(3.5) substitute $q(x,t) = \Pi(x)w(t)$ and $u(t) = \Gamma w$, and, in (3.5), we replace $y(t)$ with $y_r(t) = w_1$ corresponding to the second regulator equation condition, then it follows

$$-\alpha^2 \Pi w - \frac{d}{dx}\left(a(x)\frac{d}{dx}\Pi w\right) + \beta \Pi S w = 0, \tag{3.6}$$

$$\Pi(\pi)w = 0, \tag{3.7}$$

$$-a(0)\frac{d\Pi(0)}{dx}w - \alpha^2 M\Pi(0)w = -\beta\Pi(0)Sw + \Gamma w, \tag{3.8}$$

$$\frac{d\Pi(0)}{dx}w = C\Pi w = w_1. \tag{3.9}$$

Let us define the operator A by

$$A\varphi = \alpha^2\varphi + \frac{d}{dx}\left(a(x)\frac{d\varphi}{dx}\right),$$

with associated homogeneous boundary conditions given by

$$-a(0)\frac{d\varphi(0)}{dx}w - \alpha^2 M\varphi(0)w = 0, \quad \varphi(\pi) = 0.$$

Using this definition of A in equations (3.6)–(3.9), we derive the two regulator equations in the usual operator form

$$A\begin{bmatrix}\Pi_1\\\Pi_2\end{bmatrix} - \beta\begin{bmatrix}-\alpha\Pi_2\\\alpha\Pi_1\end{bmatrix} + B_{\text{in}}\left(\beta\begin{bmatrix}\alpha\Pi_2(0)\\-\alpha\Pi_1(0)\end{bmatrix} + \begin{bmatrix}\Gamma_1\\\Gamma_2\end{bmatrix}\right) = 0, \tag{3.10}$$

$$C\begin{bmatrix}\Pi_1\\\Pi_2\end{bmatrix} = Q = \begin{bmatrix}1\\0\end{bmatrix}, \tag{3.11}$$

where we have already dropped the dependence on w, and where B_{in} is the unbounded operator corresponding to the boundary condition (3.8). Note that the equation system (3.10)–(3.11) has been written in column-wise rather than row-wise form.

Now we proceed as usual. First we solve the system

$$AX + B_{\text{in}} = 0, \tag{3.12}$$

and set

$$G = CX = -CA^{-1}B_{\text{in}}. \tag{3.13}$$

Next we introduce the auxiliary state variable $\widetilde{\Pi}$ satisfying

$$A\begin{bmatrix}\widetilde{\Pi}_1\\\widetilde{\Pi}_2\end{bmatrix} - \beta\begin{bmatrix}-\alpha\Pi_2\\\alpha\Pi_1\end{bmatrix} + B_{\text{in}}\left(\beta\begin{bmatrix}\alpha\Pi_2(0)\\-\alpha\Pi_1(0)\end{bmatrix}\right) = 0. \tag{3.14}$$

Combining the first regulator equation (Equation (3.10)), the definition of X (Equation (3.12)) and the definition of $\widetilde{\Pi}$ (Equation (3.14)) yields

$$\begin{bmatrix} \Pi_1 \\ \Pi_2 \end{bmatrix} = -A^{-1} \left(-\beta \begin{bmatrix} -\alpha\Pi_2 \\ \alpha\Pi_1 \end{bmatrix} + B_{\text{in}} \left(\beta \begin{bmatrix} \alpha\Pi_2(0) \\ -\alpha\Pi_1(0) \end{bmatrix} \right) \right) - A^{-1} B_{\text{in}} \begin{bmatrix} \Gamma_1 \\ \Gamma_2 \end{bmatrix}$$

$$= \begin{bmatrix} \widetilde{\Pi}_1 \\ \widetilde{\Pi}_2 \end{bmatrix} + X \begin{bmatrix} \Gamma_1 \\ \Gamma_2 \end{bmatrix}.$$

Finally, using this last result together with the second regulator equation (Equation (3.11)) and the definition of G (Equation (3.13)), it follows that

$$C \begin{bmatrix} \Pi_1 \\ \Pi_2 \end{bmatrix} = C \begin{bmatrix} \widetilde{\Pi}_1 \\ \widetilde{\Pi}_2 \end{bmatrix} + CX \begin{bmatrix} \Gamma_1 \\ \Gamma_2 \end{bmatrix} = C \begin{bmatrix} \widetilde{\Pi}_1 \\ \widetilde{\Pi}_2 \end{bmatrix} + G \begin{bmatrix} \Gamma_1 \\ \Gamma_2 \end{bmatrix} = \begin{bmatrix} 1 \\ 0 \end{bmatrix},$$

or

$$\begin{bmatrix} \Gamma_1 \\ \Gamma_2 \end{bmatrix} = G^{-1} \begin{bmatrix} 1 - C\widetilde{\Pi}_1 \\ 0 - C\widetilde{\Pi}_2 \end{bmatrix}. \tag{3.15}$$

To solve for Γ, we need to solve the algebro-differential system consisting of equations (3.10), (3.14) and (3.15). Once Γ is found we solve the closed loop system (3.1)–(3.5) with $u = \Gamma_1 w_1 + \Gamma_2 w_2$.

In a specific computational example we set the reference signal to be $y_r = M_r \sin(\alpha t) = \sin(t)$ and choose the remaining parameters as

$$\beta = 1, \quad a(x) = x^2 + 1, \quad M = 0.1.$$

With these values, the numerically evaluated gains for the controller Γ are given by

$$\Gamma = [\Gamma_1, \Gamma_2] = [-2.00716, \, -0.62575].$$

Then we solve the closed loop system for $t \in (0, 6\pi)$.

Fig. 3.1 depicts the reference signal $y_r(t)$, and the controlled output $y(t) = Cq$ for the closed loop system. Fig. 3.2 depicts the error $e(t) = y(t) - y_r(t)$, and finally Fig. 3.3 contains the numerical solution for the displacement q for $x \in (0, \pi)$ and $t \in (0, 6\pi)$.

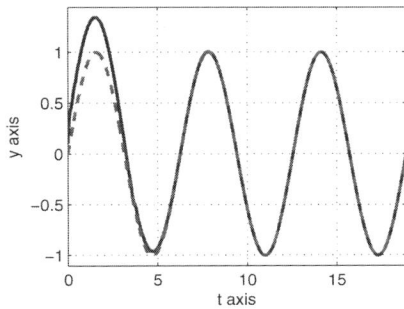

Fig. 3.1: Plot of $y(t)$, $y_r(t)$.

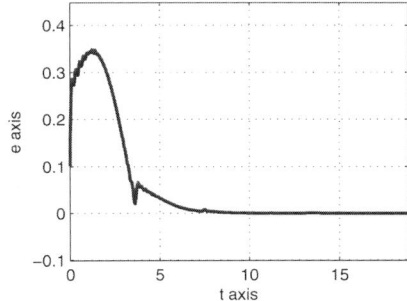

Fig. 3.2: Error $e(t)$.

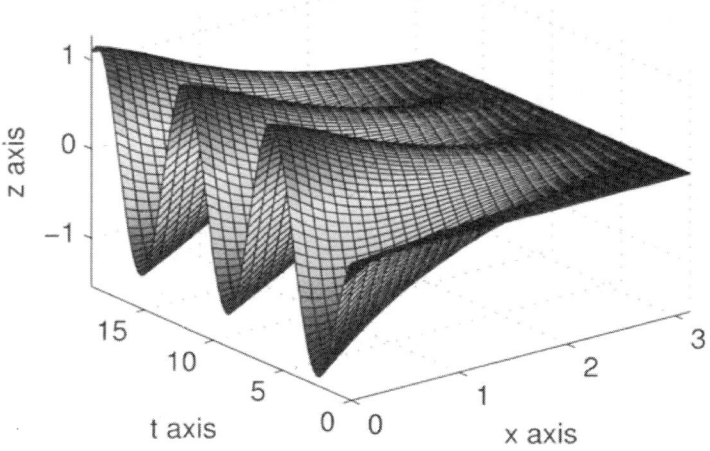

Fig. 3.3: Plot of Solution Surface.

3.3   Control of a Damped Rayleigh Beam

We consider a hinged elastic beam described by a Rayleigh beam equation. This example is adapted from a similar example considered in [89]. Let $q(x, t)$ represent the transverse displacement of the beam at the position $x \in [0, \pi]$ and the time $t \geq 0$. The system is governed by the following equations:

$$q_{tt} - aq_{xxtt} - bq_{xxt} + cq_{xxxx} = 0, \tag{3.16}$$

$$q(0, t) = 0, \quad q(\pi, t) = 0, \tag{3.17}$$

$$-q_{xx}(0, t) = u(t), \quad q_{xx}(\pi, t) = 0, \tag{3.18}$$

$$q(x, 0) = \varphi_1(x), \quad q_t(x, 0) = \varphi_2(x), \tag{3.19}$$

$$y(t) = q_{xt}(0, t). \tag{3.20}$$

Equation (3.16) has been normalized with respect to the inertial term q_{tt}. The positive constants a, b, c multiply the rotational Rayleigh inertial term $(-q_{xxtt})$, the structural rotational damping term $(-q_{xxt}$, see [48]) and the elastic bending reaction term (q_{xxxx}), respectively. Zero displacement boundary conditions are imposed on both ends (Equation (3.17)), while zero torque is imposed only on the right end (Equation (3.18)). For this single input single output boundary control system, the input u is applied as a torque at $x = 0$ (Equation (3.18)), and the measured output is the angular velocity at the same point (Equation (3.20)).

This problem can be written in the (A, B, C) state space form in a, more or less, standard way by introducing new state space variables in vector form with components given as the displacement q and its velocity q_t. For detailed discussion of the lengthy mathematical formulation of this problem in state space form together with a discussion of well-posedness and regularity see the details found in [89]. Such a discussion is not the purpose of this chapter. Rather, here we introduce different state variables (with respect to [89]), that make the derivation of the regulator equations straightforward, and their solution easy to compute.

Let us introduce the notation $(\cdot)_a$ for a new auxiliary state variable such that

$$\varphi_a = -\varphi_{xx}.$$

With this notation, the system (3.16)–(3.20) can be rewritten in its equivalent form as

$$q_{xx} + q_a = 0, \qquad (3.21)$$

$$q_{tt} + a q_{att} + b q_{at} - c q_{axx} = 0, \qquad (3.22)$$

$$q(0, t) = 0, \quad q(\pi, t) = 0, \qquad (3.23)$$

$$q_a(0, t) = u(t), \quad q_a(\pi, t) = 0, \qquad (3.24)$$

$$q(x, 0) = \varphi_1(x), \quad q_t(x, 0) = \varphi_2(x), \qquad (3.25)$$

$$y(t) = q_{xt}(0, t). \qquad (3.26)$$

In this example we consider a harmonic tracking problem in which we want the output $y(t)$ in (3.26) to track a prescribed sinusoidal trajectory $y_r(t) = M_r \sin(\alpha t)$. We want to design a control u so that the error

$$e(t) = y(t) - y_r(t) = q_{xt}(0, t) - M_r \sin(\alpha t) \xrightarrow{t \to \infty} 0.$$

Let $w = [w_1, w_2]^\mathsf{T} = [M_r \sin(\alpha t), M_r \cos(\alpha t)]^\mathsf{T}$ be the solution of the exosystem

$$w_t = Sw = \begin{bmatrix} 0 & \alpha \\ -\alpha & 0 \end{bmatrix} w, \quad w(0) = \begin{bmatrix} 0 \\ M_r \end{bmatrix}.$$

Clearly, just as in the previous example, $w_{tt} = S^2 w = -\alpha^2 w$. Let $\Pi = [\Pi_1, \Pi_2]$, $\Pi_a = [\Pi_{a1}, \Pi_{a2}]$, and $\Gamma = [\Gamma_1, \Gamma_2]$. In Equations (3.21)–(3.26) substitute $q(x, t) = \Pi(x)w(t)$, $q_a(x, t) = \Pi_a(x)w(t)$ and $u(t) = \Gamma w$, and replace $y(t)$ with $y_r(t) = w_1$. Then it follows that

$$\frac{d^2 \Pi}{dx^2} w + \Pi_a w = 0, \qquad (3.27)$$

$$- \alpha^2 \Pi w - a\alpha^2 \Pi_a w + b\Pi_a Sw - c\frac{d^2\Pi_a}{dx^2}w = 0, \tag{3.28}$$

$$\Pi(0)w = 0, \quad \Pi(\pi)w = 0, \tag{3.29}$$

$$\Pi_a(0)w = \Gamma w, \quad \Pi_a(\pi)w = 0, \tag{3.30}$$

$$\frac{d\Pi}{dx}(0)w_t = \frac{d\Pi}{dx}(0)Sw = w_1. \tag{3.31}$$

Let the operator A be defined by

$$A\begin{bmatrix} \varphi \\ \varphi_a \end{bmatrix} = \begin{bmatrix} \dfrac{d^2}{dx^2} & I \\ \alpha^2 & a\alpha^2 I + c\dfrac{d^2}{dx^2} \end{bmatrix} \begin{bmatrix} \varphi \\ \varphi_a \end{bmatrix},$$

with associated homogeneous boundary conditions given by

$$\varphi(0) = 0, \ \varphi(\pi) = 0, \ \varphi_a(0) = 0, \ \varphi_a(\pi) = 0.$$

With this, we can rewrite the system of equations (3.27)–(3.31) in the more compact form

$$0 = A\begin{bmatrix} \Pi \\ \Pi_a \end{bmatrix} w - b\begin{bmatrix} 0 \\ \Pi_a \end{bmatrix} Sw + B_{\text{in}}\begin{bmatrix} 0 \\ \Gamma \end{bmatrix} w,$$

$$\frac{d\Pi}{dx}(0)Sw = C\Pi Sw = [1, 0]w,$$

where B_{in} is the unbounded operator corresponding to the boundary control input (3.30), and the operator C is defined as $C\varphi = d\varphi(0)/dx$. Dropping w and taking into account the definition of S, Π, Π_a and Γ, we get the two regulator equations

$$0 = A\begin{bmatrix} \Pi_1 & \Pi_2 \\ \Pi_{a1} & \Pi_{a2} \end{bmatrix} - b\begin{bmatrix} 0 & 0 \\ -\alpha\Pi_{a2} & \alpha\Pi_{a1} \end{bmatrix} + B_{\text{in}}\begin{bmatrix} 0 & 0 \\ \Gamma_1 & \Gamma_2 \end{bmatrix}, \tag{3.32}$$

$$C[-\alpha\Pi_2, \ \alpha\Pi_1] = [1, 0]. \tag{3.33}$$

From here we formally proceed as usual. We first solve the system

$$A\begin{bmatrix} X \\ X_a \end{bmatrix} + B_{\text{in}}\begin{bmatrix} 0 \\ 1 \end{bmatrix} = 0, \tag{3.34}$$

and set

$$G = CX. \tag{3.35}$$

Let the auxiliary state variables $\widetilde{\Pi}$ and $\widetilde{\Pi}_a$ solve

$$A\begin{bmatrix} \widetilde{\Pi}_1 & \widetilde{\Pi}_2 \\ \widetilde{\Pi}_{a1} & \widetilde{\Pi}_{a2} \end{bmatrix} - b\begin{bmatrix} 0 & 0 \\ -\alpha\Pi_{a2} & \alpha\Pi_{a1} \end{bmatrix} = 0. \tag{3.36}$$

Combining the first regulator equation (Equation (3.32)), the definition of X (Equation (3.34)) and the definition of $\widetilde{\Pi}$ (Equation (3.36)) yields

$$\begin{bmatrix} \Pi_1 & \Pi_2 \\ \Pi_{a1} & \Pi_{a2} \end{bmatrix} = -A^{-1}\left(-b\begin{bmatrix} 0 & 0 \\ -\alpha\Pi_{a2} & \alpha\Pi_{a1} \end{bmatrix}\right) - A^{-1}B_{\text{in}}\begin{bmatrix} 0 & 0 \\ \Gamma_1 & \Gamma_2 \end{bmatrix}$$

$$= \begin{bmatrix} \widetilde{\Pi}_1 & \widetilde{\Pi}_2 \\ \widetilde{\Pi}_{a1} & \widetilde{\Pi}_{a2} \end{bmatrix} + \begin{bmatrix} X\Gamma_1 & X\Gamma_2 \\ X_a\Gamma_1 & X_a\Gamma_2 \end{bmatrix}.$$

Finally, using this last result together with the second regulator equation (Equation (3.33)) and the definition of G (Equation (3.35)), it follows

$$C\left[\Pi_1, \Pi_2\right] = C\left[\widetilde{\Pi}_1, \widetilde{\Pi}_2\right] + CX\left[\Gamma_1, \Gamma_2\right]$$

$$= C\left[\widetilde{\Pi}_1, \widetilde{\Pi}_2\right] + G\left[\Gamma_1, \Gamma_2\right] = \left[0, \frac{1}{-\alpha}\right],$$

or, equivalently

$$\begin{bmatrix} \Gamma_1 \\ \Gamma_2 \end{bmatrix} = G^{-1}\begin{bmatrix} -\dfrac{1}{\alpha} - C\widetilde{\Pi}_1 \\ 0 - C\widetilde{\Pi}_2 \end{bmatrix}. \tag{3.37}$$

To solve for Γ, we need to solve the algebro-differential system consisting of equations (3.32), (3.36) and (3.37). Once Γ is found we solve the closed loop system (3.21)–(3.26) with $u = \Gamma_1 w_1 + \Gamma_2 w_2$.

We demonstrate the above using the following specific reference signal $y_r = 3\sin(2\,t)$ and choose the remaining parameters as

$$a = 3, \ b = 1, \ c = 4.$$

With these values, the numerically evaluated controller gains given in Γ are

$$\Gamma = [\Gamma_1, \Gamma_2] = [0.33059, -0.51683].$$

Finally we solve the closed loop system for $t \in (0, 50)$ with initial conditions $\varphi_1(x) = 0$ and $\varphi_2(x) = 0$.

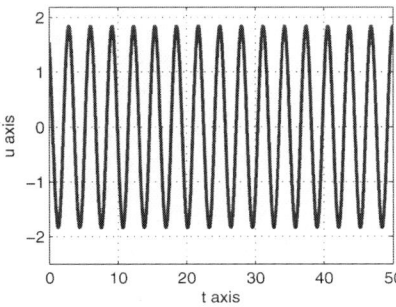

Fig. 3.4: Control $u(t) = \Gamma w$.

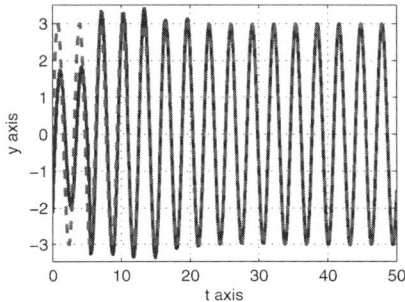

Fig. 3.5: $y(t)$, $y_r(t)$.

Fig. 3.4 depicts the control input $u(t) = \Gamma w$. Fig. 3.5 depicts the reference signal $y_r(t)$, and the controlled output $y(t)$ for the closed loop system. Fig. 3.6 depicts the error $e(t) = y(t) - y_r(t)$, and finally Fig. 3.7 contains the numerical solution for the displacement q for $x \in (0, \pi)$ and $t \in (0, 50)$.

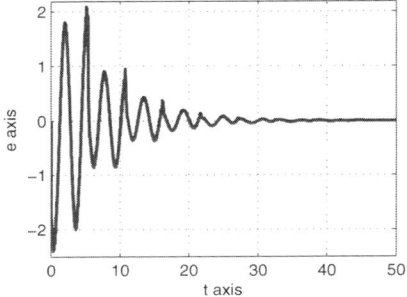
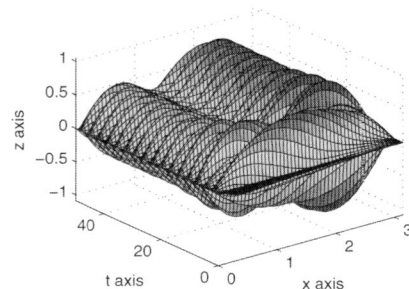

Fig. 3.6: Error $e(t)$.   **Fig. 3.7**: Solution Surface.

3.4   Vibration Regulation of a 2D Plate

In this example we consider a two-dimensional vibrating plate driven by a time-dependent disturbance acting across the entire surface of the plate.

The equation of motion for the vertical displacement q of a circular vibrating plate, whose spatial domain Ω (depicted in Fig. 3.8) is a circular disk of radius R centered at the origin, is given by

$$\rho \frac{\partial^2 q}{\partial t^2}(x, t) - \delta \frac{\partial}{\partial t} \Delta q(x, t) + EI \Delta^2 q(x, t)$$
$$= B_{\text{in}}^1(x) u_1(t) + B_{\text{in}}^2(x) u_2(t) + d(x, t), \tag{3.38}$$

$$q(x, t) = 0, \ \Delta q(x, t) = 0, \quad \mathcal{S}_0 : |x| = R, \tag{3.39}$$

$$q(x, 0) = 0, \ q_t(x, 0) = 0, \tag{3.40}$$

Here ρ is the density, EI is the stiffness constant and δ is the damping coefficient. We assume that the disturbance is a distributed load, i.e., the disturbance is a function of both the spatial variable x and the time t. We also assume the disturbance is radially symmetric and given by the following traveling wave expression

$$d(x, t) = d(r, t) = M_d \cos\left(\frac{4\pi r}{R} - \beta t\right),$$

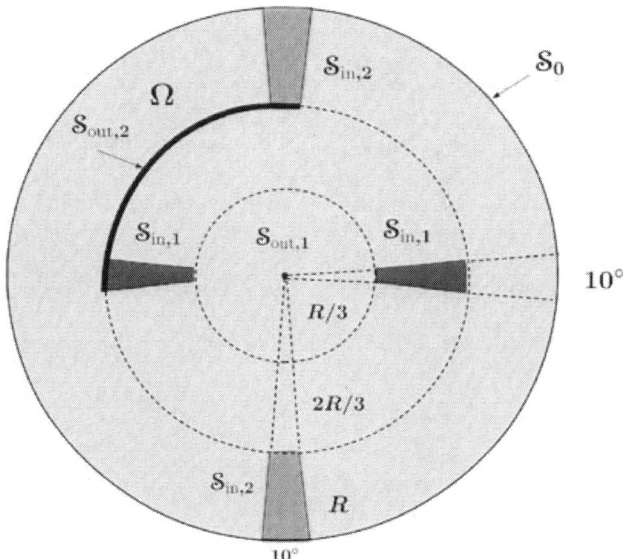

Fig. 3.8: Vibrating 2D Plate.

where $r = \sqrt{x_1^2 + x_2^2}$ is the polar radius, β is the angular frequency and $\frac{4\pi}{R}$ is the wave number.

We consider a MIMO control problem in which two measured outputs y_1 and y_2 are given by

$$y_1(t) = q|_{S_{\text{out},1}} = q(0, t) = C_1(q),$$

$$y_2(t) = \frac{1}{|S_{\text{out},2}|} \int_{S_{\text{out},2}} q(x, t)\, dx = C_2(q),$$

where $S_{\text{out},1}$ is the origin in $\mathbb{R}^2$ and $S_{\text{out},2}$ is a circular arc

$$S_{\text{out},2} = \left\{ (r, \theta) \; : \; r = \frac{2R}{3}, \quad 85° \leq \theta \leq 185° \right\},$$

with θ being the polar angle, i.e., $\tan(\theta) = \frac{x_2}{x_1}$. The time-dependent controls u_1, u_2 act through the interior regions $S_{\text{in},1}$ and $S_{\text{in},2}$, respectively, given by

$$S_{\text{in},1} = \{(r, \theta) \; : \; R/3 \leq r < 2R/3, \quad -5° \leq \theta \leq 5°, \quad 175° \leq \theta \leq 185°\},$$

and

$$S_{\text{in},2} = \{(r, \theta) \; : \; 2R/3 \leq r < R, \quad 85° \leq \theta \leq 95°, \quad 265° \leq \theta \leq 275°\}.$$

The control input actions, for $j = 1, 2$, are given by

$$B_{\text{in}}^j u_j(t) = (-1 + 2s_j(r))\, u_j(t),$$

where $s_j \in [0,1]$ is the normalized arclength measured in the radial direction

$$s_j(x) = s_j(r) = \frac{3}{R}\left(r - \frac{jR}{3}\right), \text{ for } x \in \mathcal{S}_{\text{in},j}, \ j = 1, 2.$$

The input operators, $B_{\text{in}}^1(r)$ and $B_{\text{in}}^2(r)$, act as butterfly distributed loads in the radial direction.

The objective in this example is to reject the distributed disturbance $d(x,t)$, by driving the outputs, y_1 and y_2, to the set-point values of 0. Thus we want

$$|y_j(t)| \xrightarrow{\ t \to \infty\ } 0, \ j = 1, 2.$$

In order to solve this MIMO regulator problem we first rewrite the disturbance $d(r,t)$ as

$$d(r,t) = M_d \sin\left(\frac{4\pi r}{R}\right) \sin(\beta t) + M_d \cos\left(\frac{4\pi r}{R}\right) \cos(\beta t)$$
$$= B_{\text{d}}^1(r)w_1(t) + B_{\text{d}}^2(r)w_2(t) = [B_{\text{d}}^1, \ B_{\text{d}}^2][w_1, \ w_2]^{\mathsf{T}} = B_{\text{d}}\, w$$

where $w = [M_d \sin(\beta t), \ M_d \cos(\beta t)]^{\mathsf{T}}$ and

$$B_{\text{d}} = [B_{\text{d}}^1, \ B_{\text{d}}^2] = \left[\sin\left(\frac{4\pi r}{R}\right), \cos\left(\frac{4\pi r}{R}\right)\right].$$

As in the previous example we introduce new dependent variables, an operator A and an exosystem to generate the disturbance.

1. Let $(\cdot)_{\mathfrak{a}}$ denote a new set of auxiliary state variables such that

$$\varphi_{\mathfrak{a}} = -\Delta\varphi;$$

2. Define the operator A by

$$A\begin{bmatrix} \varphi \\ \varphi_{\mathfrak{a}} \end{bmatrix} = \begin{bmatrix} \Delta & I \\ \beta^2 & EI\Delta \end{bmatrix}\begin{bmatrix} \varphi \\ \varphi_{\mathfrak{a}} \end{bmatrix},$$

   with associated homogeneous boundary conditions

$$\varphi(x) = 0, \ \varphi_{\mathfrak{a}}(x) = 0, \text{ for all } x \in \mathcal{S}_0;$$

3. w satisfies

$$w_t = Sw = \begin{bmatrix} 0 & \beta \\ -\beta & 0 \end{bmatrix} w, \quad w(0) = \begin{bmatrix} 0 \\ 1 \end{bmatrix},$$
$$w_{tt} = S^2 w = -\beta^2 w;$$

4. $\Pi = [\Pi_1, \Pi_2], \quad \Pi_{\mathfrak{a}} = [\Pi_{\mathfrak{a}1}, \Pi_{\mathfrak{a}2}], \quad$ and $\Gamma_j = [\Gamma_{j,1}, \Gamma_{j,2}],$ for $j = 1, 2.$

Substitute $q(x,t) = \Pi(x)w(t)$, $q_a(x,t) = \Pi_a(x)w(t)$ and $u_j(t) = \gamma_j(t) = \Gamma_j w$, and replace $y_j(t)$ with $y_{r_j} = 0$, then it follows

$$0 = A \begin{bmatrix} \Pi \\ \Pi_a \end{bmatrix} w - \delta \begin{bmatrix} 0 \\ \Pi_a \end{bmatrix} Sw + \begin{bmatrix} 0 \\ B_d \end{bmatrix} w + B_{in}^1 \begin{bmatrix} 0 \\ \Gamma_1 \end{bmatrix} w + B_{in}^2 \begin{bmatrix} 0 \\ \Gamma_2 \end{bmatrix} w,$$

$$C\Pi w = \begin{bmatrix} C_1 \Pi w \\ C_2 \Pi w \end{bmatrix} = \begin{bmatrix} 0 \\ 0 \end{bmatrix}.$$

Dropping w and taking into account the definitions of S, Π, Π_a, B_d and Γ_j, we obtain the two regulator equations

$$0 = A \begin{bmatrix} \Pi_1 & \Pi_2 \\ \Pi_{a1} & \Pi_{a2} \end{bmatrix} - \delta \begin{bmatrix} 0 & 0 \\ -\beta\Pi_{a2} & \beta\Pi_{a1} \end{bmatrix} + \begin{bmatrix} 0 & 0 \\ B_d^1 & B_d^2 \end{bmatrix}$$
$$+ B_{in}^1 \begin{bmatrix} 0 & 0 \\ \Gamma_{1,1} & \Gamma_{1,2} \end{bmatrix} + B_{in}^2 \begin{bmatrix} 0 & 0 \\ \Gamma_{2,1} & \Gamma_{2,2} \end{bmatrix}, \tag{3.41}$$

$$\begin{bmatrix} C_1(\Pi_1) & C_1(\Pi_2) \\ C_2(\Pi_1) & C_2(\Pi_2) \end{bmatrix} = \begin{bmatrix} 0 & 0 \\ 0 & 0 \end{bmatrix}. \tag{3.42}$$

Now we formally proceed as usual taking into account that we have two inputs and two outputs. Our first step is to solve the systems

$$A \begin{bmatrix} X_1 \\ X_{a1} \end{bmatrix} + B_{in}^1 \begin{bmatrix} 0 \\ 1 \end{bmatrix} = 0, \quad A \begin{bmatrix} X_2 \\ X_{a2} \end{bmatrix} + B_{in}^2 \begin{bmatrix} 0 \\ 1 \end{bmatrix} = 0, \tag{3.43}$$

and set the matrix

$$G = \begin{bmatrix} G_{11} & G_{12} \\ G_{21} & G_{22} \end{bmatrix} = \begin{bmatrix} C_1(X_1) & C_1(X_2) \\ C_2(X_1) & C_2(X_2) \end{bmatrix}.$$

Let the auxiliary state variables $\widetilde{\Pi}$ and $\widetilde{\Pi}_a$ be defined by

$$A \begin{bmatrix} \widetilde{\Pi}_1 & \widetilde{\Pi}_2 \\ \widetilde{\Pi}_{a1} & \widetilde{\Pi}_{a2} \end{bmatrix} - \delta \begin{bmatrix} 0 & 0 \\ -\beta\Pi_{a2} & \beta\Pi_{a1} \end{bmatrix} + \begin{bmatrix} 0 & 0 \\ B_d^1 & B_d^2 \end{bmatrix} = 0. \tag{3.44}$$

Combining the first regulator equation (in Equation (3.41)), the definitions of X_1 and X_2 (in Equation (3.43)), and the definition of $\widetilde{\Pi}$ (in Equation (3.44)) yields

$$\begin{bmatrix} \Pi_1 & \Pi_2 \\ \Pi_{a1} & \Pi_{a2} \end{bmatrix} = -A^{-1} \left(-\delta \begin{bmatrix} 0 & 0 \\ -\beta\Pi_{a2} & \beta\Pi_{a1} \end{bmatrix} + \begin{bmatrix} 0 & 0 \\ B_d^1 & B_d^2 \end{bmatrix} \right)$$
$$- A^{-1} B_{in}^1 \begin{bmatrix} 0 & 0 \\ \Gamma_{1,1} & \Gamma_{1,2} \end{bmatrix} - A^{-1} B_{in}^2 \begin{bmatrix} 0 & 0 \\ \Gamma_{2,1} & \Gamma_{2,2} \end{bmatrix}$$
$$= \begin{bmatrix} \widetilde{\Pi}_1 & \widetilde{\Pi}_2 \\ \widetilde{\Pi}_{a1} & \widetilde{\Pi}_{a2} \end{bmatrix} + \begin{bmatrix} X_1\,\Gamma_{1,1} & X_1\,\Gamma_{1,2} \\ X_{a1}\,\Gamma_{1,1} & X_{a1}\,\Gamma_{1,2} \end{bmatrix} + \begin{bmatrix} X_2\,\Gamma_{2,1} & X_2\,\Gamma_{2,2} \\ X_{a2}\,\Gamma_{2,1} & X_{a2}\,\Gamma_{2,2} \end{bmatrix}.$$

Using this last result together with the second regulator equation (Equation (3.42)) and the definition of G (Equation (3.35)), it follows

$$\begin{bmatrix} 0 & 0 \\ 0 & 0 \end{bmatrix} = \begin{bmatrix} C_1(\Pi_1) & C_1(\Pi_2) \\ C_2(\Pi_1) & C_2(\Pi_2) \end{bmatrix}$$

$$= \begin{bmatrix} C_1(\widetilde{\Pi}_1) & C_1(\widetilde{\Pi}_2) \\ C_2(\widetilde{\Pi}_1) & C_2(\widetilde{\Pi}_2) \end{bmatrix} + G \begin{bmatrix} \Gamma_{1,1} & \Gamma_{1,2} \\ \Gamma_{2,1} & \Gamma_{2,2} \end{bmatrix},$$

or, equivalently

$$\begin{bmatrix} \Gamma_{1,1} & \Gamma_{1,2} \\ \Gamma_{2,1} & \Gamma_{2,2} \end{bmatrix} = \begin{bmatrix} 0 & 0 \\ 0 & 0 \end{bmatrix} - G^{-1} \begin{bmatrix} C_1(\widetilde{\Pi}_1) & C_1(\widetilde{\Pi}_2) \\ C_2(\widetilde{\Pi}_1) & C_2(\widetilde{\Pi}_2) \end{bmatrix}$$

$$= - \begin{bmatrix} g_{11} & g_{12} \\ g_{21} & g_{22} \end{bmatrix} \begin{bmatrix} C_1(\widetilde{\Pi}_1) & C_1(\widetilde{\Pi}_2) \\ C_2(\widetilde{\Pi}_1) & C_2(\widetilde{\Pi}_2) \end{bmatrix}. \qquad (3.45)$$

To solve for Γ_j, $j = 1, 2$, we need to solve the algebro-differential system consisting of equations (3.41), (3.44) and (3.45). Once the Γ_j are found we solve the closed loop system (3.38)–(3.40) with $u_1 = \gamma_1 = \Gamma_{1,1} w_1 + \Gamma_{1,2} w_2$ and $u_2 = \gamma_2 = \Gamma_{2,1} w_1 + \Gamma_{2,1} w_2$.

For our numerical simulation we have chosen as parameter values $R = 2$, $\rho = 1$, $\delta = 1$, $EI = 1$, $\beta = \pi$, $M_d = 1$. With these values, the numerically evaluated controller gains are given by

$$\Gamma_1 = [1.44513, -7.26252], \quad \Gamma_2 = [-0.80521, -13.90966].$$

Finally then, we solve the closed loop system for $t \in (0, 15)$.

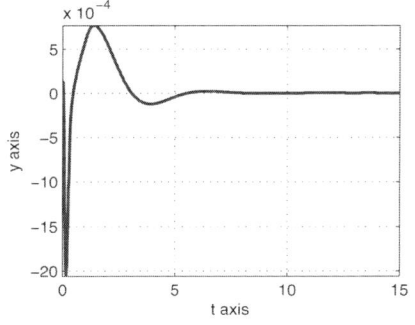

Fig. 3.9: $y_1(t)$.

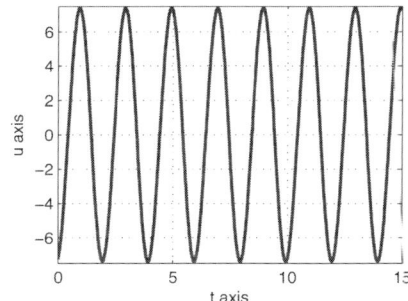

Fig. 3.10: $u_1(t)$.

Fig. 3.9 depicts the output $y_1(t)$ for the closed loop system. Fig. 3.10 depicts the control input $u_1(t) = \Gamma_1 w$.

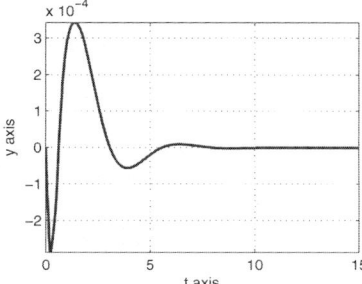

Fig. 3.11: $y_2(t)$.

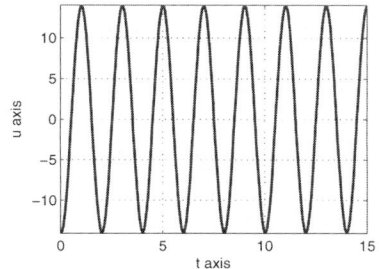

Fig. 3.12: $u_2(t)$.

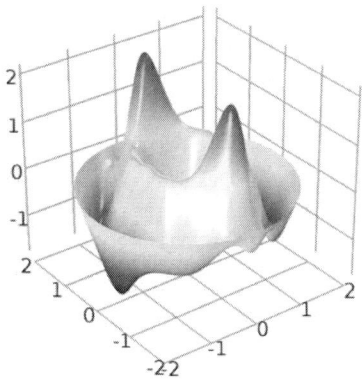

Fig. 3.13: Surf. $t = 0.1$.

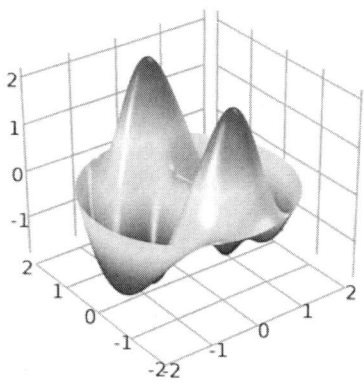

Fig. 3.14: Surf. $t = 2.5$.

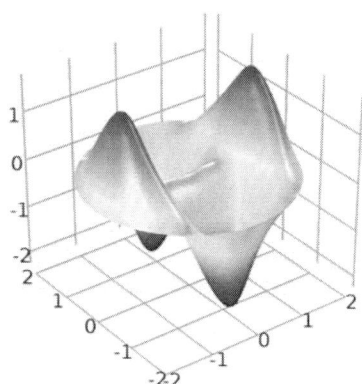

Fig. 3.15: Surf. $t = 5$.

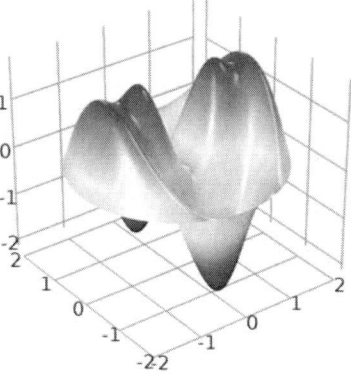

Fig. 3.16: Surf. $t = 7.5$.

Fig. 3.11 depicts the controlled output $y_2(t)$ for the closed loop system. Fig. 3.12 depicts the control input $u_2(t) = \Gamma_2 w$. Finally, Fig.s 3.13, 3.14, 3.15 and 3.16 contain the numerical solution for the displacement q profiles at the times $t = 0.1$, $t = 0.25$, $t = 0.5$ and $t = 0.75$, respectively.

3.5   Control of a Linearized Stokes Flow in 2 Dimensions

In this section we present an example of control for a linearized two-dimensional Stokes flow in a forked channel. Note that in Chapter 5 we will return to this example and examine the full nonlinear Navier-Stokes version.

Consider the two-dimensional forked channel (domain Ω) depicted in Fig. 3.17 (see also Fig. 3.18). We use the notation $v = [v_1, \, v_2]^{\mathsf{T}}$ for the velocity vector field, p for the pressure, ρ for the density and μ for the dynamic viscosity. The stress vector τ on the generic boundary S is given by

$$\tau = (-pI + \mu\,[(\nabla v) + (\nabla v)^{\mathsf{T}}]) \cdot n_S,$$

where $n_S = \begin{pmatrix} n_x \\ n_y \end{pmatrix}$ is the exterior normal on the boundary S and I is the identity tensor. We also denote by $\tau_S = \begin{pmatrix} n_y \\ -n_x \end{pmatrix}$ the unit tangent vector to the surface S. The normal and tangential components of the stress vector are given by $\tau_{n_S} = \tau \cdot n_S$ and $\tau_{\tau_S} = \tau \cdot \tau_S$, respectively.

An incompressible Stokes flow system is considered with disturbance $d(t)$ entering through a normal stress on the boundary S_1 and boundary control $u(t)$ entering through a normal stress on S_2. We also assume the output is given by the total flux of volume on S_3 where a zero stress boundary condition is applied.

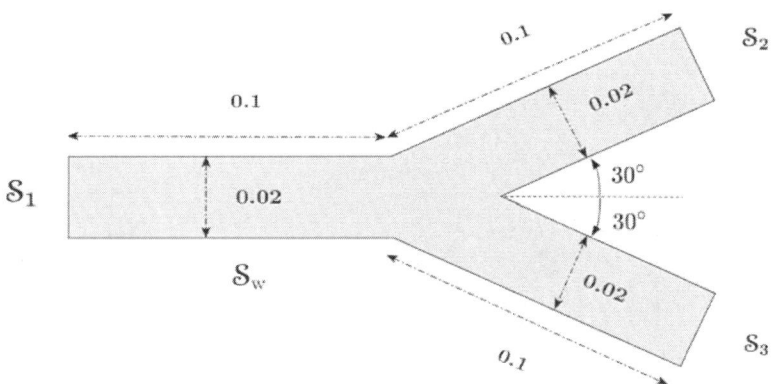

Fig. 3.17: Forked Channel with Boundaries in Meters.

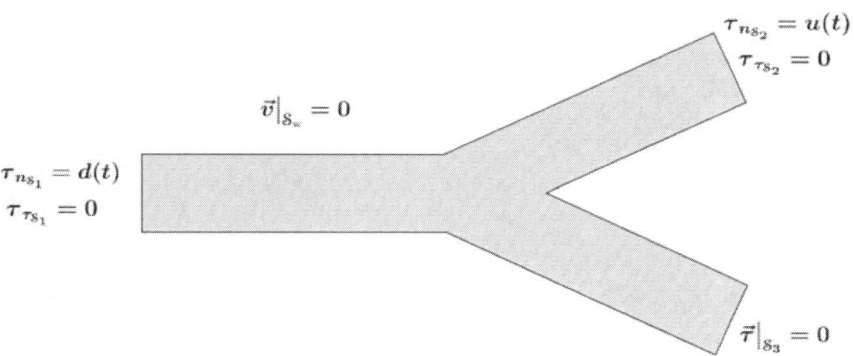

Fig. 3.18: Forked Channel Stokes Flow.

The mathematical formulation for the linear SISO Stokes control problem is given by

$$\rho \frac{\partial \boldsymbol{v}}{\partial t} = \nabla \cdot (\mu \left[(\nabla \boldsymbol{v}) + (\nabla \boldsymbol{v})^{\mathsf{T}} \right]) - \nabla p, \qquad \text{(Momentum balance)} \qquad (3.46)$$

$$\nabla \cdot \boldsymbol{v} = 0, \qquad \text{(Mass continuity)} \qquad (3.47)$$

$$\boldsymbol{v}(\boldsymbol{x}, 0) = \boldsymbol{\varphi}(\boldsymbol{x}), \; \boldsymbol{x} \in \Omega \qquad \text{(Initial data)} \qquad (3.48)$$

$$\tau_{n_{S_1}} = d(t), \quad \tau_{\tau_{S_1}} = 0, \qquad \text{(Disturbance)} \qquad (3.49)$$

$$\tau_{n_{S_2}} = u(t), \quad \tau_{\tau_{S_2}} = 0, \qquad \text{(Boundary control)} \qquad (3.50)$$

$$\boldsymbol{\tau}|_{S_3} = 0, \qquad \text{(Outflow)} \qquad (3.51)$$

$$\boldsymbol{v}|_{S_w} = 0, \qquad \text{(Wall)} \qquad (3.52)$$

$$y(t) = C\boldsymbol{v} = \gamma \int_{S_3} \boldsymbol{v} \cdot n_{S_3} ds. \qquad \text{(Measured output)} \qquad (3.53)$$

In Equation (3.53) γ is a scaling constant.

Due to the presence of the pressure variable p in Equation (3.46) and of the incompressibility constraint (3.47) (neither of which contain pressure time derivative terms), the system (3.46)–(3.53) does not seem to be a dynamical system as considered in Chapters 1 and 2. However, it is well known that on a suitable state space of divergence free vector fields, we have $\nabla \cdot \boldsymbol{v} = 0$ and the pressure gradient in (3.46) drops out. This space is called the divergence-free space $\mathcal{Z}$, which can be expressed in terms of a projection P as $\mathcal{Z} = PL^2(\Omega, \mathbb{R}^2)$.

With this understanding, we define the state operator A by

$$A = P\nabla \cdot \left(\frac{\mu}{\rho} \left[(\nabla \boldsymbol{v}) + (\nabla \boldsymbol{v})^{\mathsf{T}} \right] \right),$$

with associated homogeneous boundary conditions

$$\boldsymbol{\tau}|_{S_i} = 0 \; \text{(with } i = 1, 2 \text{ and } 3\text{)}, \quad \text{and} \quad \boldsymbol{v}|_{S_w} = 0.$$

The operator A is well known to possess properties similar to the Laplace operator. We do not pursue the details here but refer the reader to the many articles on this subject, see for example [77, 47]. Some finite element software packages, including COMSOL, have built-in solvers for incompressible fluid flows that automatically solve the problem on the appropriate projected subspaces so that we do not need to be concerned with the details of obtaining the linear Stokes operator or the special subspaces.

Let us replace the boundary disturbance on S_1 by the corresponding unbounded operator $B_{\mathrm{d}} = \delta_{S_1} n_{S_1}$, and the boundary control on S_2 by the corresponding unbounded operator $B_{\mathrm{in}} = \delta_{S_2} n_{S_2}$. Then the incompressible linear Stokes flow control problem in (3.46)–(3.53) for the state variable $Z \in \mathcal{Z}$ can be rewritten as

$$Z_t = AZ + B_{\mathrm{d}} d(t) + B_{\mathrm{in}} u(t),$$
$$Z(0) = \varphi,$$
$$CZ = y(t),$$

which is now formally equivalent to the dynamical systems considered in Chapters 1 and 2.

We are interested in solving a SISO tracking problem, where the signal to be tracked and disturbance to be rejected are generated by a finite-dimensional neutrally stable exogenous system

$$w_t = Sw, \quad w(0) = w_0 \in \mathcal{W} = \mathbb{R}^k,$$
$$y_r = Qw, \quad d = Pw. \tag{3.54}$$

In the specific numerical example we consider the problem of tracking a reference signal in the form $y_r(t) = M_r + A_r \sin(\alpha t)$ on S_3, while rejecting a normal stress disturbance in the form $d(t) = M_d + A_d \sin(\beta t)$ on S_1.

Let $\alpha_1 = \alpha_2 = 0$, $\alpha_3 = \beta_1$, $\alpha_4 = \beta_2$ and define $w = [w_1, w_2, w_3, w_4]^{\mathsf{T}}$ as the following:

$$w_1 = \begin{bmatrix} w_{1,1} \\ w_{1,2} \end{bmatrix} = \begin{bmatrix} M_1 \sin(\alpha_1 t) \\ M_1 \cos(\alpha_1 t) \end{bmatrix}, \quad w_2 = \begin{bmatrix} w_{2,1} \\ w_{2,2} \end{bmatrix} = \begin{bmatrix} M_2 \sin(\alpha_2 t) \\ M_2 \cos(\alpha_2 t) \end{bmatrix},$$

$$w_3 = \begin{bmatrix} w_{3,1} \\ w_{3,2} \end{bmatrix} = \begin{bmatrix} A_1 \sin(\alpha_3 t) \\ A_1 \cos(\alpha_3 t) \end{bmatrix}, \quad w_4 = \begin{bmatrix} w_{4,1} \\ w_{4,2} \end{bmatrix} = \begin{bmatrix} A_2 \sin(\alpha_4 t) \\ A_2 \cos(\alpha_4 t) \end{bmatrix}.$$

With these, in Equation (3.54) we have $S = \mathrm{diag}(S_1, S_2, S_3, S_4)$ with

$$S_1 = \begin{bmatrix} 0 & \alpha_1 \\ -\alpha_1 & 0 \end{bmatrix}, \ S_2 = \begin{bmatrix} 0 & \alpha_2 \\ -\alpha_2 & 0 \end{bmatrix}, \ S_3 = \begin{bmatrix} 0 & \alpha_3 \\ -\alpha_3 & 0 \end{bmatrix}, \ S_4 = \begin{bmatrix} 0 & \alpha_4 \\ -\alpha_4 & 0 \end{bmatrix},$$

$$w_0 = [0, M_1, 0, M_2, 0, A_1, 0, A_2]^{\mathsf{T}}, \ Q = [Q_1, Q_2, Q_3, Q_4] \text{ with}$$

$$Q_1 = [Q_{1,1}, Q_{1,2}] = [0, 1], \quad Q_2 = [Q_{2,1}, Q_{2,2}] = [0, 0],$$

$$Q_3 = \begin{bmatrix} Q_{3,1},\, Q_{3,2} \end{bmatrix} = \begin{bmatrix} 1,\, 0 \end{bmatrix}, \quad Q_4 = \begin{bmatrix} Q_{4,1},\, Q_{4,2} \end{bmatrix} = \begin{bmatrix} 0,\, 0 \end{bmatrix},$$

and $P = \begin{bmatrix} P_1,\, P_2,\, P_3,\, P_4 \end{bmatrix}$ with

$$P_1 = \begin{bmatrix} P_{1,1},\, P_{1,2} \end{bmatrix} = \begin{bmatrix} 0,\, 0 \end{bmatrix}, \quad P_2 = \begin{bmatrix} P_{2,1},\, P_{2,2} \end{bmatrix} = \begin{bmatrix} 0,\, 1 \end{bmatrix},$$
$$P_3 = \begin{bmatrix} P_{3,1},\, P_{3,2} \end{bmatrix} = \begin{bmatrix} 0,\, 0 \end{bmatrix}, \quad P_4 = \begin{bmatrix} P_{4,1},\, P_{4,2} \end{bmatrix} = \begin{bmatrix} 1,\, 0 \end{bmatrix}.$$

Remark 3.1. Notice that we have introduced the harmonic oscillators w_1 and w_2 with zero frequency to track and reject the constant signal M_1 and M_2. Clearly we have $w_1 = [0, M_1]$, $w_2 = [0, M_2]$ and $S_1 = S_2 = \mathbf{0}_{2\times2}$. Although this is not necessary, we will follow this approach to automate the solution process strategy. From now on this trick will be used extensively and it will be fully exploited later in Section 3.8, when tracking and rejecting generic periodic signals.

Problem 3.1 (Output Regulation Problem). Find a control $u(t) \in \mathcal{U}$ in the input space, so that

$$e(t) = y(t) - y_r(t) \to 0 \text{ as } t \to \infty,$$

for any initial data.

3.5.1   Solving the Regulator Equations

Let

$$\Pi = \begin{bmatrix} \Pi_1,\, \Pi_2,\, \Pi_3,\, \Pi_4 \end{bmatrix}, \quad \text{with } \Pi_k = \begin{bmatrix} \Pi_{k,1},\, \Pi_{k,2} \end{bmatrix} \quad \text{for } k = 1, \cdots, 4,$$

and

$$\Gamma = \begin{bmatrix} \Gamma_1,\, \Gamma_2,\, \Gamma_3,\, \Gamma_4 \end{bmatrix}, \quad \text{with } \Gamma_k = \begin{bmatrix} \Gamma_{k,1},\, \Gamma_{k,2} \end{bmatrix} \quad \text{for } k = 1, \cdots, 4,$$

where $\Pi_{k,m} \in \mathcal{Z}$ and $\Gamma_{k,m} \in \mathbb{R}$ for $m = 1$ or 2.

As we have seen in Chapters 1 and 2 the regulator equations are given in terms of Π and Γ by a pair of equations for Πw and Γw, with

$$\Pi w = \sum_{k=1}^{4} \Pi_k w_k \in \mathcal{Z} \quad \text{and} \quad \Gamma w = \sum_{k=1}^{4} \Gamma_k w_k \in \mathcal{U},$$

as

$$\Pi S w = A \Pi w + B_{\mathrm{d}} P w + B_{\mathrm{in}} \Gamma w, \tag{3.55}$$

$$0 = C \Pi w - Q w. \tag{3.56}$$

Since the equations (3.55) and (3.56) are linear and assumed to hold for all w we can write the system as an operator equation

$$\Pi S = A \Pi + B_{\mathrm{d}} P + B_{\mathrm{in}} \Gamma,$$

$$0 = C \Pi - Q,$$

which should be solved for Π and Γ. Because of the particular structure of S, Q and P the previous system is decoupled in Π_k and Γ_k, and can be split and solved separately as

$$\Pi_k S_k = A\Pi_k + B_\mathrm{d}P_k + B_\mathrm{in}\Gamma_k,$$
$$0 = C\Pi_k - Q_k, \tag{3.57}$$

for $k = 1, \cdots, 4$.

To solve the equation system (3.57) we proceed formally. First we solve the unit input problem for the auxiliary state variable $X \in \mathcal{Z}$

$$0 = AX + B_\mathrm{in},$$

and, as usual, we use the notation G to denote the transfer function $G(s) = C(sI - A)^{-1}B_\mathrm{in}$ evaluated at $s = 0$, i.e., $G = G(0)$. Thus we have

$$G = CX = -A^{-1}B_\mathrm{in}.$$

Then we solve the coupled algebro-differential systems

$$0 = A\Pi_k - \Pi_k S_k + B_\mathrm{d}P_k + B_\mathrm{in}\Gamma_k,$$
$$0 = A\widetilde{\Pi}_k - \Pi_k S_k + B_\mathrm{d}P_k,$$
$$\Gamma_k = G^{-1}\left[Q_k - C\widetilde{\Pi}_k\right],$$

for $k = 1, \cdots, 4$, where we have introduced a new set of auxiliary state variables

$$\widetilde{\Pi}_k = \left[\widetilde{\Pi}_{k,1}, \widetilde{\Pi}_{k,2}\right].$$

Once the gains Γ_k are found, we set

$$u(t) = \Gamma w = \Gamma_1 w_1 + \Gamma_2 w_2 + \Gamma_3 w_3 + \Gamma_4 w_4,$$

and solve the closed loop system

$$Z_t = AZ + B_\mathrm{d}d(t) + B_\mathrm{in}\Gamma w,$$
$$Z(0) = \varphi,$$
$$CZ = y(t).$$

In our specific numerical simulation we have taken the boundary S_1 (i.e., the height of the first channel) to be $2\,\mathrm{cm}$ wide and the length of the main channel to be $10\,\mathrm{cm}$ long. The same dimensions are taken for the two forked branches, which are each inclined by $\pm 30°$ with respect to the main channel. We have set the following parameters for the problem

$$M_1 = 10, \quad A_1 = 10, \quad \beta_1 = 1,$$
$$M_2 = 25, \quad A_2 = 12, \quad \beta_2 = 2,$$

$$\rho = 1000, \quad \mu = 0.1, \quad \gamma = 10^4,$$

and we have solved the problem for $t \in (0, 20)$.

The results from our numerical simulations are depicted in the following figures.

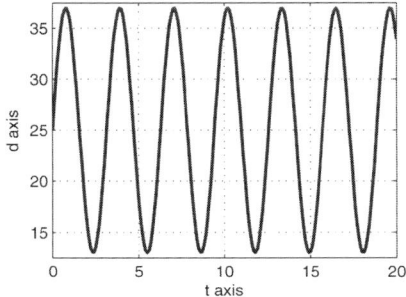

Fig. 3.19: $d(t)$.

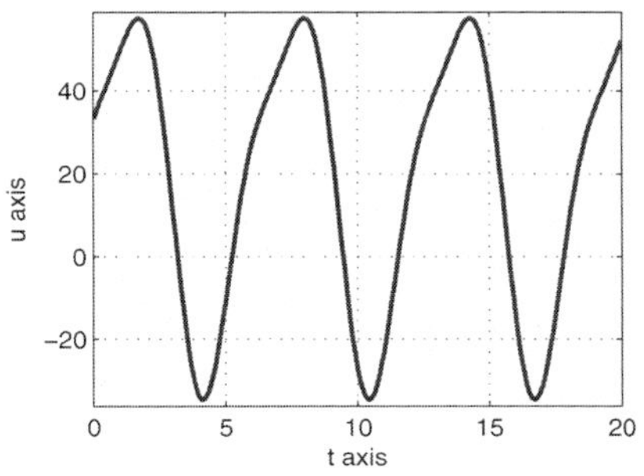

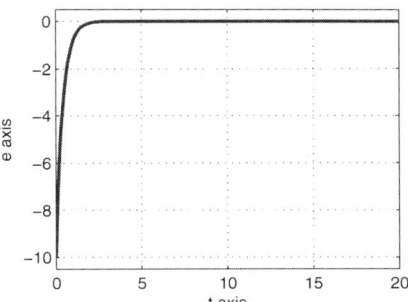

Fig. 3.20: $u(t)$.

Fig. 3.19 depicts the disturbance $d(t)$. Fig. 3.20 depicts the control input $u(t) = \Gamma w$.

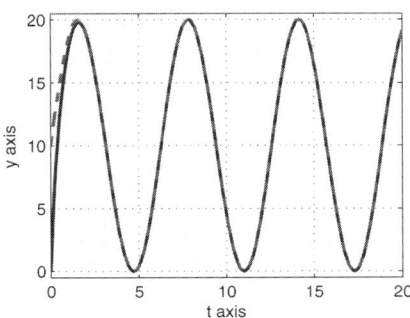

Fig. 3.21: $y(t)$ and $y_r(t)$.

Fig. 3.22: $e(t)$.

Fig. 3.21 depicts the output $y(t)$ and the signal to be tracked $y_r(t)$ for the closed loop system. Fig. 3.22 depicts the error $e(t) = y(t) - y_r(t)$.

Finally, Figs. 3.23, 3.24, 3.25 and 3.26 contain the numerical solution for the velocity vector field and its magnitude (grayscale) color map at times $t = 0.5$, $t = 6$, $t = 12$ and $t = 12.5$, respectively.

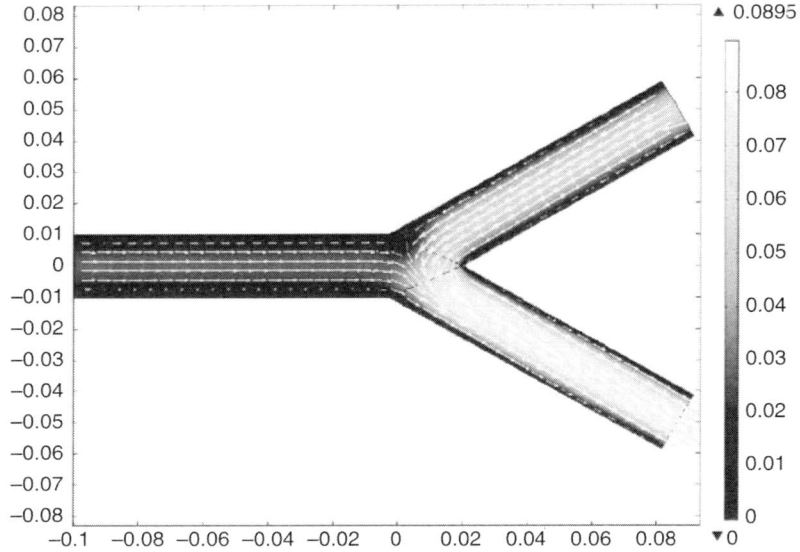

Fig. 3.23: Solution $t = 0.5$.

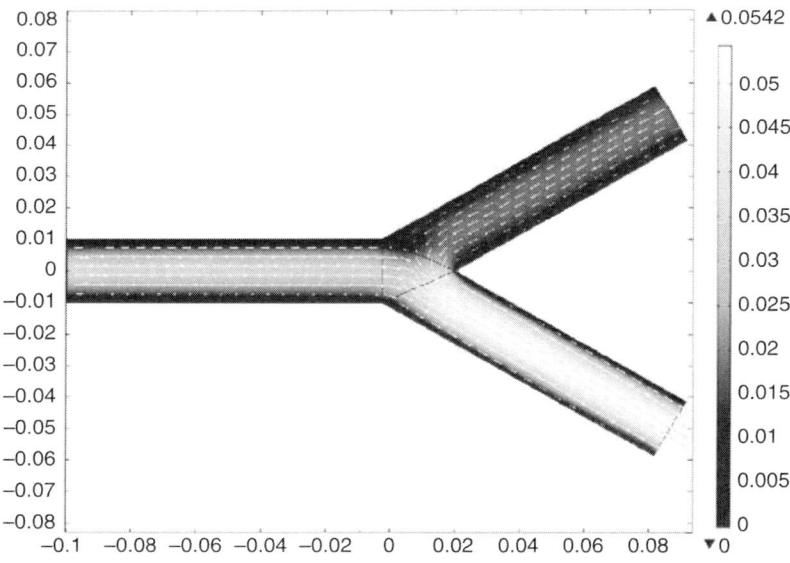

Fig. 3.24: Solution $t = 6$.

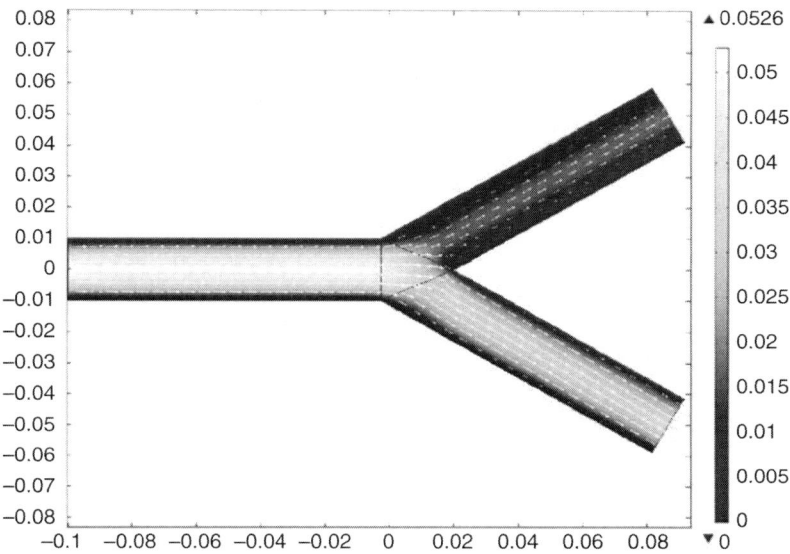

Fig. 3.25: Solution $t = 12$.

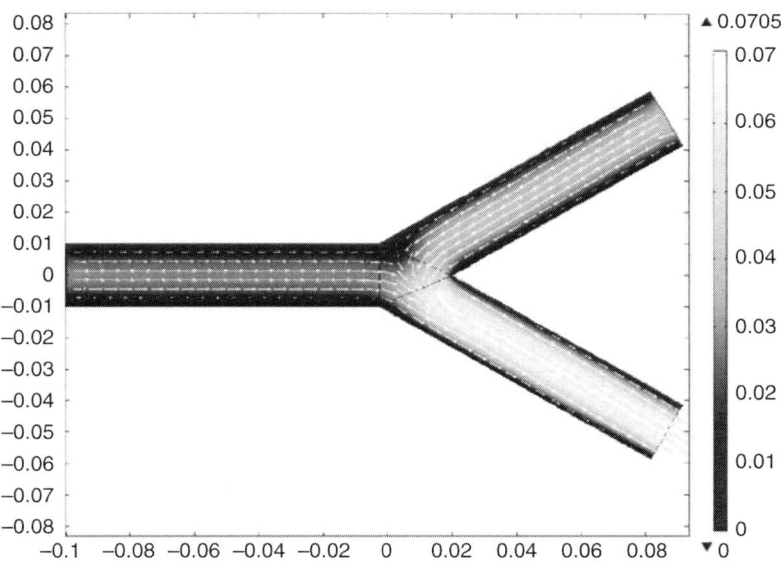

Fig. 3.26: Solution $t = 12.5$.

3.6　Thermal Control of a 2D Fluid Flow

In this section we consider a linear problem of thermo-fluid mechanics regulation. A steady-state incompressible Navier-Stokes fluid flow passes through a two-dimensional region in which two desired temperature targets are tracked. The control is achieved by changing the temperature of the inlet flow and through a heat flux in part of the boundary. Two different disturbances are also rejected on different parts of the boundary.

A more general example of thermo-fluid mechanics regulation will be discussed in Chapter 5, where a full nonlinear Boussinesq flow will be considered. These kind of examples are motivated by recent research directed in the design of energy efficient buildings.

The domain Ω in Figs. 3.27 and 3.28 consists of a main rectangular region of size $20D \times 10D$ for a fixed positive number D. Inlet and outlet square regions, of side length D, are located on the left-bottom and right-bottom part of the room, respectively.

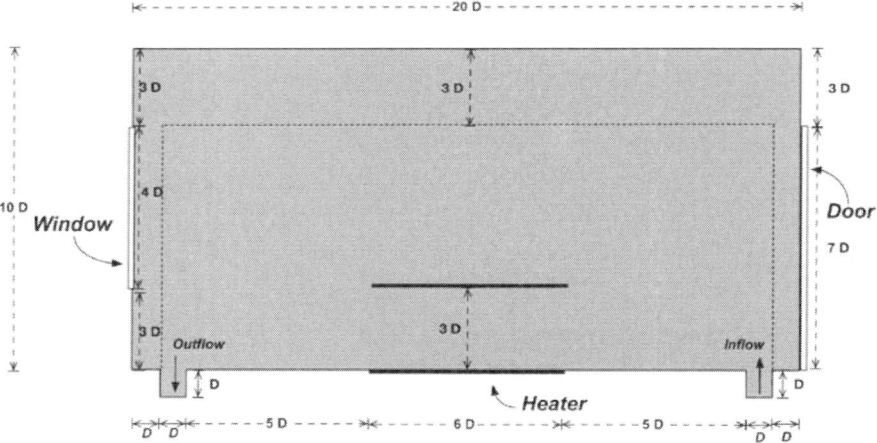

Fig. 3.27: Two-Dimensional Box Domain Ω.

The boundary of the region Ω consists of an inflow boundary $S_{in,1}$, an outflow boundary S_o, two disturbances entering through a window on $S_{d,1}$ and a door on $S_{d,2}$. Insulated walls are considered on the rest of the boundary S_w. Two distinct average temperature outputs are considered on the two interior lines $S_{out,1}$ and $S_{out,2}$ in Fig. 3.28.

The mathematical formulation of the above fluid mechanics problem is given by the steady-state Navier-Stokes equation for incompressible fluids

$$\rho(\boldsymbol{v} \cdot \nabla)\boldsymbol{v} = \nabla \cdot (\mu\,[(\nabla\boldsymbol{v}) + (\nabla\boldsymbol{v})^{\mathsf{T}}]) - \nabla p, \qquad (3.58)$$

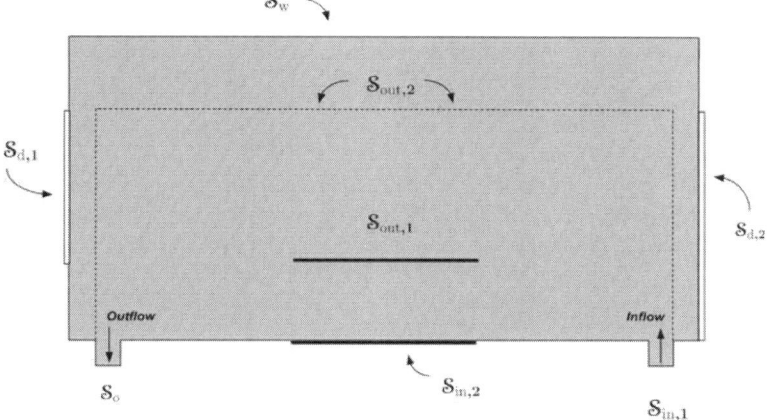

Fig. 3.28: Two-Dimensional Room with Boundaries.

$$\nabla \cdot \boldsymbol{v} = 0, \tag{3.59}$$

$$\boldsymbol{v} \cdot \boldsymbol{n} = f(s) \quad \text{on } \mathcal{S}_{\text{in},1}, \tag{3.60}$$

$$\boldsymbol{\tau} = \boldsymbol{0} \quad \text{on } \mathcal{S}_o, \tag{3.61}$$

$$\boldsymbol{v} = \boldsymbol{0} \quad \text{on } \mathcal{S}_{\text{w}} \cup \mathcal{S}_{\text{in},2} \cup \mathcal{S}_{\text{d},1} \cup \mathcal{S}_{\text{d},2}. \tag{3.62}$$

Here, we have used the notation $\boldsymbol{v} = \begin{bmatrix} u, & v \end{bmatrix}^{\mathsf{T}}$ for the velocity vector field, p for the pressure, ρ for the density and μ for the dynamic viscosity. The stress vector $\boldsymbol{\tau}$ on the generic boundary $\mathcal{S}$ is given by

$$\boldsymbol{\tau} = (-pI + \mu \left[(\nabla \boldsymbol{v}) + (\nabla \boldsymbol{v})^{\mathsf{T}} \right]) \cdot \boldsymbol{n},$$

where $\boldsymbol{n}$ is the external unit normal vector on the boundary $\mathcal{S}$ and I is the identity tensor. We also use the function

$$f(s) = s(1 - s)$$

for the parabolic inflow profile with maximum amplitude 0.25, where s is the arclength normalized between 0 and 1. A zero stress outflow boundary condition is considered on $\mathcal{S}_o$, while zero velocity is considered on $\mathcal{S}_{\text{w}}$.

The mathematical formulation of the heat control problem is given through the temperature equation

$$\rho c_p \frac{\partial T}{\partial t} + \rho c_p (\boldsymbol{v} \cdot \nabla) T = K \Delta T, \tag{3.63}$$

$$T(x, 0) = \phi(x), \tag{3.64}$$

$$\mathcal{B}_{\text{in}}^1 T = T = u_1(t) \text{ on } \mathcal{S}_{\text{in},1}, \tag{3.65}$$

$$\mathcal{B}_{\text{in}}^2 T = q(T) = u_2(t) \text{ on } \mathcal{S}_{\text{in},2}, \tag{3.66}$$

$$q(T) = 0 \text{ on } \mathcal{S}_{\mathrm{o}}, \tag{3.67}$$

$$q(T) = 0 \text{ on } \mathcal{S}_{\mathrm{w}}, \tag{3.68}$$

$$\mathcal{B}_{\mathrm{d}}^1 T = q(T) + h_1 T = h_1 d_1(t) \text{ on } \mathcal{S}_{\mathrm{d},1}, \tag{3.69}$$

$$\mathcal{B}_{\mathrm{d}}^2 T = q(T) + h_2 T = h_2 d_2(t) \text{ on } \mathcal{S}_{\mathrm{d},2}. \tag{3.70}$$

Here we have used the notation T for the temperature, c_p for heat capacity at constant pressure, K for the thermal conductivity and h_1 and h_2 for the heat transfer coefficients. The normal heat flux q on $\mathcal{S}$ is given by

$$q(T) = K \nabla T \cdot \boldsymbol{n}.$$

Thus we have two boundary controls u_j, for $j = 1, 2$, entering through a given temperature on $\mathcal{S}_{\mathrm{in},1}$ and through a heat flux input on $\mathcal{S}_{\mathrm{in},2}$. There are also two disturbances d_j, for $j = 1, 2$, corresponding to mixed-thermal boundary conditions through a window at $\mathcal{S}_{\mathrm{d},1}$ and a door at $\mathcal{S}_{\mathrm{d},2}$. Outflow boundary conditions are considered on $\mathcal{S}_{\mathrm{o}}$, while an insulated boundary is considered on the remaining boundary $\mathcal{S}_{\mathrm{w}}$.

We have two measured outputs on $\mathcal{S}_{\mathrm{out},1}$ and on $\mathcal{S}_{\mathrm{out},2}$ given by the average temperature

$$y_1 = C_1(T) = \frac{1}{|\mathcal{S}_{\mathrm{out},1}|} \int_{\mathcal{S}_{\mathrm{out},1}} T \, ds,$$

$$y_2 = C_2(T) = \frac{1}{|\mathcal{S}_{\mathrm{out},2}|} \int_{\mathcal{S}_{\mathrm{out},2}} T \, ds.$$

Again, in this example, our objective is to find control inputs $u_1(t)$ and $u_2(t)$ so that the measured outputs $y_1(t)$ and $y_2(t)$ track given reference signals y_{r_1} and y_{r_2}, while rejecting the disturbances $d_1(t)$ and $d_2(t)$.

We reiterate that in this example we do not consider any nonlinear coupling for the non-isothermal Navier-Stokes flow. Rather, we first solve the nonlinear steady Navier-Stokes equation system (3.58)–(3.62) and then focus our attention on the solution of the regulation heat problem (3.63)–(3.70) in which the velocity vector field $\boldsymbol{v}$ in the convective term is assumed to have already been evaluated.

Remark 3.2. Equation (3.63) is linear in T, since the convective velocity $\boldsymbol{v}$ in the advection term is given. Then we consider the linear homogeneous operator A applied to the state variable φ to be given by $K\Delta\varphi - \rho c_p (\boldsymbol{v} \cdot \nabla)\varphi$ rather than $K\Delta\varphi$, only. We also assume $\boldsymbol{v}$ to be so that A is always invertible. Here, of course, the definition of the linear operator A is also supplemented by the homogeneous boundary conditions on the boundary of the domain Ω.

Under this assumption we can rewrite the control problem for the state variable $T = Z \in \mathcal{Z}$ in the form

$$Z_t = AZ + B_{\mathrm{d}}^1 d_1(t) + B_{\mathrm{d}}^2 d_2(t) + B_{\mathrm{in}}^1 u_1(t) + B_{\mathrm{in}}^2 u_2(t),$$

$$Z(0) = \varphi,$$
$$C_1 Z = y_1(t),$$
$$C_2 Z = y_2(t),$$

where B_d^1, B_d^2, B_{in}^1 and B_{in}^2 are the unbounded operators corresponding to the boundary conditions $\mathcal{B}_d^1$, $\mathcal{B}_d^2$, $\mathcal{B}_{in}^1$ and $\mathcal{B}_{in}^2$, respectively. As we have seen throughout the examples given in Chapter 2, as well as those already presented in this chapter, we do not need to know the exact form or action of these unbounded operators.

Our regulation problem involves solving a MIMO tracking problem, where the signals to be tracked and disturbances to be rejected are generated by a finite-dimensional neutrally stable exogenous system

$$w_t = Sw, \quad w(0) = w_0 \in W = \mathbb{R}^k,$$
$$y_{r_1} = Q_1 w, \quad y_{r_1} = Q_2 w, \tag{3.71}$$
$$d_1 = P_1 w, \quad d_2 = Q_2 w,$$

In the specific numerical example we consider the problem of tracking reference signals in the form $y_{r_1}(t) = M_1 + A_1 \sin(\beta_1 t)$ on $S_{out,1}$ and $y_{r_2}(t) = M_2 + A_2 \sin(\beta_2 t)$ on $S_{out,2}$, while rejecting disturbances in the form $d_1(t) = M_3 + A_3 \sin(\beta_3 t)$ on $S_{d,1}$ and $d_2(t) = M_4 + A_4 \sin(\beta_4 t)$ on $S_{d,2}$.

Let $\alpha_1 = \alpha_2 = \alpha_3 = \alpha_4 = 0$, $\alpha_5 = \beta_1$, $\alpha_6 = \beta_2$, $\alpha_7 = \beta_3$, $\alpha_8 = \beta_4$ and define $w = \begin{bmatrix} w_1, w_2, w_3, w_4, w_5, w_6, w_7, w_8 \end{bmatrix}^{\mathsf{T}}$ as follows

$$w_1 = \begin{bmatrix} w_{1,1} \\ w_{1,2} \end{bmatrix} = \begin{bmatrix} M_1 \sin(\alpha_1 t) \\ M_1 \cos(\alpha_1 t) \end{bmatrix}, \quad w_2 = \begin{bmatrix} w_{2,1} \\ w_{2,2} \end{bmatrix} = \begin{bmatrix} M_2 \sin(\alpha_2 t) \\ M_2 \cos(\alpha_2 t) \end{bmatrix},$$

$$w_3 = \begin{bmatrix} w_{3,1} \\ w_{3,2} \end{bmatrix} = \begin{bmatrix} M_3 \sin(\alpha_3 t) \\ M_3 \cos(\alpha_3 t) \end{bmatrix}, \quad w_4 = \begin{bmatrix} w_{4,1} \\ w_{4,2} \end{bmatrix} = \begin{bmatrix} M_4 \sin(\alpha_4 t) \\ M_4 \cos(\alpha_4 t) \end{bmatrix},$$

$$w_5 = \begin{bmatrix} w_{5,1} \\ w_{5,2} \end{bmatrix} = \begin{bmatrix} A_1 \sin(\alpha_5 t) \\ A_1 \cos(\alpha_5 t) \end{bmatrix}, \quad w_6 = \begin{bmatrix} w_{6,1} \\ w_{6,2} \end{bmatrix} = \begin{bmatrix} A_2 \sin(\alpha_6 t) \\ A_2 \cos(\alpha_6 t) \end{bmatrix},$$

$$w_7 = \begin{bmatrix} w_{7,1} \\ w_{7,2} \end{bmatrix} = \begin{bmatrix} A_3 \sin(\alpha_7 t) \\ A_3 \cos(\alpha_7 t) \end{bmatrix}, \quad w_8 = \begin{bmatrix} w_{8,1} \\ w_{8,2} \end{bmatrix} = \begin{bmatrix} A_4 \sin(\alpha_8 t) \\ A_4 \cos(\alpha_8 t) \end{bmatrix}.$$

With these, in Equation (3.71) we have $S = \mathrm{diag}(S_1, S_2, \cdots, S_8)$ with

$$S_i = \begin{bmatrix} 0 & \alpha_i \\ -\alpha_i & 0 \end{bmatrix}, \text{ for } i = 1, 2, \cdots, 8,$$

$$w_0 = \begin{bmatrix} 0, M_1, 0, M_2, 0, M_3, 0, M_4, 0, A_1, 0, A_2, 0, A_3, 0, A_4 \end{bmatrix}^{\mathsf{T}},$$
$$Q_i = \begin{bmatrix} Q_1^i, Q_2^i, \cdots, Q_8^i \end{bmatrix}, \text{ for } i = 1, 2,$$

where

$$Q_k^i = \begin{bmatrix} Q_{k,1}^i, Q_{k,2}^i \end{bmatrix} = \begin{bmatrix} 0, 0 \end{bmatrix},$$

and with the following exceptions

$$Q_1^1 = \left[Q_{1,1}^1, Q_{1,2}^1\right] = \left[0, 1\right], \quad Q_5^1 = \left[Q_{5,1}^1, Q_{5,2}^1\right] = \left[1, 0\right],$$
$$Q_2^2 = \left[Q_{2,1}^2, Q_{2,2}^2\right] = \left[0, 1\right], \quad Q_6^2 = \left[Q_{6,1}^2, Q_{6,2}^2\right] = \left[1, 0\right],$$

and $P_i = \left[P_1^i, P_2^i, \cdots, P_8^i\right]$, for $i = 1, 2$, where

$$P_k^i = \left[P_{k,1}^i, P_{k,2}^i\right] = \left[0, 0\right],$$

with the following exceptions

$$P_3^1 = \left[P_{3,1}^1, P_{3,2}^1\right] = \left[0, 1\right], \quad P_7^1 = \left[P_{7,1}^1, P_{7,2}^1\right] = \left[1, 0\right],$$
$$P_4^2 = \left[P_{4,1}^2, P_{4,2}^2\right] = \left[0, 1\right], \quad P_8^2 = \left[P_{8,1}^2, P_{8,2}^2\right] = \left[1, 0\right].$$

Here the terms Q_1^1 and Q_5^1 correspond to the set-point and harmonic components of the signal y_{r_1}; the terms Q_2^2 and Q_6^2 correspond to the set-point and harmonic components of the signal y_{r_2}; the terms P_3^1 and P_7^1 correspond to the set-point and harmonic components of the disturbance d_1; and the terms P_4^2 and P_8^2 correspond to the set-point and harmonic components of the disturbance d_2.

Remark 3.3. Again as in the previous example we have introduced the harmonic oscillators associated with the variables w_1, w_2, w_3, w_4 with zero frequency to track and reject the constant signals M_1, M_2, M_3 and M_4. Clearly, we have $w_1 = [0, M_1]$, $w_2 = [0, M_2]$, $w_3 = [0, M_2]$, $w_4 = [0, M_2]$ and $S_1 = S_2 = S_3 = S_4 = \mathbf{0}_{2 \times 2}$.

We also introduce the corresponding state variables and control gains for the regulator equations. Let

$$\Pi = \left[\Pi_1, \Pi_2, \cdots, \Pi_8\right], \quad \text{with } \Pi_k = \left[\Pi_{k,1}, \Pi_{k,2}\right] \quad \text{for } k = 1, \cdots, 8,$$

and

$$\Gamma_i = \left[\Gamma_1^i, \Gamma_2^i, \cdots, \Gamma_8^i\right], \quad \text{with } \Gamma_k^i = \left[\Gamma_{k,1}^i, \Gamma_{k,2}^i\right] \quad \text{for } i = 1, 2 \text{ and } k = 1, \cdots, 8.$$

As we have seen in Chapters 1 and 2 the regulator equations are given in terms of Π and Γ_i by a pair of equations for Πw and $\Gamma_i w$, with

$$\Pi w = \sum_{k=1}^{8} \Pi_k w_k \in \mathcal{Z} \quad \text{and} \quad \Gamma_i w = \sum_{k=1}^{8} \Gamma_k^i w_k \in \mathcal{U}^i, \text{ for } i = 1, 2,$$

as

$$\Pi S w = A \Pi w + B_{\mathrm{d}}^1 P_1 w + B_{\mathrm{d}}^2 P_2 w + B_{\mathrm{in}}^1 \Gamma_1 w + B_{\mathrm{in}}^2 \Gamma_2 w, \tag{3.72}$$

$$0 = C_1 \Pi w - Q_1 w, \tag{3.73}$$
$$0 = C_2 \Pi w - Q_2 w. \tag{3.74}$$

Since the equations (3.72), (3.73) and (3.74) are linear and assumed to hold for all w we can write the system as an operator equation

$$\Pi S = A\Pi + B_d^1 P_1 + B_d^2 P_2 + B_{in}^1 \Gamma_1 + B_{in}^2 \Gamma_2,$$
$$0 = C_1 \Pi - Q_1, \quad 0 = C_2 \Pi - Q_2,$$

which should be solved for Π, Γ_1 and Γ_2. Because of the particular structure of S, Q and P the previous system is decoupled in Π_k and Γ_k^i, and can be split and solved separately as

$$\Pi_k S_k = A\Pi_k + B_d^1 P_k^1 + B_d^2 P_k^2 + B_{in}^1 \Gamma_k^1 + B_{in}^2 \Gamma_k^2,$$
$$0 = C_1 \Pi_k - Q_k^1, \quad 0 = C_2 \Pi_k - Q_k^2,$$

for $k = 1, \cdots, 8$.

The solutions of the previous systems are obtained formally following the same strategy adopted in Section 1.5.2 for the harmonic oscillator.

Once the gains Γ_k are found, we set

$$u_i(t) = \Gamma_i w = \Gamma_1^i w_1 + \Gamma_2^i w_2 + \cdots + \Gamma_8^i w_8, \quad \text{for } i = 1, 2,$$

and solve the closed loop system

$$Z_t = AZ + B_d^1 d_1(t) B_d^2 d_2(t) + B_{in}^1 \Gamma_1 w + B_{in}^2 \Gamma_2 w,$$
$$Z(0) = \phi,$$
$$C_1 Z = y_1(t), \quad C_2 Z = y_2(t).$$

In our specific numerical simulation we have taken the reference length D to be $0.2\,\text{m}$ and we have set the remaining parameters for the problem to be

$$M_1 = 22°\text{C}, \quad A_1 = 3°\text{C}, \quad \beta_1 = \frac{2\pi}{86400\text{s}},$$
$$M_2 = 18°\text{C}, \quad A_2 = 1°\text{C}, \quad \beta_2 = \frac{2\pi}{86400\text{s}},$$
$$M_3 = 5°\text{C}, \quad A_3 = 10°\text{C}, \quad \beta_3 = \frac{2\pi}{86400\text{s}},$$
$$M_4 = 8°\text{C}, \quad A_4 = 5°\text{C}, \quad \beta_4 = \frac{2\pi}{86400\text{s}},$$
$$\rho = 1\frac{\text{kg}}{\text{m}^3}, \quad \mu = 0.001\text{Pa s}, \quad K = 0.001\frac{\text{W}}{\text{m}\,°\text{C}},$$
$$c_p = \frac{1\text{J}}{\text{kg}\,°\text{C}}, \quad h_1 = 0.01\frac{\text{W}}{\text{m}^2\,°\text{C}}, \quad h_2 = 0.02\frac{\text{W}}{\text{m}^2\,°\text{C}}.$$

and we have solved the problem for $t \in (0, 2 \times 86400s) = (0, 2\,\text{days})$. The initial data for the energy equation is chosen to be $\phi(x) = 18°\text{C}$.

The time evolution $y_1(t)$ is given in Fig. 3.29 for all time. The asymptotic convergence to zero of the error $e_1(t) = y_1(t) - y_{r,1}$ is given in Fig. 3.30.

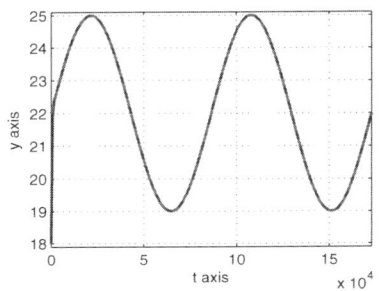
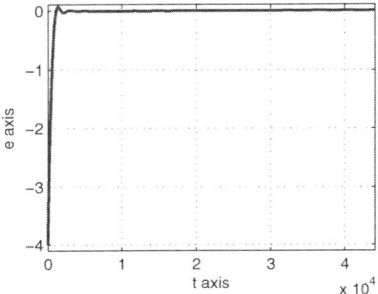

Fig. 3.29: $y_1(t)$.   **Fig. 3.30**: $e_1(t)$.

Similar results for $y_2(t)$ and $e_2(t) = y_2(t) - y_{r,2}(t)$ are shown in Figs. 3.31 and 3.32.

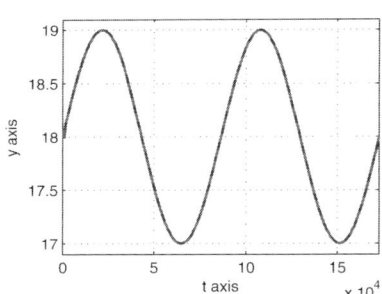
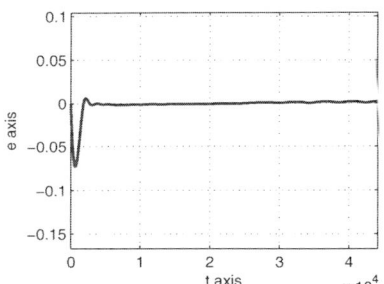

Fig. 3.31: $y_2(t)$ and $y_{r,2}(t)$.   **Fig. 3.32**: $e_2(t)$.

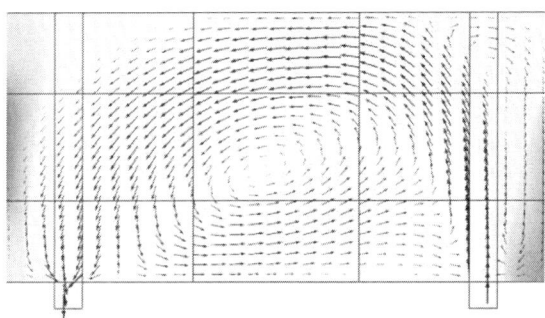

Fig. 3.33: Solution $t = 18$ Hours.

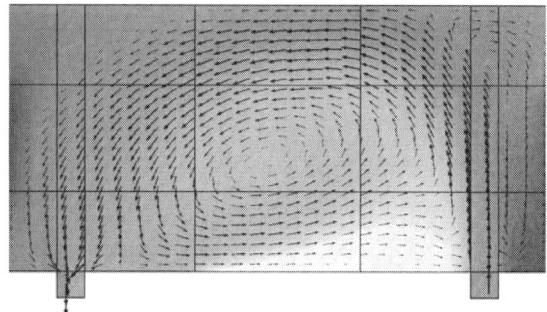

Fig. 3.34: Solution $t = 48$ Hours.

In Figs. 3.33 and 3.34 a screenshot of the velocity vector field v and the temperature profile T are given at times $t = 18$ hours and $t = 48$ hours, respectively.

3.7 Thermal Regulation in a 3D Room

In this example we again consider a thermal control problem but for a three-dimensional room. In order to keep things a little bit simpler we consider a SISO example in which we have one control input on $\mathcal{S}_{\text{in}}$, one measured output $\mathcal{S}_{\text{out}}$ and a single disturbance on $\mathcal{S}_{\text{d}}$. We assume there is a steady fluid flow velocity field in the room generated by a three-dimensional steady Navier-Stokes flow with velocity denoted by v.

The computational domain is given in Fig. 3.35. The main room, a box with a corner placed at the origin, extends 4m long in the positive x direction, 4m wide in the positive y direction and 2m tall in the positive z direction. An inlet box region of size $(0.4, 0.6, 0.4)$ is located on top of the ceiling with sides parallel to the reference planes. The lower surface of the inlet region is centered at the point $(0.4, 2, 2)$. An output box region of size $(0.4, 0.6, 0.4)$ is located under the floor with sides parallel to the reference planes. The upper surface of the output region is centered at the point $(3.6, 2, 0)$. A rectangular window of size 2×1 is centered on the side wall with $y = 4$ with edges parallel to the x and z axes. The output region $\mathcal{S}_{\text{out}}$ inside the domain is the horizontal rectangle of size 2×0.6, centered at $(2, 2, 0.5)$ and with edges parallel to the x and y axes.

Our measured output $C(T)$ is given by the average temperature on $\mathcal{S}_{\text{out}}$

$$y(t) = C(T) = \frac{1}{|\mathcal{S}_{\text{out}}|} \int_{\mathcal{S}_{\text{out}}} T \, dS,$$

and is designed to track the prescribed temperature profile $y_r = M_1 + A_1 \sin(\beta_1 t)$.

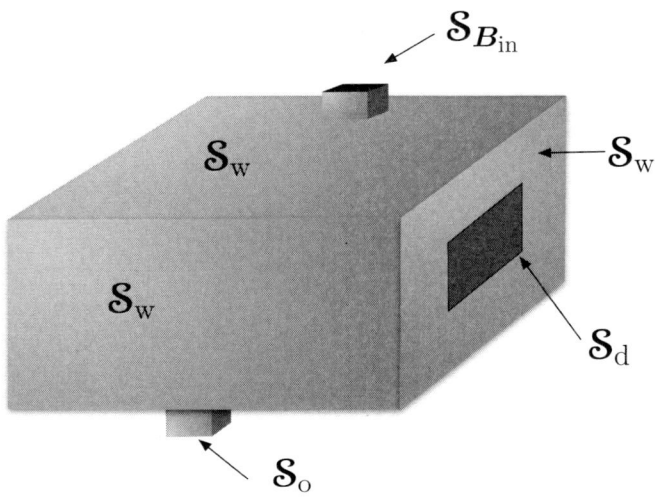

Fig. 3.35: 3D Room.

The control is achieved by changing the temperature of the inlet flow on S_{in}. We assume there is a disturbance entering the window surface S_d through a thermal mixed boundary condition with external temperature given by $d(t) = M_2 + A_2 \sin(\beta_2 t)$. Outflow boundary conditions are considered on S_o, while insulated boundary conditions are considered on the rest of the wall.

Just as in the previous example the fluid flow satisfies the steady state incompressible Navier-Stokes equation system given by

$$\rho(\boldsymbol{v} \cdot \nabla)\boldsymbol{v} = \nabla \cdot (\mu\,[(\nabla\boldsymbol{v}) + (\nabla\boldsymbol{v})^{\intercal}]) - \nabla p, \tag{3.75}$$

$$\nabla \cdot \boldsymbol{v} = 0, \tag{3.76}$$

$$\boldsymbol{\tau} = 1\boldsymbol{n} \text{ on } S_{\text{in}}, \tag{3.77}$$

$$\boldsymbol{\tau} = \boldsymbol{0} \text{ on } S_o, \tag{3.78}$$

$$\boldsymbol{v} = \boldsymbol{0} \quad \text{on } S_w \cup S_d. \tag{3.79}$$

Once again we have used the notation $\boldsymbol{v} = [v_1,\, v_2,\, v_3]^{\intercal}$ for the velocity vector field, p for the pressure, μ for the dynamic viscosity, ρ for the density, $\boldsymbol{\tau}$ for the surface stress on a surface S

$$\boldsymbol{\tau}(\boldsymbol{v}, p) = (-pI + \mu\,[(\nabla\boldsymbol{v}) + (\nabla\boldsymbol{v})^{\intercal}]) \cdot \boldsymbol{n}$$

and $\boldsymbol{n} = [n_x,\, n_y,\, n_z]^{\intercal}$ for the outward normal on the boundary S.

Note that in the inflow boundary condition (3.77) we have used a normal stress rather than an inflow velocity profile.

The mathematical formulation of the heat control problem is given by the temperature equation

$$\rho c_p \frac{\partial T}{\partial t} + \rho c_p (\boldsymbol{v} \cdot \nabla) T = K \Delta T, \tag{3.80}$$

$$T(x, 0) = \phi(x), \tag{3.81}$$

$$\mathcal{B}_{\text{in}} T = T = u(t) \text{ on } \mathcal{S}_{\text{in}}, \tag{3.82}$$

$$q(T) = 0 \text{ on } \mathcal{S}_o, \tag{3.83}$$

$$q(T) = 0 \text{ on } \mathcal{S}_w, \tag{3.84}$$

$$\mathcal{B}_d T = q(T) + h T = h d(t) \text{ on } \mathcal{S}_d. \tag{3.85}$$

Again, we used the notation T for the temperature, c_p for heat capacity at constant pressure, K for the thermal conductivity and h for the heat transfer coefficient. The heat normal flux q on $\mathcal{S}$ is given by

$$q(T) = K \nabla T \cdot \boldsymbol{n}.$$

First we solve the Navier-Stokes equation system (3.75)–(3.79) for the velocity $\boldsymbol{v}$. Then we solve the control problem (3.80)–(3.85), which is linear in T, in the same way as we solved the control thermal problem in the previous example.

We have set the following parameters for the problem:

$$M_1 = 22°\text{C}, \quad A_1 = 3°\text{C}, \quad \beta_1 = \frac{2\pi}{86400\text{s}},$$

$$M_2 = 5°\text{C}, \quad A_2 = 10°\text{C}, \quad \beta_3 = \frac{2\pi}{86400\text{s}},$$

$$\rho = 1\frac{\text{kg}}{\text{m}^3}, \quad \mu = 0.001\text{Pa s}, \quad K = 0.001\frac{\text{W}}{\text{m °C}},$$

$$c_p = \frac{1\text{J}}{\text{kg °C}} \quad h = 0.01\frac{\text{W}}{\text{m}^2 °\text{C}}$$

and we have solved the problem for $t \in (0, 2 \times 86400s) = (0, 2\,\text{days})$. The initial data for the energy equation is chosen to be $\phi(x) = 18°\text{C}$.

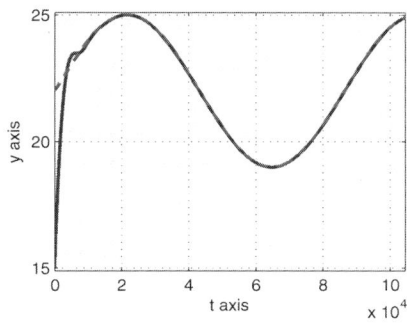

Fig. 3.36: $y(t)$ and $y_r(t)$.

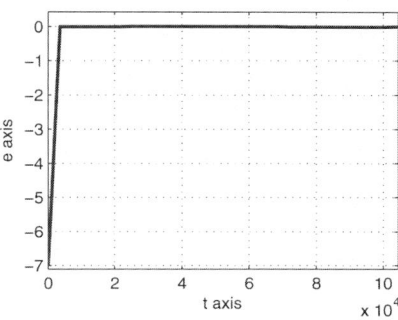

Fig. 3.37: $e(t)$.

The time evolution $y(t)$ and $y_r(t)$ is given in Fig. 3.36 for all time. The asymptotic convergence to zero of the error $e(t) = y(t) - y_r$ is given in Fig. 3.37. The disturbance $d(t)$ and the control $u(t)$ are given in Figs. 3.38 and 3.39 for all times.

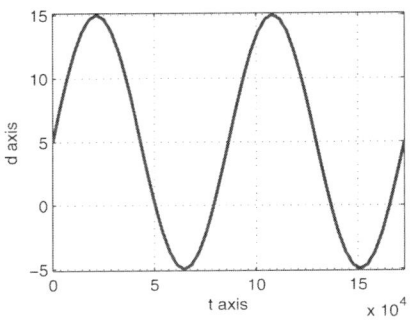

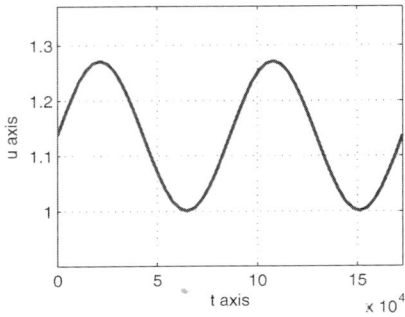

Fig. 3.38: $d(t)$.   **Fig. 3.39**: $u(t)$.

In Figs. 3.40 a screenshot of the velocity stream lines and the temperature map T is given at time $t = 48$ hours.

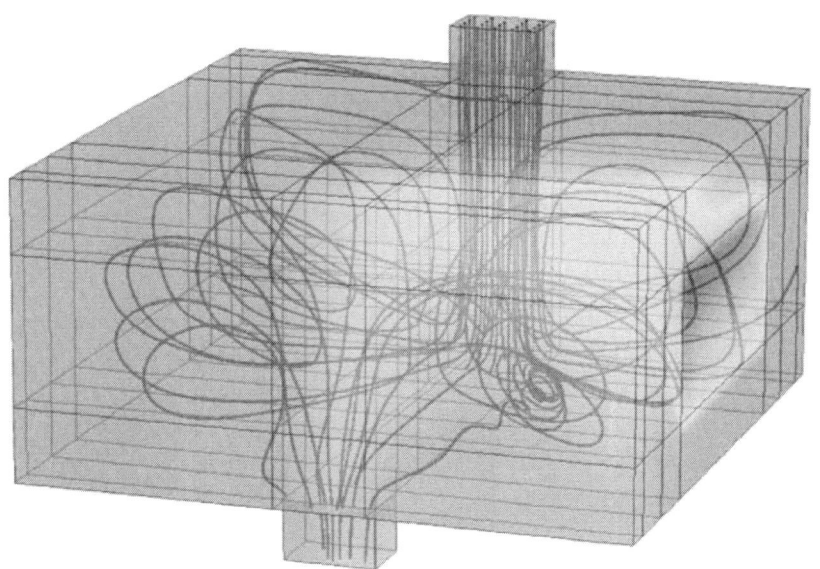

Fig. 3.40: Solution $t = 48$ Hours.

3.8   Using Fourier Series for Tracking Periodic Signals

We consider a MIMO linear regulator problem

$$z_t(t) = Az(t) + B_\mathrm{d}d + B_\mathrm{in}u, \quad t > 0,$$
$$z(0) = z_0,$$
$$y = Cz,$$

where the objective is to find a vector of controls u so that the vector of outputs $y(t)$ tracks a vector of periodic reference signals $y_r(t)$, while rejecting a vector of periodic disturbances d, as t tends to infinity. Here we assume there are n_in inputs, n_out outputs and n_d disturbances. In the most common case we will have $n_\mathrm{in} = n_\mathrm{out}$. So we have

$$B_\mathrm{in}u = B_\mathrm{in}^1 u_1 + B_\mathrm{in}^2 u_2 + \cdots B_\mathrm{in}^{n_\mathrm{in}} u_{n_\mathrm{in}},$$
$$B_\mathrm{d}d = B_\mathrm{d}^1 u_1 + B_\mathrm{d}^2 u_2 + \cdots B_\mathrm{d}^{n_\mathrm{d}} u_{n_\mathrm{d}},$$
$$Cz = \begin{bmatrix} C_1 z, & C_2 z, \cdots, C_{n_\mathrm{out}} z \end{bmatrix}.$$

In order to compute the controls $u_j(t)$ we will proceed by solving a sequence of simpler problems and enforce the superposition principle to obtain the desired control laws. Namely, we will solve a sequence of $n_\mathrm{out} + n_\mathrm{d}$ problems for controls that can be added together to obtain the desired control.

In our first sequence of problems we will solve, for each $\ell = 1, \cdots, n_\mathrm{out}$, a tracking problem in which there is a single signal to be tracked and no disturbance. Thus we consider for each fixed ℓ

$$y_{r,\ell} \neq 0 \quad \text{while} \quad y_{r,j} = 0 \text{ for all } j \neq \ell \text{ and } d_j = 0 \text{ for all } j.$$

Then we will solve a sequence of n_d disturbance rejection problems for $\ell = 1, \cdots, n_\mathrm{d}$. In this case, for each fixed ℓ we assume that there is a single disturbance to be rejected and no signals to be tracked. In particular, for each ℓ we will have

$$d_\ell \neq 0 \quad \text{while} \quad d_j = 0 \text{ for all } j \neq \ell \text{ and } y_{r,j} = 0 \text{ for all } j.$$

Thus for each $\ell = 1, \cdots, n_\mathrm{out} + n_\mathrm{d}$, as described above, we will consider the control system

$$z_t(x,t) = Az(x,t) + \sum_{j=1}^{n_\mathrm{d}} (B_\mathrm{d}^j d_j)(x,t) + \sum_{j=1}^{n_\mathrm{in}} (B_\mathrm{in}^j u_j)(x,t),$$

in which there is either only one nonzero reference signal $y_{r,\ell}$ to be tracked or only one nonzero disturbance d_ℓ to be rejected.

We also assume that $y_{r,\ell}$ is a periodic function with period $2P^{r,\ell}$, possessing the following Fourier series representation:

$$y_{r,\ell}(t) = \frac{a_0^{r,\ell}}{2} + \sum_{k=1}^{\infty} \left\{ a_k^{r,\ell} \cos\left(\frac{k\pi t}{P^{r,\ell}}\right) + b_k^{r,\ell} \sin\left(\frac{k\pi t}{P^{r,\ell}}\right) \right\}, \quad \ell = 1, \cdots, n_{\text{out}},$$

and we assume that $y_{r,\ell}$ is in the class $C^{(p)}[0, 2P^{r,\ell}]$, for some $p \geq 1$.

With these assumptions the coefficients in the series satisfy

$$\sum_{k=1}^{\infty} |k^p a_k^{r,\ell}| < \infty \quad \text{and} \quad \sum_{k=1}^{\infty} |k^p b_k^{r,\ell}| < \infty,$$

and in particular this implies that

$$k^p a_k^{r,\ell} \xrightarrow{k \to \infty} 0 \quad \text{and} \quad k^p b_k^{r,\ell} \xrightarrow{k \to \infty} 0.$$

It follows that given any desired level of tracking accuracy, we can choose a truncation level $N^{r,\ell}$ for the infinite Fourier series that will generate a finite-dimensional exosystem and at the same time achieve the desired approximation for the control law.

Thus, we truncate the infinite series (if it is indeed infinite) at the value $N^{r,\ell}$ to obtain

$$y_{r,\ell}(t) \simeq y_{r,\ell}^{N^{r,\ell}}(t) = \frac{a_0^{r,\ell}}{2} + \sum_{k=1}^{N^{r,\ell}} \left\{ a_k^{r,\ell} \cos\left(\frac{k\pi t}{P^{r,\ell}}\right) + b_k^{r,\ell} \sin\left(\frac{k\pi t}{P^{r,\ell}}\right) \right\},$$

for $\ell = 1, \cdots, n_{\text{out}}$.

Making the same assumptions for the disturbance d_ℓ, we end up with the following finite term Fourier series approximation

$$d_\ell(t) \simeq d_\ell^{N^{d,\ell}}(t) = \frac{a_0^{d,\ell}}{2} + \sum_{k=1}^{N^{d,\ell}} \left\{ a_k^{d,\ell} \cos\left(\frac{k\pi t}{P^{d,\ell}}\right) + b_k^{d,\ell} \sin\left(\frac{k\pi t}{P^{d,\ell}}\right) \right\},$$

for $\ell = 1, \cdots, n_{\text{d}}$.

At this point we depart somewhat from the developments presented in Chapters 1 and 2. The idea remains exactly the same but we try to organize things a bit differently to accommodate the many Fourier series representations for the reference signals, disturbances and controls.

Let us first introduce some new notation by defining the set of functions f_ℓ which incorporates all the approximate signals to be tracked and all the approximate disturbances to be rejected.

$$f_\ell(t) = \sum_{k=0}^{N^\ell} \left(a_k^\ell \cos(\alpha_k^l t) + b_k^\ell \sin(\alpha_k^\ell t) \right), \quad \text{for } \ell = 1, \cdots, n_{\text{out}} + n_{\text{d}}, \quad (3.86)$$

where we define

$$
N^\ell = \begin{cases} N^{r,\ell} & \text{for } \ell = 1, \cdots, n_{\text{out}}, \\ N^{d,m} & \text{for } \ell = (n_{\text{out}} + 1), \cdots, (n_{\text{out}} + n_{\text{d}}) \text{ and } m = (\ell - n_{\text{out}}), \end{cases}
$$

$$
\alpha_k^\ell = \begin{cases} \dfrac{k\pi}{P^{r,\ell}} & \text{for } \ell = 1, \cdots, n_{\text{out}}, \\ \dfrac{k\pi}{P^{d,m}} & \text{for } \ell = (n_{\text{out}} + 1), \cdots, (n_{\text{out}} + n_{\text{d}}) \text{ and } m = (\ell - n_{\text{out}}), \end{cases}
$$

$$
a_k^\ell = \begin{cases} \begin{cases} \dfrac{a_0^{r,\ell}}{2} & \text{for } k = 0, \\ a_k^{r,\ell} & \text{for } k \geq 1, \end{cases} & \text{for } \ell = 1, \cdots, n_{\text{out}}, \\[2em] \begin{cases} \dfrac{a_0^{d,m}}{2} & \text{for } k = 0, \\ a_k^{d,m} & \text{for } k \geq 1, \end{cases} & \begin{aligned} &\text{for } \ell = n_{\text{out}} + 1, \cdots, n_{\text{out}} + n_{\text{d}}, \\ &\text{and } m = \ell - n_{\text{out}}, \end{aligned} \end{cases}
$$

$$
b_k^\ell = \begin{cases} \begin{cases} 0 & \text{for } k = 0, \\ b_k^{r,\ell} & \text{for } k \geq 1, \end{cases} & \text{for } \ell = 1, \cdots, n_{\text{out}}, \\[2em] \begin{cases} 0 & \text{for } k = 0, \\ b_k^{d,m} & \text{for } k \geq 1, \end{cases} & \begin{aligned} &\text{for } \ell = n_{\text{out}} + 1, \cdots, n_{\text{out}} + n_{\text{d}}, \\ &\text{and } m = \ell - n_{\text{out}}. \end{aligned} \end{cases}
$$

Remark 3.4. Note that in Equation (3.86) all the constant terms in the Fourier series have been moved into the sum of harmonic signals. This did not change the nature of the signals due to $\alpha_0^\ell = 0$ and $b_0^\ell = 0$ for any ℓ. This choice is motivated by the idea of building a general solver that can take into account either constant or harmonic signals without making any distinction. Generally speaking we consider a constant signal to be a harmonic signal with frequency zero.

As already mentioned we will solve only one problem at a time. Clearly for $\ell = 1, \cdots, n_{\text{out}}$ we solve a tracking problem, while for $\ell = n_{\text{out}} + 1, \cdots, n_{\text{out}} + n_{\text{d}}$ we solve a disturbance rejection problem. Furthermore, for any signal ℓ we will consider $N^\ell + 1$ separate problems, each of them corresponding to the tracking (or rejection) of the response from a single harmonic oscillator, i.e,

$$
a_k^\ell \cos(\alpha_k^\ell t) + b_k^\ell \sin(\alpha_k^\ell t), \quad \text{for } k = 0, \cdots, N^\ell, \quad \text{for } \ell = 1, \cdots, n_{\text{out}} + n_{\text{d}}.
$$

Once all these problems are solved we apply the superposition principle to find the desired control vector u.

Fix ℓ and k, and consider the tracking (or rejecting) of the single pair of exosystem variables denoted by

$$
w_k^\ell = \begin{bmatrix} w_{k_1}^\ell, & w_{k_2}^\ell \end{bmatrix}^\mathsf{T} = \begin{bmatrix} \cos\left(\alpha_k^\ell t\right), & \sin\left(\alpha_k^\ell t\right) \end{bmatrix}^\mathsf{T},
$$

so that we track (or reject) the signal given by

$$a_k^\ell w_{k_1}^\ell + b_k^\ell w_{k_2}^\ell.$$

For this we solve the two-dimensional regulator equations

$$\Pi_k^\ell S_k^\ell = A\Pi_k^\ell + B_{\mathrm{d}} P_k^\ell + B_{\mathrm{in}} \Gamma_k^\ell,$$
$$C\Pi_k^\ell = Q_k^\ell,$$

where

$$\Pi_k^\ell = \begin{bmatrix} \Pi_{k_1}^\ell & \Pi_{k_2}^\ell \end{bmatrix},$$

$$S_k^\ell = \begin{bmatrix} 0 & -\alpha_k^\ell \\ \alpha_k^\ell & 0 \end{bmatrix},$$

$$\Gamma_k^\ell = \begin{bmatrix} \Gamma_{k_1}^\ell \\ \Gamma_{k_2}^\ell \\ \vdots \\ \Gamma_{k_{n_{in}}}^\ell \end{bmatrix} = \begin{bmatrix} \Gamma_{k_{1,1}}^\ell & \Gamma_{k_{1,2}}^\ell \\ \Gamma_{k_{2,1}}^\ell & \Gamma_{k_{2,2}}^\ell \\ \vdots & \vdots \\ \Gamma_{k_{n_{in},1}}^\ell & \Gamma_{k_{n_{in},2}}^\ell \end{bmatrix},$$

$$Q_k^\ell = \begin{cases} \begin{bmatrix} Q_{k_1}^\ell \\ \vdots \\ Q_{k_\ell}^\ell \\ \vdots \\ Q_{n_{\mathrm{out}}}^\ell \end{bmatrix} = \begin{bmatrix} 0 & 0 \\ \vdots & \vdots \\ a_k^\ell & b_k^\ell \\ \vdots & \vdots \\ 0 & 0 \end{bmatrix}, & \text{for } \ell = 1, \cdots, n_{\mathrm{out}}, \\[4em] \begin{bmatrix} Q_{k_1}^\ell \\ \vdots \\ Q_{n_{\mathrm{out}}}^\ell \end{bmatrix} = \begin{bmatrix} 0 & 0 \\ \vdots & \vdots \\ 0 & 0 \end{bmatrix}, & \text{for } \ell = n_{\mathrm{out}} + 1, \cdots, n_{\mathrm{out}} + n_{\mathrm{d}}, \end{cases}$$

and

$$P_k^\ell = \begin{cases} \begin{bmatrix} P_{k_1}^\ell \\ \vdots \\ P_{n_{\mathrm{d}}}^\ell \end{bmatrix} = \begin{bmatrix} 0 & 0 \\ \vdots & \vdots \\ 0 & 0 \end{bmatrix}, & \text{for } \ell = 1, \cdots, n_{\mathrm{out}}, \\[4em] \begin{bmatrix} P_{k_1}^\ell \\ \vdots \\ P_{k_m}^\ell \\ \vdots \\ P_{n_{\mathrm{d}}}^\ell \end{bmatrix} = \begin{bmatrix} 0 & 0 \\ \vdots & \vdots \\ a_k^\ell & b_k^\ell \\ \vdots & \vdots \\ 0 & 0 \end{bmatrix}, & \begin{aligned} &\text{for } \ell = n_{\mathrm{out}} + 1, \cdots, n_{\mathrm{out}} + n_{\mathrm{d}}, \\ &\text{and } m = \ell - n_{\mathrm{out}}. \end{aligned} \end{cases}$$

Just as we have seen many times, in order to solve these regulator equations, we introduce the matrix G defined as

$$G = -CA^{-1}B_{\mathrm{in}},$$

and the new auxiliary variable $\widetilde{\Pi}_k^\ell$ defined as

$$\widetilde{\Pi}_k^\ell = -A^{-1}(-\Pi_k^\ell S_k^\ell + B_{\mathrm{d}}^\ell P_k^\ell).$$

With these, the gains are given by

$$\Gamma_k^\ell = G^{-1}\left(Q_k^\ell - C\widetilde{\Pi}_k^\ell\right).$$

For $j = 1, \cdots, n_{\mathrm{in}}$ we combine all the tracking and disturbance rejection controls to finally obtain

$$u_j(t) = \sum_{\ell=1}^{n_{\mathrm{out}}+n_{\mathrm{d}}} \sum_{k=0}^{N^\ell} \Gamma_{k_j}^\ell w_k^\ell.$$

Example 3.1. In this example we consider a MIMO tracking, disturbance rejection problem for a two-dimensional heat equation in a rectangle of height one unit and width two units. Here, as usual we use the coordinates $\boldsymbol{x} = (x_1, x_2)$ where x_1 denotes the horizontal direction and x_2 the vertical one. Our rectangle is oriented so that the lower left corner is at $(-1, 0)$ and the upper right corner is at $(1, 1)$ so that $-1 < x_1 < 1$ and $0 < x_2 < 1$.

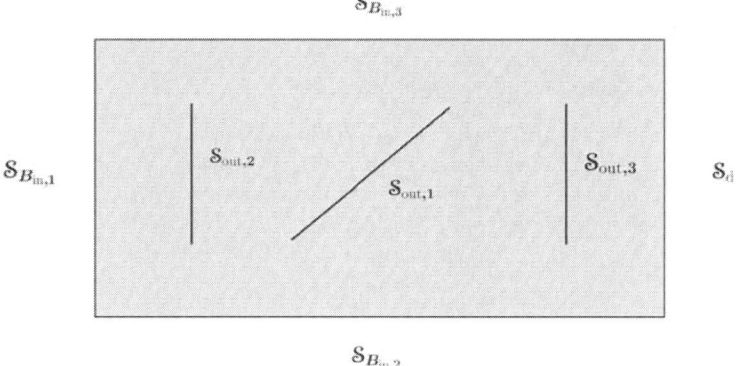

Fig. 3.41: Two-Dimensional Domain with Boundaries.

For this example we have three inputs, three outputs and one disturbance. The control inputs take place through Robin (or Neumann) boundary operators $\mathcal{S}_{B_{\mathrm{in},j}}$ (for $j = 1, 2, 3$) on three sides of the boundary with a disturbance entering through the remaining side $\mathcal{S}_{\mathrm{d}}$. The measured outputs consist of the average temperatures over three lines inside the rectangular domain, $\mathcal{S}_{\text{out},i}$ for $i = 1, 2, 3$. The problem can be written as a boundary control system with $z = z(\boldsymbol{x}, t)$ as the state variable in the form

$$z_t(\boldsymbol{x}, t) = c\Delta z(\boldsymbol{x}, t),$$
$$z(\boldsymbol{x}, 0) = \phi(\boldsymbol{x}),$$
$$\mathcal{B}_{\text{in}}^1(z)(t) = (q_1 + c\partial_{\boldsymbol{n}})\, z = u_1(t) \quad \text{on } \mathcal{S}_{\text{in},1},$$
$$\mathcal{B}_{\text{in}}^2(z)(t) = (q_2 + c\partial_{\boldsymbol{n}})\, z = u_2(t) \quad \text{on } \mathcal{S}_{\text{in},2},$$
$$\mathcal{B}_{\text{in}}^3(z)(t) = (q_3 + c\partial_{\boldsymbol{n}})\, z = u_3(t) \quad \text{on } \mathcal{S}_{\text{in},3},$$
$$\mathcal{B}_{\text{d}}(z)(t) = z = d(t) \quad \text{on } \mathcal{S}_{\text{d}},$$

where $\partial_{\boldsymbol{n}}$ denotes the normal derivative and u_j are the desired control inputs. Notice in this notation we are assuming that the scalar control inputs and the disturbance enter along the entire boundary surface $\mathcal{S}$.

The measured outputs are given by

$$y_1(t) = C_1(z)(t) = \frac{1}{|\mathcal{S}_{\text{out},1}|} \int_{\mathcal{S}_{\text{out},1}} z(s, t)\, ds,$$

$$y_2(t) = C_2(z)(t) = \frac{1}{|\mathcal{S}_{\text{out},2}|} \int_{\mathcal{S}_{\text{out},2}} z(s, t)\, ds,$$

$$y_3(t) = C_3(z)(t) = \frac{1}{|\mathcal{S}_{\text{out},3}|} \int_{\mathcal{S}_{\text{out},3}} z(s, t)\, ds.$$

In our specific numerical example we set $c = 0.1$, $q_1 = 0$, $q_2 = 5$ and $q_3 = 1$. The slanted line $\mathcal{S}_{\text{out},1}$ is the line between the points $(-.25, .25)$ and $(.25, .75)$. The line $\mathcal{S}_{\text{out},2}$ has end points $(-.5, .25)$ and $(-.5, .75)$ and $\mathcal{S}_{\text{out},3}$ has end points $(.5, .25)$ and $(.5, .75)$.

The desired reference signals are given by

$$y_{r,1}(t) = 1 + \cos(t) + \sin(2t) + \cos(5t) + \sin(10t).$$

The reference signals $y_{r,2}$ and $y_{r,3}$ are periodic sawtooth functions defined by

$$y_{r,2} = \begin{cases} \dfrac{(t - t_0)}{\sqrt{2}}, & t_0 \le t < t_0 + \sqrt{2}, \\[2mm] 2 - \dfrac{(t - t_0)}{\sqrt{2}}, & t_0 + \sqrt{2} < t < t_0 + 2\sqrt{2}, \end{cases}$$

$$y_{r,3} = \begin{cases} \dfrac{t - t_0}{\sqrt{3}}, & t_0 \le t < t_0 + \sqrt{3}, \\[2mm] 1 - \dfrac{t - t_0}{\sqrt{3}}, & t_0 + \sqrt{3} < t < t_0 + 2\sqrt{3}. \end{cases}$$

Finally we consider a disturbance in the form

$$d(t) = 2 + \sin(t) + \cos(2t) + \sin(5t) + \cos(10t).$$

Note that $P^1 = P^4 = \pi$, $P^2 = \sqrt{2}$ and $P^3 = \sqrt{3}$, so that the periods are incommensurate. We also approximate the signals with several different truncation levels for comparison: $N^1 = N^2 = N^3 = N^4 = 20$. While this gives an exact representation of the signals for $y_1(t)$ and $d(t)$, for the signals $y_2(t)$ and $y_3(t)$ this results in only an approximation.

We solve this problem for $0 < t < 14$ and zero initial data. The time evolution of $y_1(t)$ and $y_{r,1}(t)$ is given in Fig. 3.42 for all time, while the asymptotic convergence to zero of the error $e_1(t) = y_1(t) - y_{r,1}$ is given in Fig. 3.43.

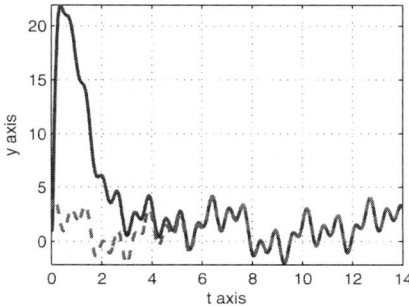
Fig. 3.42: $y_1(t)$ and $y_{r,1}(t)$.

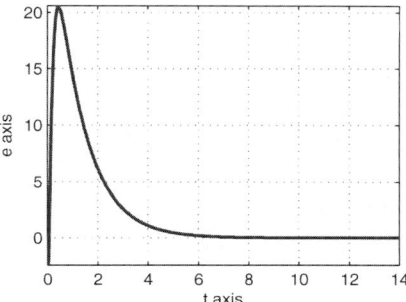
Fig. 3.43: $e_1(t)$.

Similarly, the time evolution of $y_2(t)$ and $y_{r,2}(t)$ is given in Fig. 3.44 for all time, and the asymptotic convergence to zero of the error $e_2(t) = y_2(t) - y_{r,2}$ is given in Fig. 3.45.

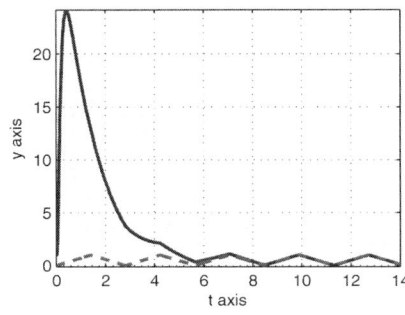
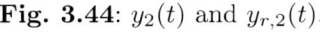
Fig. 3.44: $y_2(t)$ and $y_{r,2}(t)$.

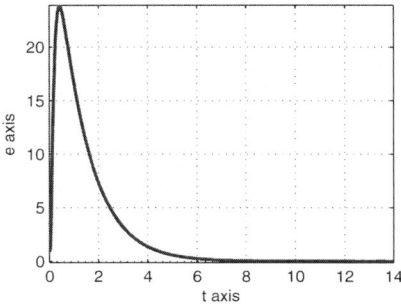
Fig. 3.45: $e_2(t)$.

Finally, the time evolution of $y_3(t)$ and $y_{r,3}(t)$ is given in Fig. 3.46 for all time, and the asymptotic convergence to zero of the error $e_3(t) = y_3(t) - y_{r,3}$ is given in Fig. 3.47.

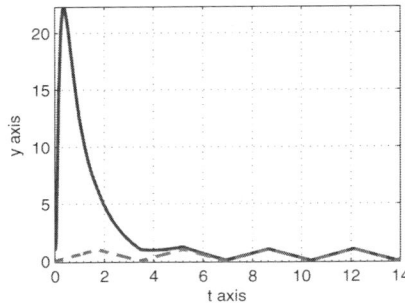

Fig. 3.46: $y_3(t)$ and $y_{r,3}(t)$.

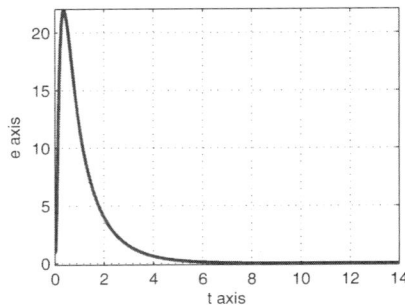

Fig. 3.47: $e_3(t)$.

3.9   Zero Dynamics Inverse Design

Up to now, in all of our examples, the measured outputs and control laws were functions of time t alone. That is, we have never considered an example in which we had a distributed output in the form $y = C(z) = z(x, t)$ for x on a hypersurface inside or on the boundary of the spatial domain. Furthermore, the signals to be tracked have not been distributed, i.e., in the form $y_r(x, t)$. The main difficulty in applying the geometric methodology to regulation problems for systems with infinite-dimensional input and output spaces is that, in order to obtain the control $u(x, t)$, one needs to invert the transfer function $G(x, s) = C(sI - A)^{-1}B_{\text{in}}$ which is given in terms of an integral operator with a kernel $K(x, \xi; s)$, i.e.,

$$G(x, s) = \int_S K(x, \xi; s)\, d\xi$$

for some surface S. It is far more difficult to invert such an operator than it is for all the examples considered so far in the book. The authors have applied such a numerical scheme to obtain approximate solutions problems with infinite-dimensional input and output spaces in which they obtain very accurate approximate solutions of the corresponding regulator equations. This idea was pursued in a Texas Tech University Masters Thesis [73] (available online through the Texas Tech Library) in connection with geometric methods for optimal control problems. Rather than side track to investigate these topics here we present a detailed discussion of the so-called "zero dynamics" design method which works very well for boundary control problems with infinite-dimensional input and output spaces.

In particular, for co-located boundary and interior control systems the zero dynamics method provides a design methodology that is equivalent to the regulator theory approach but avoids the explicit inversion of the infinite-dimensional transfer function. In particular, in this section we discuss the zero dynamics inverse design method for distributed parameter systems governed by linear parabolic equations, with co-located boundary control and sensing as it was developed in [18].

The idea of zero dynamics presented in [18] is motivated by the classical root-locus construct which has been an important tool in finite-dimensional linear systems theory for many years. In particular for a scalar proportional output feedback the trajectories of the closed loop system converge to those of the zero dynamics system as the feedback gain tends to infinity. The main contribution of [18] is to provide a modeling and design mechanism for solving certain tracking and disturbance rejection problems. This mechanism provides what appears to be an alternative to the geometric approach involving regulator equations. As we will show below, it is actually just a very special case of the geometric method. The approach is based on the fact that for certain boundary control systems it is easy to model the system's "zero dynamics", which, in turn, provides an easy to understand systematic methodology for solving many interesting problems of output regulation.

Historically the utility of zero dynamics design for linear and nonlinear lumped systems is well documented. Indeed there is a considerable literature devoted to this topic (see, e.g., [42, 40, 91, 55, 28]). The classical development and definitions of zero dynamics for finite-dimensional systems involves normal forms, invariant output zeroing subspaces and so on. We conclude that it would not be of practical interest to include the formal definition of zero dynamics found in the literature of finite-dimensional systems. Rather, since we are mostly interested in infinite-dimensional systems, we direct the reader to the discussions found in [29, 63, 64].

In classical automatic control there is a vast literature concerned with system zeros, e.g., root-locus design. Along these lines there are also a few research articles devoted to zero dynamics and (transmission and invariant) zeros for SISO relative degree one linear distributed parameter systems [93, 29, 63, 64]. Most of the papers in the literature including these are concerned with systems having bounded input and output operators.

In what follows we give a very brief discussion of root-locus methods and its relationship to transmission zeros. Then we will provide a formal argument motivating our definition of zero dynamics for co-located boundary control systems.

For finite-dimensional scalar-input scalar-output (SISO) linear systems, there are several classical graphical techniques, such as the root-locus method, Bode-plot representations, Nyquist diagrams and Nichols charts in the frequency domain which have been used for many years in classical automatic control design to facilitate the design of stabilizing feedback control laws. In particular, the root-locus method has been widely used because it deals directly, efficiently, and visually with the problem of the location of the closed loop system poles, which govern the stability and natural behavior of the system. It provides a valuable aid in system design, especially in showing stability margins and in evaluating and improving the performance of a feedback system.

In its simplest form the idea of root-locus can be understood by considering a SISO finite-dimensional linear system as

$$z_t = Az + Bu,$$
$$y = Cz,$$

where $z \in \mathbb{R}^n$, $B \in \mathcal{L}(\mathbb{R}, \mathbb{R}^n)$, $C \in \mathcal{L}(\mathbb{R}^n, \mathbb{R})$ (i.e., SISO system).

In addition, for simplicity we will assume the system has relative degree one, which is equivalent to $CB \neq 0$. For this system the transfer function is a rational function given by

$$G_0(s) = C(sI - A)^{-1}B,$$

which we can write in the form

$$G_0(s) = \frac{N(s)}{D(s)} = \frac{b_0 + b_1 s + \cdots + s^{n-1}}{a_0 + a_1 s + \cdots + s^n}.$$

The zeros of the numerator are called the *open loop zeros* and the zeros of the denominator are called the *open loop poles*. For a proportional error feedback law

$$u(t) = -ky(t) + v(t),$$

the closed loop transfer function becomes

$$G_k(s) = \frac{N(s)}{D(s) + kN(s)}.$$

We see that the zeros of the closed loop system are the same zeros as the open loop system (up to possible cancellations). Those zeros which are not in the spectrum of A are called the invariant zeros or transmission zeros. The closed loop poles are given by the zeros of the denominator (i.e., the roots of the return difference equation),

$$D(s) + kN(s) = 0.$$

The sign of $cb \neq 0$ is referred to as the "instantaneous gain" which determines whether for $k > 0$ or $k < 0$ the single infinite branch of the root-locus goes to $\pm\infty$. This is also called the high frequency gain, since

$$G_0(s) = C(sI - A)^{-1}B = \sum_{i=1}^{\infty} \frac{CA^{i-1}B}{s^i} \sim \frac{CB}{s}, \quad s \gg 1.$$

Perhaps the simplest and best known result from finite-dimensional SISO root-locus theory is that *"The closed loop poles vary from the open loop poles to the open loop zeros (and infinity) as the gain k varies from 0 to ∞"*.

The idea is that the closed loop poles converge to the eigenvalues of a system governed by the open loop zeros. In other worlds, the trajectories of the closed loop system converge to the trajectories of a system determined by the open loop zeros. This system, which is a bit difficult to describe precisely without a fairly lengthy excursion, is the zero dynamics system. Thus we say that trajectories of the closed loop system converge to trajectories of the "zero dynamics" as the gain k tends to infinity. We reiterate, the zero dynamics is a system whose eigenvalues are the transmission zeros.

For infinite-dimensional linear systems it is typically impossible to apply directly the root-locus techniques developed for finite-dimensional linear systems. Nevertheless, due to the geometric appeal of such a theory and with the considerable interest in systems governed by partial differential equations, it is reasonable and desirable to develop a root-locus methodology, retaining as much as possible the important features of the finite-dimensional case. There has been some work in this direction either explicit or implicit. For example, in the special case of a distributed parameter control system

$$z_t = Az + Bu,$$
$$y = Cz,$$

where A is a self-adjoint operator on a Hilbert space with compact resolvent, B is bounded, C is A-bounded and the input and output spaces are finite-dimensional with the same dimensions. S. Pohjolainen [75] was able to obtain some information for a simple scalar output feedback $u = ky$, for $k \in \mathbb{R}$. In particular, in [75] it was observed that the closed loop dynamics is governed by the system

$$z_t = (A + kBC)z,$$

and it was shown that if $\det[C(\lambda I - A)^{-1}B] \not\equiv 0$ on the resolvent set $\rho(A)$ then a number $\lambda^* \notin \sigma(A)$ (i.e., is not the spectrum of A) is a transmission zero if and only if $\det[C(A - \lambda^* I)^{-1}B] = 0$. In particular, the limit eigenvalues of the operator $(A + kBC)$ for $k \to \infty$ are transmission zeros for the system (A, B, C) (see also [34], Exercise 4.28).

In [24], C. I. Byrnes, D. S. Gilliam and J. He presented a general root-locus theory for a large class of single-input single-output distributed parameter parabolic boundary feedback control problems in one space dimension.

As we have mentioned a state space analog of root-locus plots is that as the gain parameters tend to infinity, the trajectories of the closed loop system converge to the trajectories of the "zero dynamics" system. Here the authors have presented a new definition of zero dynamics suitable for co-located boundary control systems. The basic idea is very simple. Consider a SISO boundary control system

$$z_t = A_0 z, \quad z_0 = z(0), \quad \mathcal{B}z = u,$$

with measured output $y = Cz$. Then considering a proportional output feedback $u = -ky$ for $k \in \mathbb{R}$ we obtain a closed loop system

$$z_t = A_0 z, \quad z_0 = z(0), \quad (\mathcal{B} + kC)z = 0,$$

which, on dividing by k, we can also write in the form

$$z_t = A_0 z, \quad z_0 = z(0), \quad \left(C + \frac{1}{k}\mathcal{B}\right) z = v.$$

At least formally, as the gain parameter k tends to infinite we obtain a system whose dynamics are driven by a system in which we have constrained the output of the control system to zero, a main property of the classical zero dynamics from finite-dimensional system theory. Further for a large class of infinite-dimensional systems we can prove that the trajectories of the closed loop system approach the trajectories of the system

$$z_t = A_0 z, \quad z_0 = z(0), \quad Cz = 0,$$

which we call the "zero dynamics". Thus at least formally this system has the main desired properties of the classical zero dynamics system. Certainly, for exponentially stable Riesz spectral systems one can prove this to be exactly the same as for finite-dimensional systems.

In this section we are interested in linear distributed parameter systems governed by a parabolic partial differential equation. Unfortunately for general distributed parameter systems, there is no well defined classical zero dynamics (see [94]) due to a variety of factors including the nonequivalence of different forms of invariance, nonequivalence of transmission and invariant zeros and, even more serious, difficulties that arise due to the presence of unbounded densely defined operators (see [34, 94, 93]). As an example, and as we have already mentioned, even in the simplest cases it may not make sense to write expressions like AB, CA, etc, for unbounded operators A, B and C in the standard Hilbert state space.

However, for systems governed by partial differential equations with co-located boundary inputs and outputs the above described concept of zero dynamics provides a very useful analog of the classical definition. To our knowledge this definition of zero dynamics was first introduced in the work in [9] and has been exploited and developed by the authors for linear and nonlinear distributed parameter systems in several papers, see, for example, [25, 26]. As we will demonstrate in this section this definition of zero dynamics proves to be very useful in developing regulator theory for distributed parameter systems with unbounded controls.

Let us try to explain the design methodology by way of a general parabolic boundary control model. The notation follows in part the discussion in Section 2.2 from Chapter 2. We begin with a linear elliptic partial differential operator (as given in (1.8)) L acting on a domain $D(L)$ in a Hilbert state space $\mathcal{Z}$. The operator L is a state operator without any boundary conditions imposed. From this operator we define an additional densely defined operator A_0 obtained by restricting the operator L by imposing boundary conditions on some (but not all) parts of the physical domain boundary. Namely we have $A_0 = L$ with $\mathcal{D}(A_0) \supset \mathcal{D}(L)$ plus some homogeneous boundary conditions. Typically A_0 does not generate a C_0-semigroup.

We obtain an abstract boundary control system from a pair $(A_0, \mathcal{B})$ where $\mathcal{B}$ is a boundary input control operator through which the system can be influenced via an external force.

$$z_t = A_0 z + B_d d, \quad z(0) = \varphi, \tag{3.87}$$

$$\mathcal{B}_{\text{in}} z(t) = u_{\text{in}}(t), \qquad \text{(Boundary input)} \tag{3.88}$$

$$y(t) = Cz(t). \qquad \text{(Boundary output)} \tag{3.89}$$

For an operator M, we will use the notation $\ker(M)$ to denote the null space of M. In our intended applications the operator A_0 is a densely defined operator with the following additional properties (see the discussion of Hilbert scales in Section 2.2.1. In particular, see the definitions of $\mathcal{Z}_1$ in (2.13) and $\mathcal{Z}_{-1}$ in (2.14)).

Assumption 3.1.

1. *The operator $A = A_0$ with $\mathcal{D}(A) = \ker(\mathcal{B}_{in}) \cap \mathcal{D}(A_0)$ is the generator of an exponentially stable C_0-semigroup $T(t)$ in $\mathcal{Z}$ and in $\mathcal{Z}_1$ (see Lemma 2.2 in Section 2.2.1).*

$$\|T(t)\varphi\|_{\mathcal{Z}_1} \le M_1(t_0)e^{-\alpha t}\|\varphi\|, \quad \forall \ t \ge t_0 > 0, \tag{3.90}$$

where $\varphi \in \mathcal{Z}$, $\|\cdot\|$ denotes the norm in $\mathcal{Z}$ and $\lim_{t_0 \to 0} M_1(t_0) = \infty$.

2. *The operator $\widehat{A} = A_0$ with $\mathcal{D}(\widehat{A}) = \ker(\mathcal{C}) \cap \mathcal{D}(A_0)$ is also the generator of an exponentially stable C_0-semigroup $\widehat{T}(t)$ in $\mathcal{Z}_1$.*

$$\|\widehat{T}(t)\varphi\|_{\mathcal{Z}_1} \le C_1(t_0)e^{-\alpha t}\|\varphi\| \quad \forall \ t \ge t_0 > 0, \tag{3.91}$$

where $\varphi \in \mathcal{Z}$ and $\lim_{t_0 \to 0} C_1(t_0) = \infty$.

3. $\mathcal{Z}_1 \subset \mathcal{D}(C)$, $\mathcal{Z}_1 \subset \mathcal{D}(\mathcal{B}_{in})$ *and* $C \in \mathcal{L}(\mathcal{Z}_1, \mathcal{Y})$ *and* $\mathcal{B}_{in} \in \mathcal{L}(\mathcal{Z}_1, \mathcal{U})$.

The zero dynamics controller is obtained from the zero dynamics system by choosing the desired control as an output given by the boundary control operator $\mathcal{B}_{\text{in}}$ applied to the solution of the zero dynamics. Here the zero dynamics system is obtained by constraining the error defined by $e(t) = \mathcal{C}\xi(t) - y_r(t)$ to be identically zero (the analog of constraining the output to zero by passing to the limit as the gain parameter passes to infinity as in our remarks concerning the zero dynamics for boundary control systems).

Definition 3.1 (Zero Dynamics Controller).

$$\xi_t = A_0\xi + B_d d, \quad \xi(0) = \psi, \tag{3.92}$$

$$\mathcal{C}\xi(t) = y_r, \tag{3.93}$$

$$u_{\text{out}}(t) = \mathcal{B}_{\text{in}}\xi(t), \tag{3.94}$$

where y_r (here considered as an input) is the original signal to be tracked.

Remark 3.5. It is well known (see [61, 60]) that for problems governed by initial boundary value problems for parabolic partial differential equations the initial boundary problem (3.87)–(3.89) has a unique strong solution z. More precisely, for any $T > 0$ we have

$$z \in C([0,T], L^2(\Omega)) \cap L^2([0,T], H^1(\Omega)).$$

Moreover, for $t > 0$ this solution is instantly classical, which means in particular, that $z(\cdot, t) \in H^s(\Omega)$ for any $s > 0$.

Proposition 3.1. *For the system* (3.87)–(3.89), *the problem of output regulation (for all initial data φ) is solved by the controller given in* (3.92)–(3.94) *with $u_{in}(t) = u_{out}(t)$ in* (3.94).

The proof of Proposition 3.1 is very simple due to our strong assumptions but, from a practical point of view, the proposition provides a powerful tool for constructing controllers for co-located boundary controlled distributed parameter systems.

Proof. Let $u(t) = \mathcal{B}_{\text{in}}\xi(t)$ as in (3.94) and define $\eta(t) = z(t) - \xi(t)$ where z is the solution to (3.87)–(3.89) and ξ the solution to (3.92)–(3.94). Then it is easy to show that η satisfies

$$\eta_t = A\eta, \tag{3.95}$$
$$\eta(0) = \varphi - \psi \equiv \eta_0, \tag{3.96}$$
$$e(t) = \mathcal{C}\eta(t), \tag{3.97}$$

where A is defined in Assumption 3.1. The result now follows from Remark 3.5. Indeed, it is well known (cf., [61], [60]) that the solution η of (3.95)–(3.96) is a continuous (in x and t for $t > 0$) globally defined solution which satisfies

$$\sup_{x \in \overline{\Omega}} |\eta(x,t)| \xrightarrow{t \to \infty} 0.$$

More precisely, $\eta(\cdot, t) \in H^s(\Omega)$ for $s > [n/2] + 1$ which implies $\eta(\cdot, t) \in C(\overline{\Omega})$ and our result follows from (3.90). Namely, we have

$$\|\eta(\cdot, t)\|_{H^s(\Omega)} \le M_s(t_0)e^{-\alpha t}\|\eta_0\|, \quad \text{for some} \ \alpha > 0, \quad t \ge t_0 > 0,$$

and the result follows from (3.97), the estimate (3.91) and the Sobolev Embedding Theorem. $\qquad\square$

Under our strong assumptions on the systems (3.87)–(3.89) and (3.92)–(3.94) given as *Abstract Boundary Control Systems*, it is well known that there are operators B and C which, as we have already mentioned in Chapter 2, may not be easy to describe explicitly and, in practice, involve complicated spaces and relations. For our purposes it is sufficient to know that such mappings exist. Their utility is that they allow us to write the systems (3.87)–(3.89) and (3.92)–(3.94) in standard systems form. For a detailed discussion of this process see M. Tucsnak and G. Weiss [84]. Namely system (3.87)–(3.89) can be rewritten as

$$z_t = Az + B_{\mathrm{d}}d + B_{\mathrm{in}}u,$$
$$y = \mathcal{C}z,$$

where $A = A_0$ with $D(A) = D(A_0) \cap \ker(\mathcal{B}_{\mathrm{in}})$, and the system (3.92)–(3.94) can be written as

$$\xi_t = \widetilde{A}\xi + Cy_r,$$
$$u = \mathcal{B}_{\mathrm{in}}\xi,$$

where $\widetilde{A} = A_0$ with $D(\widetilde{A}) = D(A_0) \cap \ker(\mathcal{C})$.

Formally, we can apply the Laplace transform to these systems to obtain

$$s\widehat{z} = A\widehat{z} + B_{\mathrm{in}}\widehat{u},$$
$$\widehat{y} = \mathcal{C}\widehat{z},$$

so that

$$(sI - A)\widehat{z} = B_{\mathrm{in}}\widehat{u} \;\Rightarrow\; \widehat{z} = (sI - A)^{-1}B_{\mathrm{in}}\widehat{u} \;\Rightarrow\; \widehat{y} = \mathcal{C}(sI - A)^{-1}B_{\mathrm{in}}\widehat{u} = G(s)\widehat{u}.$$

Similarly we have

$$s\widehat{\xi} = \widetilde{A}\widehat{\xi} + C\widehat{w},$$
$$\widehat{u} = \mathcal{B}_{\mathrm{in}}\widehat{\xi},$$

$$(sI - \widetilde{A})\widehat{\xi} = C\widehat{w} \;\Rightarrow\; \widehat{\xi} = (sI - \widetilde{A})^{-1}C\widehat{w} \;\Rightarrow\; \widehat{u} = \mathcal{B}_{\mathrm{in}}(sI - \widetilde{A})^{-1}C\widehat{w} = K(s)\widehat{w}.$$

Thus we have

$$G(s) = \mathcal{C}(sI - A)^{-1}B_{\mathrm{in}}, \qquad K(s) = \mathcal{B}_{\mathrm{in}}(sI - \widetilde{A})^{-1}C.$$

Notice that, at least formally, in order to have $\widehat{y} = \widehat{w}$, we need

$$\widehat{y} = G(s)\widehat{u} = G(s)K(s)\widehat{w} \;\Rightarrow\; G(s)K(s) = I.$$

From this we see that, at least in this setting, finding the inverse of $G(s)$ is very easy as it is obtained as the transfer function of the zero dynamics.

Thankfully we can verify the above relation between G and K without explicitly knowing B_{in} and C. In particular, the desired result can be seen by noting that if $\widehat{u}_{out}$ (given in (3.94)) is used as the control input $\widehat{u}_{in}$ in (3.88), then $\widehat{y} = \widehat{w}$. This amounts to the same as showing that the transfer function of the zero dynamics $K(s)$ is the inverse of the transfer function, $G(s)$, of the control system. In other words we have $G(s)K(s) = I$.

This can be seen by considering the two problems (3.87)–(3.89) and (3.92)–(3.94) which we repeat here

$$s\widehat{z} = A_0\widehat{z} + B_d d, \qquad\qquad s\widehat{\xi} = A_0\widehat{\xi} + B_d d,$$
$$\mathcal{B}_{in}\widehat{z} = \widehat{u}_{in}, \qquad\qquad\qquad \mathcal{C}\widehat{\xi} = \widehat{w},$$
$$\widehat{y} = \mathcal{C}\widehat{z}, \qquad\qquad\qquad \widehat{u}_{out} = \mathcal{B}_{in}\widehat{\xi}.$$

Given that $\widehat{z}$ and $\widehat{\xi}$ are solutions of these problems for any $\widehat{u}_{in}$ and $\widehat{u}_{out}$ we now suppose that $\widehat{u}_{in} = \widehat{u}_{out}$ and we want to show that $\widehat{y} = \widehat{w}$.

To this end set $\widehat{\eta} = \widehat{z} - \widehat{\xi}$ and we see that

$$\mathcal{B}_{in}(\widehat{\eta}) = \mathcal{B}_{in}(\widehat{z}) - \mathcal{B}_{in}(\widehat{\xi}) = \widehat{u}_{in} - \widehat{u}_{out} = 0,$$

which implies $\widehat{\eta}$ satisfies

$$s\widehat{\eta} = A_0\widehat{\eta},$$
$$\mathcal{B}_{in}\widehat{\eta} = 0.$$

In other words we have $\eta \in D(A)$ so the above is equivalent to

$$(sI - A)\widehat{\eta} = 0.$$

But due to Assumption 3.1 the resolvent set of A contains the entire right half of the complex plane. Therefore, for every $s \in \mathbb{C}_0^+ = \{\zeta \in \mathbb{C} : \text{Re}\,(\zeta) \geq 0\}$ the operator $(sI - A)$ is invertible so that $\widehat{\eta} = 0$ but this implies

$$\widehat{y} - \widehat{w} = \mathcal{C}\widehat{z} - \mathcal{C}\widehat{\xi} = \mathcal{C}(\widehat{z} - \widehat{\xi}) = \mathcal{C}(\widehat{\eta}) = 0.$$

3.9.1 Zero Dynamics and the Regulator Equations

Let us now return to the linear state feedback regulator problem as considered in Chapters 1 and 2. Under suitable assumptions such as that the references and disturbances are generated by a neutrally stable exosystem we recall that the problem is solvable if and only if there exist mappings $\Pi \in \mathcal{L}(W, Z)$ with $\text{Ran}(\Pi) \subset \mathcal{D}(\mathcal{A})$ and $\Gamma \in \mathcal{L}(W, U)$ satisfying the "regulator equations,"

$$\Pi S = A\Pi + B_d P + B_{in}\Gamma, \qquad\qquad (3.98)$$
$$\mathcal{C}\Pi = Q. \qquad\qquad\qquad\qquad (3.99)$$

In this case a feedback law solving the state feedback regulator problem is given by

$$u(t) = \Gamma w(t).$$

As we have seen in Remark 1.5 (which also holds in the case of unbounded operators considered in Chapter 2) assuming the exosystem state matrix S is diagonalizable in the k-dimensional vector space $\mathcal{W}$ with eigenvalues λ_j, eigenvectors Φ_j and bi-orthogonal sequence Ψ_j, then we have

$$S\Phi_j = \lambda_j \Phi_j, \quad S^* \Psi_i = \overline{\lambda_i} \Psi_i, \quad \langle \Phi_i, \Psi_j \rangle_{\mathrm{w}} = \delta_{ij}.$$

In this case the operator Π in the regulator equations has the representation

$$\Pi w = \sum_{j=1}^{k} \langle w, \Psi_j \rangle (\lambda_j I - A)^{-1}(B\Gamma + P)\Phi_j. \tag{3.100}$$

So, as we have already seen, the most challenging computation is finding $G(\lambda_j)^{-1}$. But for co-located boundary control problems we can use the relation that $K(\lambda_j) = G(\lambda_j)^{-1}$ where $K(\lambda_j)$ is easily obtained by solving the systems

$$(\lambda_j I - A_0)\xi = 0, \quad \mathcal{C}\xi = Q\Psi_j \quad \Rightarrow \quad K(\lambda_j)Q\Psi_j = \mathcal{B}_{\mathrm{in}}\xi,$$

for each $j = 1, \cdots k$ which does not involve computing an inverse.

In particular, suppose that Π is given by (3.100) (see also (1.24)). Then, in order to find Γ so that (3.98)–(3.99) are satisfied, we can take $w = \Phi_\ell$ for each ℓ and apply C to Π, and use the second regulator equation, to obtain the equation

$$Q\Phi_\ell = C(\lambda_\ell I - A)^{-1}B\Gamma\Phi_\ell + C(\lambda_\ell I - A)^{-1}B_{\mathrm{d}}P\Phi_\ell,$$

for $\Gamma\Phi_\ell$, for $\ell = 1, \cdots, k$.

This equation can be rewritten as

$$G(\lambda_\ell)\Gamma\Phi_\ell = Q\Phi_\ell - C(\lambda_\ell I - A)^{-1}B_{\mathrm{d}}P\Phi_\ell,$$

where $G(\lambda_\ell)$ is the transfer function $G(s) = C(sI - A)^{-1}B$ evaluated at λ_ℓ.

In order to solve for $\Gamma\Phi_\ell$ for all ℓ we need to invert $G(\lambda_\ell)$ for $\ell = 1, \cdots, k$. Under the assumption of invertibility of $G(\lambda_\ell)$ (which we called the non-resonance assumption) and replacing $G(\lambda_\ell)^{-1}$ by $K(\lambda_\ell)$, summing over all $\ell = 1, \cdots, k$ we arrive at

$$\Gamma w = \sum_{\ell=1}^{K} \langle w, \Psi_\ell \rangle K(\lambda_\ell) \left[Q\Phi_\ell - C(\lambda_\ell I - A)^{-1}P\Phi_\ell \right].$$

So clearly the zero dynamics system can be used to obtain the desired control $u = \Gamma w$.

Remark 3.6 (Dynamic Regulator Equations). We have already observed that the zero dynamics controller given in (3.92)–(3.94), with arbitrary initial condition ψ, asymptotically solves the regulation problem. But we would like to draw attention to the relation between the zero dynamics controller and the so-called dynamic regulator equations. The regulator equations (3.98)–(3.99) were derived in Section 1.2, based on the existence of an attractive invariant subspace $\mathcal{V}^+$ in the composite space $\mathfrak{X} = \mathcal{Z} \times \mathcal{W}$. This subspace was defined by a mapping $\Pi \in \mathcal{L}(\mathcal{W}, \mathcal{Z})$ and the invariance of this subspace implies that if an initial condition starts in $\mathcal{V}^+$ then it must remain in $\mathcal{V}^+$ for all time. In other words, given a reference signal $y_r(t) = Qw(t)$ where w is given as a solution of $w_t = Sw$ determined by an initial condition $w(0)$. The invariance of the subspace $\mathcal{V}^+$ can be expressed in terms of $\Pi(t) = \Pi(w(t))$ as follows

$$\Pi_t(t) = A_0\Pi(t) + B_\mathrm{d}Pw(t), \tag{3.101}$$

$$\mathcal{B}_\mathrm{in}(\Pi(t)) = \Gamma w, \tag{3.102}$$

$$\xi(0) = \Pi(w(0)), \tag{3.103}$$

and the error zeroing condition given by the second regulator equation is

$$C(\Pi(t)) = y_r(t). \tag{3.104}$$

If we introduce the notation $\widetilde{\xi}(t) = \Pi(w(t))$ then Equations (3.101)–(3.104) can easily be interpreted as the zero dynamics controller, i.e., we have $\widetilde{\xi}$ satisfies

$$\widetilde{\xi}_t(t) = A_0\widetilde{\xi}(t) + B_\mathrm{d}Pw(t), \tag{3.105}$$

$$C(\widetilde{\xi}(t)) = y_r(t), \tag{3.106}$$

$$\widetilde{\xi}(0) = \Pi(w(0)), \tag{3.107}$$

where now with the following equation is considered a constraint

$$\Gamma w(t) = \mathcal{B}_\mathrm{in}(\widetilde{\xi}(t)) \tag{3.108}$$

defining Γ.

But we must notice that the initial condition $\Pi(w(0))$ requires an *a priori* knowledge of the value of Π which we do not have. Indeed, in the zero dynamics controller we do not claim that we obtain exactly the control gain Γ obtained from the regulator equations. Rather we obtain a different controller gain for every different initial condition ψ in (3.87).

Let us investigate the relation between the control gain obtained from the zero dynamics in (3.92)–(3.94) and the one obtained from the regulator equations expressed in (3.105)–(3.108) where we have assumed that the initial condition involving $\Pi(w(0))$ is known. We will denote by $\widetilde{\Gamma}$ the gain obtained from the regulator equations and use the symbol Γ to denote the gain obtained from the zero dynamics controller (3.92)–(3.94), i.e.,

$$\xi_t = A_0\xi + B_\mathrm{d}d,$$

$$\xi(0) = \psi,$$
$$\mathcal{C}\xi = y_r,$$
$$\Gamma = \mathcal{B}_{\text{in}}\xi(t),$$

Then, defining $\eta = \xi - \widetilde{\xi}$ we see from (3.105)–(3.108) that η satisfies

$$\eta_t = A_0\eta,$$
$$\eta(0) = \psi - \Pi(w(0)),$$
$$\mathcal{C}\eta = 0,$$
$$\Gamma - \widetilde{\Gamma} = \mathcal{B}_{\text{in}}\eta(t).$$

Notice that looking back at the Assumption 3.1 we see that the two conditions $\eta_t = A_0\eta$ and $\mathcal{C}\eta = 0$ can be combined into the single condition $\eta_t = \widetilde{A}\eta$ so the above system for η can be written as

$$\eta_t = \widetilde{A}\eta,$$
$$\eta(0) = \psi - \Pi(w(0)),$$
$$\Gamma - \widetilde{\Gamma} = \mathcal{B}_{\text{in}}\eta(t).$$

But according to our assumption on $\widetilde{A}$ (see (3.91)) we have

$$\|\Gamma w(t) - \widetilde{\Gamma}w(t)\|_{\mathcal{U}} = \|\mathcal{B}_{\text{in}}(\eta(t))\|_{\mathcal{U}} \leq C_2\|\eta\|_{\mathcal{Z}_1} \leq C_3 e^{-\alpha t}\|\psi - \Pi(w(0))\|$$

for some constants C_2 and C_3 whose exact values are irrelevant. Here we have used the assumption that $\mathcal{Z}_1 \subset \mathcal{D}(\mathcal{B}_{\text{in}})$ and $\mathcal{B}_{\text{in}} \in \mathcal{L}(\mathcal{Z}_1, \mathcal{U})$.

Therefore, even though we obtain a slightly different feedback gain by choosing different initial conditions ψ the difference of the gains applied to a trajectory from the exosystem converge exponentially fast to zero. This point becomes more important when we consider the zero dynamics method applied to nonlinear systems in the next chapters.

Example 3.2. As we have seen, in the case of co-located boundary control systems the dynamic regulator equations can be interpreted in terms of the zero dynamics inverse design [18]. In this numerical example we consider a boundary controlled heat equation with state variable $z(x, t)$ in $\mathcal{Z} = L^2(\Omega)$, where $x = (x_1, x_2) \in \Omega$ and Ω is the rectangular region

$$\Omega = \{(x_1, x_2) : 0 \leq x_1 \leq 2, \ -0.5 \leq x_2 \leq 0.5\}.$$

We assume that the boundary of Ω, denoted $\partial\Omega$, is partitioned into

$$\mathcal{S}_{B_{\text{in},1}} = \{(x_1, x_2) : 0 < x_1 < 2, \ x_2 = -0.5\},$$

$$\mathcal{S}_{B_{\text{in},2}} = \{(x_1, x_2) : 0 < x_1 < 2, \ x_2 = 0.5\},$$

$$\mathcal{S}_{\text{d}} = \{(x_1, x_2) : x_1 = 2, \ -0.5 < x_2 < 0.5\},$$

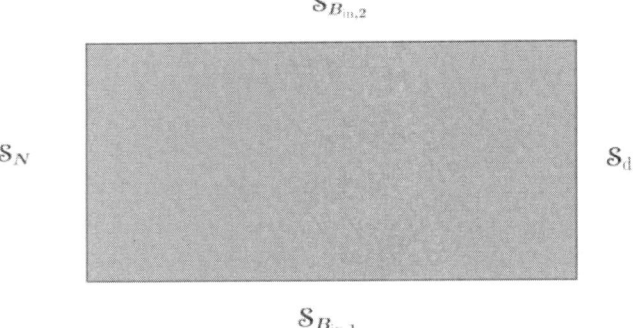

Fig. 3.48: Computational Domain for the MIMO Co-located Example.

$$\mathcal{S}_{\mathrm{N}} = \{(x_1, x_2) : x_1 = 0, \quad -0.5 < x_2 < 0.5\}$$

(as shown in Fig. 3.48). We assume homogeneous Neumann boundary conditions on $\mathcal{S}_{\mathrm{N}}$. A disturbance enters through a non-homogeneous Robin boundary condition on $\mathcal{S}_{\mathrm{d}}$, i.e.,

$$\mathcal{B}_{\mathrm{d}}(z) = (c\partial_{\mathbf{n}} + k_d)\, z\Big|_{\mathcal{S}_{B_{\mathrm{d}}}} = q(x_2)\sin(t),$$

where $q(x_2)$ is a smooth approximation (with a continuous second derivative) to a characteristic function, which is one on the interval $|x_2| < 1/16$, zero for $|x_2| > 3/16$ and $1/2$ at $|x_2| = 1/8$. We have two boundary control inputs given by

$$\mathcal{B}_{\mathrm{in}}^{1} z(x, t) = c\partial_{\mathbf{n}} z\Big|_{\mathcal{S}_{B_{\mathrm{in},1}}},$$

$$\mathcal{B}_{\mathrm{in}}^{2} z(x, t) = c\partial_{\mathbf{n}} z\Big|_{\mathcal{S}_{B_{\mathrm{in},2}}}.$$

Recall that $\partial_{\mathbf{n}}$ denotes the normal derivative. Notice in this case we have infinite-dimensional input space defined by $\mathcal{U}_1 \oplus \mathcal{U}_2$ where

$$\mathcal{U}_1 = L^2(\mathcal{S}_{B_{\mathrm{in},1}}) \text{ and } \mathcal{U}_2 = L^2(\mathcal{S}_{B_{\mathrm{in},2}}).$$

Denote by γ_j the trace operator $\gamma_j : H^s(\Omega) \to H^{s-1/2}(\mathcal{S}_{B_{\mathrm{in},j}})$, $j = 1, 2$. For this example we assume co-located outputs defined by

$$y_1(x, t) = C_1(z) = \gamma_1 z(x, t) \text{ for } x \in \mathcal{S}_{B_{\mathrm{in},1}}, \quad t > 0,$$

$$y_2(x, t) = C_2(z) = \gamma_2 z(x, t) \text{ for } x \in \mathcal{S}_{B_{\mathrm{in},2}}, \quad t > 0.$$

We refer to the articles [20, 21] for a discussion of the regular linear systems produced with these types of inputs and outputs and the appropriate spaces in which all the operations are well defined.

The controlled heat problem in $\mathcal{Z} = L^2(\Omega)$ can be written as

$$z_t(x,t) = c\Delta z(x,t) + B_d d(x,t), \tag{3.109}$$

$$z(x,0) = \varphi(x), \quad \varphi \in \mathcal{Z},$$

$$\mathcal{B}_{\text{in}}^j z(x,t) = u_j(x,t), \quad j = 1,2 \quad x \in \mathcal{S}_{B_{\text{in},j}}, \tag{3.110}$$

$$\partial_{\mathbf{n}} z(x,t) = 0, \quad x \in \mathcal{S}_{\text{N}}, \tag{3.111}$$

$$y_j(x,t) = \gamma_j z(x,t), \quad j = 1,2, \quad x \in \mathcal{S}_{B_{\text{in},j}}. \tag{3.112}$$

Here we have transformed the boundary disturbance term into a distributional forcing term in the equation as in Chapter 2. Recall that the exact form of this operator is of no consequence for the calculations.

Let us introduce the operator $A_0 = c\Delta$ in $\mathcal{Z}$ with domain

$$\mathcal{D}(A_0) = \left\{ \varphi \in H^2(\Omega) : c\frac{\partial}{\partial \mathbf{n}}\varphi\Big|_{\mathcal{S}_{\text{N}}} = 0, \ \left(c\frac{\partial}{\partial \mathbf{n}} + k_d\right)\varphi\Big|_{\mathcal{S}_{\text{d}}} = 0 \right\}.$$

In addition we also define

$$\mathcal{B}z = \begin{bmatrix} \mathcal{B}_{\text{in}}^1 z, & \mathcal{B}_{\text{in}}^2 z \end{bmatrix}^{\top}.$$

$$\mathcal{C}z = \begin{bmatrix} C_1 z, & C_2 z \end{bmatrix}^{\top},$$

$$u = \begin{bmatrix} u_1, & u_2 \end{bmatrix}^{\top}, \quad y = \begin{bmatrix} y_1, & y_2 \end{bmatrix}^{\top}.$$

With this notation the system (3.109)–(3.112) can be written as

$$z_t = A_0 z,$$

$$z(0) = \varphi,$$

$$\mathcal{B}z(t) = u(t),$$

$$y(t) = \mathcal{C}z(t).$$

We assume that the distributed reference signals are given as

$$y_{r,1}(x,t) = A_{r,1} + M_{r,1}\cos(\alpha_1(x - t)) \quad \text{on } \mathcal{S}_{B_{\text{in},1}},$$

$$y_{r,2}(x,t) = A_{r,2} + M_{r,2}\sin(\alpha_2(x + t)) \quad \text{on } \mathcal{S}_{B_{\text{in},2}},$$

and we define the vector of reference signals as

$$y_r = \begin{bmatrix} y_{r,1}(x,t) \\ y_{r,2}(x,t) \end{bmatrix},$$

for some constants $A_{r,j}$, $M_{r,j}$. Finally we define the errors

$$e_j(t) = \left(\int_{\mathcal{S}_{B_{\text{in},j}}} |y_j(x,t) - y_{r,j}(x,t)|^2 \, dx \right)^{1/2}, \quad j = 1,2. \tag{3.113}$$

With all these notations our objective is to find controls $u_j(x,t)$ in (3.110) so that

$$e(t) = \max_{j=1,2} e_j(t) \xrightarrow{t\to\infty} 0.$$

To this end we introduce the zero dynamics for this example given by

$$\xi_t = A_0\xi,$$
$$\xi(0) = \psi,$$
$$\mathcal{C}\xi(t) = y_r.$$

We must show that the above problem fits into the framework described in the first part of this section. Following the development above we let $\eta = z - \xi$ and introduce the control $u = \mathcal{B}\xi$ which in the present case amounts to

$$(\mathcal{B}\xi)(x,t) = \left[\mathcal{B}_{\text{in}}^1\xi, \ \mathcal{B}_{\text{in}}^2\xi\right]$$

and define the desired infinite-dimensional control input by

$$u(x,t) = \mathcal{B}\xi(x,t).$$

With this the equation for $\eta(x,t) = z(x,t) - \xi(x,t)$ becomes

$$\eta_t(x,t) = c\Delta\eta(x,t),$$
$$\eta(0,t) = \varphi(x) - \psi(x) \equiv \varphi_0(x),$$
$$\partial_{\mathbf{n}}\eta(x,t) = 0, \quad t > 0, \quad x \in \mathcal{S}_{B_{\text{in},j}}, \ j = 1,2.$$
$$(c\partial_{\mathbf{n}} + k_d)\,\eta(x,t) = 0, \quad x \in \mathcal{S}_{\text{d}}, \quad \partial_{\mathbf{n}}\eta(x,t) = 0, \quad x \in \mathcal{S}_{\text{N}}.$$

It is well known (cf., [61, 60] that there is a continuous (in x and t, for $t > 0$) globally defined solution to this problem which satisfies

$$\sup_{x\in\overline{\Omega}} |\eta(x,t)| \xrightarrow{t\to\infty} 0.$$

Moreover, $\eta(\cdot,t) \in H^s(\Omega)$ for $s > [n/2] + 1$ which implies $\eta(\cdot,t) \in C(\overline{\Omega})$ and

$$\|\eta(\cdot,t)\|_{H^s(\Omega)} \leq M\frac{e^{-\alpha t}}{t}\|\varphi_0\|, \quad \text{for some} \ \ \alpha > 0, \ \ M > 0, \ \ \text{for} \ \ t > 0.$$

Under our assumptions the operator $A = c\Delta$ with domain

$$\mathcal{D}(A) = \left\{\varphi \ : \ (c\partial_{\mathbf{n}} + k_d)\,\varphi(x)\Big|_{\mathcal{S}_{\text{d}}} = 0, \ \ \partial_{\mathbf{n}}\varphi(x) = 0 \ \text{ for } x \in \partial\Omega\backslash\mathcal{S}_{\text{d}}\right\},$$

is a strictly negative, unbounded self-adjoint operator in $L^2(\Omega)$ and generates an exponentially stable analytic semigroup, e^{At}. This semigroup is also exponentially stable in $H^s(\Omega)$ for $s > 0$ (in particular for $s > [n/2] + 1$).

In our specific numerical examples we have set the signals to be

$$y_{r,1}(x_1, t) = 1 + \cos(0.25(x_1 - t)), \quad y_{r,2}(x_1, t) = 2 + 2\sin(0.5(x_1 + t)),$$

the physical parameters have been set to

$$c = 0.1, \quad k_d = 1.$$

We used initial conditions $\varphi(x) = xy$, $\psi(x) = 0$, and we have solved for $t \in (0, 100)$.

Since the measured outputs $y_j(x_1, t)$ are infinite-dimensional we have chosen to display the trajectories of the outputs for all time but only at the points $x_1 = -0.5$ for $y_2(x_1, t)$ and $x_1 = 0.5$ for $y_2(x_1, t)$. In particular, the time evolution of $y_1(0.5, t)$ and $y_{r,1}(0.5, t)$ is given in Fig. 3.49 for all time. The asymptotic convergence to zero of the error $e_1(t)$ is given in Fig. 3.50 (see (3.113)). Similarly, the time evolution of $y_2(0.5, t)$ and $y_{r,2}(0.5, t)$ is given in Fig. 3.51 for all time, and the asymptotic convergence to zero of the error $e_2(t)$ is given in Fig. 3.52.

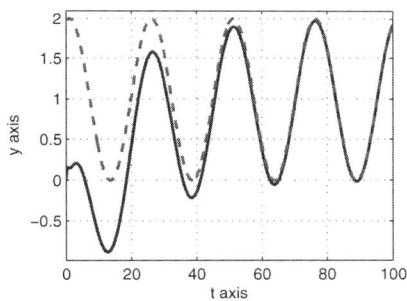

Fig. 3.49: y_1 and $y_{r,1}$ at $(0.5,t)$.

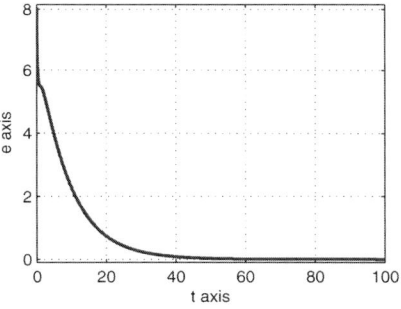

Fig. 3.50: $e_1(t)$.

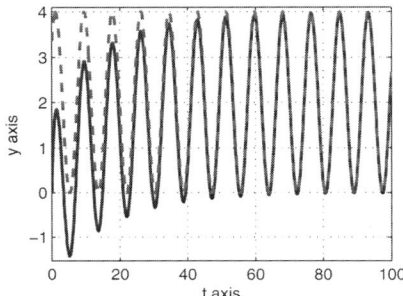

Fig. 3.51: y_2 and $y_{r,2}$ at $(0.5,t)$.

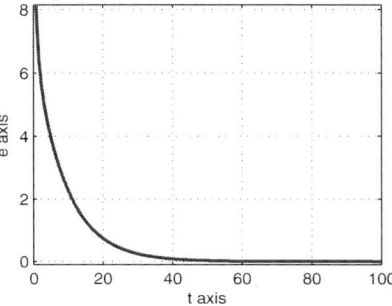

Fig. 3.52: $e_2(t)$.

Example 3.3. In this example we consider a SISO, co-located interior control tracking problem for a two-dimensional heat equation with state variable $z(x,t)$ in $\mathcal{Z} = L^2(\Omega)$, where $x = (x_1, x_2) \in \Omega$ and Ω is the square region, depicted in Fig. 3.53,

$$\Omega = \{(x_1, x_2) : -L \le x_1 \le L, \quad -L \le x_2 \le L\},$$

with boundary denoted by $\mathcal{S}_0$. Inside this domain we have a closed curve $\mathcal{S}_{B_{in}}$ forming the boundary of a square with sides $L/3$ and centered inside Ω.

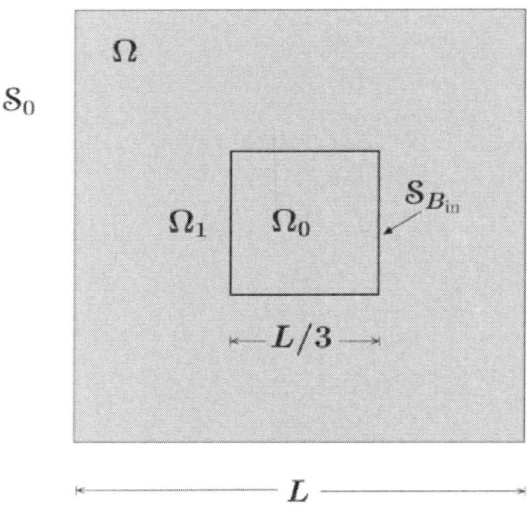

Fig. 3.53: Two-Dimensional Region.

Consider the boundary controlled heat equation in the domain Ω.

$$z_t(x,t) = c\Delta z(x,t), \quad x \in \Omega = \Omega_0 \cup \Omega_1 , \quad t > 0,$$

$$\mathcal{B}_{\mathcal{S}_0}(z)(x,t) = (c\partial_{\mathbf{n}}z + kz)\Big|_{x \in \mathcal{S}_0} (x,t) = 0,$$

$$[z]\Big|_{x \in \mathcal{S}_{B_{in}}} (x,t) = 0, \tag{3.114}$$

$$\mathcal{B}_{in}(z)(x,t) = c[\partial_{\mathbf{n}}z]\Big|_{x \in \mathcal{S}_{B_{in}}} (x,t) = u_{in}(x,t), \tag{3.115}$$

$$y(x,t) = \mathcal{C}(z) = z\Big|_{x \in \mathcal{S}_{B_{in}}} (x,t),$$

$$z(x,0) = \varphi(x), \quad x \in \Omega.$$

where the condition (3.114)

$$0 = [z]\Big|_{x \in \mathcal{S}_{B_{in}}} = z(x^+,t)\Big|_{x^+ \in \Omega_1} - z(x^-,t)\Big|_{x^- \in \Omega_0},$$

represents the continuity of the solution across the interface between Ω_0 and Ω_1 and (3.115) represents the jump in the normal derivative

$$[\partial_{\mathbf{n}} z]\Big|_{x \in S_{B_{in}}} = (\partial_{\mathbf{n}^+}) z(x^+, t)\Big|_{x^+ \in \Omega_1} - (\partial_{\mathbf{n}^-}) z(x^-, t)\Big|_{x^- \in \Omega_0}.$$

Here $\mathbf{n}^+$ is the exterior normal on Ω_0 into Ω_1, $\mathbf{n}^-$ is the exterior normal on $S_{B_{in}}$ into Ω_0. The terms

$$(\partial_{\mathbf{n}^+}) z(x^+, t)\Big|_{x^+ \in \Omega_1} \quad \text{and} \quad (\partial_{\mathbf{n}^-}) z(x^-, t)\Big|_{x^- \in \Omega_0}$$

represent the normal derivatives when approaching the boundary $S_{B_{in}}$ from Ω_1 and Ω_0, respectively. Thus, $\mathcal{B}_{S_{B_{in}}}(z)(x, t) = u_{in}$ represents a control of the flux across the interface $S_{B_{in}}$, i.e., a jump in the normal derivative.

Note that the output is co-located

$$y(x, t) = \mathcal{C}(z)(x, t) = z\Big|_{x \in S_{B_{in}}} (x, t).$$

Our design objective is to force the measured output y to track a prescribed reference trajectory on the interior interface $S_{B_{in}}$. More specifically, we assume that there is a reference signal (to be tracked) in the form

$$y_r(x, t) = w(t)Q(x), \quad x \in S_{B_{in}}, \quad w \in C_b^\infty([0, \infty]), \quad Q \in C^2(\Omega_0).$$

Here $C_b^\infty([0, \infty])$ is the class of infinitely differentiable functions with w and dw/dt bounded on $[0, \infty)$. The closed curve $S_{B_{in}}$ describing the boundary of Ω_0 has four sides and on each one the value of $y_r(x, t)$ is given in terms of a parametric function, parameterized in x by a normalized arclength s varying between 0 and 1. In particular, on each side we track the function

$$y_r = A_r + M_r 16 s^2 (1 - s)^2 \cos(\alpha t).$$

We define the point-wise error for $x \in S_{B_{in}}$ as

$$e(x, t) = (y(x, t) - y_r(x, t)). \tag{3.116}$$

Problem 3.2 (The Main Objective). *Find a control input u_{in} in (3.115) so that the error defined in (3.116) satisfies*

$$\|e(\cdot, t)\|_{H^{s-1/2}(S_{B_{in}})} \xrightarrow{t \to \infty} 0 \quad \text{for some } s > 1/2.$$

To solve this problem we introduce the zero dynamics system

$$\xi_t(x, t) = c\Delta\xi(x, t), \quad x \in \Omega = \Omega_0 \cup \Omega_1, \quad t > 0, \tag{3.117}$$

$$\mathcal{B}_0(\xi)(x, t) = (c\partial_{\mathbf{n}}\xi + k\xi)\Big|_{x \in S_0} (x, t) = 0, \tag{3.118}$$

$$\mathcal{C}(\xi)(x,t) = \xi\Big|_{x \in \mathcal{S}_{B_{\text{in}}}} (x,t) = Q(x)w(t), \tag{3.119}$$

$$\xi(x,0) = \psi(x), \quad x \in \Omega. \tag{3.120}$$

The desired control is obtained by choosing as input function $u_{\text{in}}(x,t)$ in (3.115) the following output of the zero dynamics (3.117)–(3.120):

$$u_{\text{out}}(x,t) = \mathcal{B}_{\text{in}}(\xi)(x,t) = c\,[\partial_{\mathbf{n}}\xi]\Big|_{x \in \mathcal{S}_{B_{\text{in}}}} (x,t).$$

In our specific numerical simulation we have set

$$L = 3\,, \;\; k = 1\,, \;\; c = 1\,, \;\; \alpha = \pi\,, \;\; A_r = 1\,, \;\; \text{and } M_r = 1.$$

We have taken the initial condition for the control system to be $z(x,0) = 0$. For the zero dynamics we have solved the system starting at $t = -4$, with zero initial conditions, to give time for the zero dynamics to reach the invariant manifold. Then we have solved the coupled system for $t \in (0,20)$.

Rather than attempt to visualize the evolution of y_r on the entire boundary $\mathcal{S}_{B_{\text{in}}}$, we have chosen two particular points along the lower portion of the boundary given by $\{(x_1, x_2) : \; -0.5 < x_1 < 0.5, \;\; x_2 = -0.5\}$. The selected points are the lower left corner at $(s = 0)$, and the midpoint at $(s = 0.5)$. In there we have $y_r(0,t) = 1$ and $y_r(0.5,t) = 1 + \cos(\pi t)$. The snapshots of y and y_r at these two points are depicted in Figs. 3.54 and 3.55 for all times. In Fig. 3.56 we depict the error norm computed as

$$\|e(t)\| = \left(\int_{\mathcal{S}_{B_{\text{in}}}} (y(x,t) - y_r(x,t))^2 \, dx \right)^{1/2}.$$

Finally, Figs. 3.57, 3.58, 3.59 and 3.60 are snapshots of the solution surface at times $t = 2.0$, $t = 2.5$, $t = 3.0$ and $t = 3.5$, respectively.

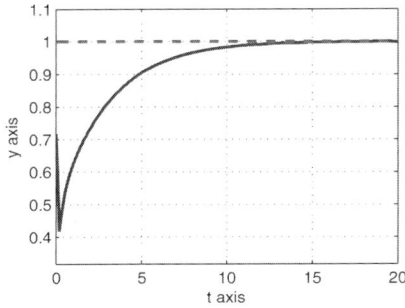

Fig. 3.54: $y(-0.5, -0.5, t)$.

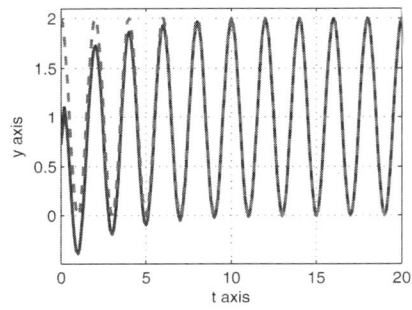

Fig. 3.55: $y(0, -0.5, t)$.

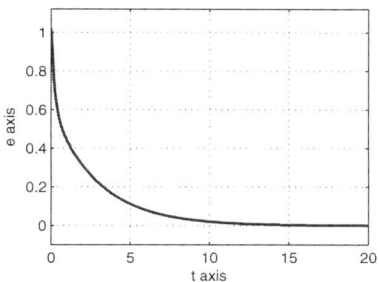

Fig. 3.56: $\|e(t)\|$.

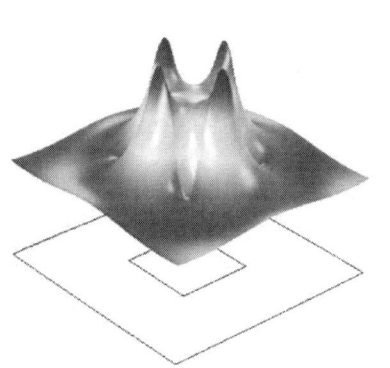

Fig. 3.57: Solution $t = 2.0$.

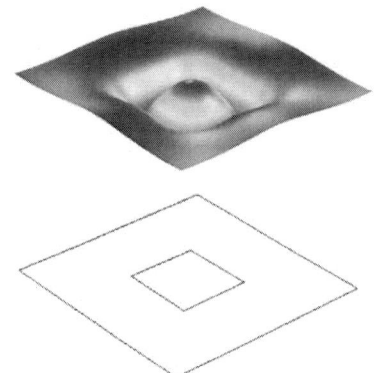

Fig. 3.58: $t = 2.5$.

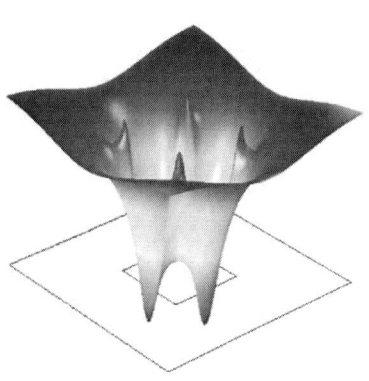

Fig. 3.59: Solution $t = 3.0$.

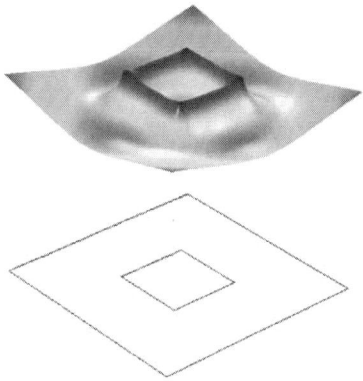

Fig. 3.60: $t = 3.5$.

Part II

Regulation for Nonlinear Systems

Chapter 4

Nonlinear Distributed Parameter Systems

4.1 Introduction

Certainly the most interesting and challenging problems in regulation of distributed parameter systems arise in applications to nonlinear systems. For finite-dimensional nonlinear systems one of the most important theoretical results in geometric regulation is the 1990 paper by C.I. Byrnes and A. Isidori [56] in which an application of center manifold theory was used to prove necessary and sufficient conditions for solvability of the state (and error) feedback regulator problem in terms of a set of regulator equations analogous to those given in (1.1) for linear systems. While there has been progress in extending the nonlinear theory to distributed parameters systems (cf. Theorem (4.1) below), at the time of writing of this book there is no complete analog of the results of Byrnes and Isidori due in large part to the technical complications that arise due to unbounded control input operators. As there are currently several research groups working towards such a theory there is little doubt that a comprehensive theory will emerge in the not-too-distant future. In this work we hope to encourage this research and demonstrate the utility of the geometric methods for nonlinear distributed parameters systems.

We choose to focus on presenting conditions, given as assumptions, under which a problem of nonlinear regulation is solvable. These conditions are essentially hypothesizing the existence of an error zeroing, attractive, invariant manifold for a given problem. Under these assumptions it is straightforward, almost by definition, that the regulation problem is solvable. For some of our examples it is possible to demonstrate that these conditions are satisfied. For example, for some one-dimensional parabolic initial boundary value problems, results such as those found in [92] can be used to prove the existence of an attractive invariant manifold. Verifying the existence of such manifolds is generally quite difficult. Nevertheless, in general, we have found that our methodology for obtaining control laws based on the assumptions of the existence of such an invariant manifold can be applied in many cases in which we are not able to prove the existence of the manifold. In other words we show, by way of many interesting examples, that the design methodology works on a remarkable number of problems. Of course there are limitations. For example, this theory is an equilibrium-based theory which means that it is a local theory and must be applied in the neighborhood of an equilibrium. In this book we always assume, without loss of generality, that the equilibrium in question is zero. Therefore we expect that our regulation results only hold for small initial conditions and small reference and disturbance signals. In general we find that, while there are some restrictions, in applications the method is quite robust.

In what follows we will explain a basic formalism for designing a control law and provide several methodologies for approximately solving the design paradigm, for example, the so-called β regularization method and truncated

Fourier series expansions. Based on the many practical examples contained in this chapter and in Chapter 5 we hope the reader is encouraged to pursue (as we will continue to do so ourselves) a more careful and rigorous mathematical justification for the validity of the methods.

We note that starting in 2003 a somewhat more general nonequilibrium theory has evolved based on omega limit sets, zero dynamics and internal models by C.I. Byrnes and A. Isidori in [28, 22], and these results have been extended in many directions in more recent years by A. Isidori, and several of his colleagues including L. Marconi and A. Serrani where the equilibrium theory is replaced by regulation on a global attractor or perhaps on an inertial manifold.

The main goal of this chapter is to examine problems of regulation for nonlinear distributed parameter systems. Once again the basis for our approach is geometric, for example based on the center manifold theorem. Even for local problems, however, one must surmount technical issues that inevitably arise in the infinite-dimensional setting. For example, it is well known that there is no complete analog of fundamental existence and uniqueness theorems available for ordinary differential equations in the case of initial, boundary value problems for partial differential equations. Of course, there are results such as the well known Cauchy–Kowalevski theorem which provides the main local existence and uniqueness theorem for analytic partial differential equations associated with Cauchy initial value problems. But this theorem is of no use in practical situations involving boundary control problems for systems of nonlinear partial differential equations. There has been a tremendous research effort over the years providing a vast literature on existence and uniqueness for various classes of equations, see for example [61, 60, 92]. The substantial literature on global in time existence and long time behavior of solutions to dynamical systems governed by nonlinear initial boundary values demonstrates the complex nature of this subject. As we have repeatedly pointed out we do not intend to add to this literature. By this we mean that we do not intend to provide detailed discussions of existence and uniqueness of globally defined solutions to the initial boundary value problems discussed in our examples. In some cases we will provide, when appropriate, references to such discussions.

On the other hand, we recall that since the theory of output regulation considered in this book is an asymptotic theory, the long time existence and behavior of solutions to is of utmost importance. Nonetheless, we will not attempt to demonstrate the existence or long term behavior for our examples but only comment that for many of our problems the results are well known.

Much of the material contained in this chapter derives from research papers published by one or both of the authors. In particular for the problem of setpoint control for nonlinear partial differential equations we will follow our work in [2]. For the discussion on zero dynamics design for nonlinear partial differential equations we will borrow extensively from [13, 14, 16, 18, 15, 17].

As in the previous chapters our main focus is regulation problems of tracking and disturbance attenuation for nonlinear systems and with exogenous systems for which center manifold methods can be used to obtain state feedback control laws for solving problems.

Our approach in the first section of this chapter will be to establish a theory parallel to that which has been established for finite-dimensional nonlinear systems ([8], [55], [56], [49], [71]). We will proceed much as we did in Part I of the book where we described the extension of the finite-dimensional linear theory to include a large class of linear distributed parameter systems problems. In particular we were able to solve problems of output regulation for stable linear systems with either bounded or unbounded inputs and outputs in an appealing systems theoretic fashion as in [76, 74], [81, 80], [30] and [66].

The results presented in this chapter are primarily concerned with output regulation with respect to signals and disturbances generated by finite-dimensional exogeneous systems (see, however, [50, 51, 53] for a discussion of infinite-dimensional exosystems). In the following section the exosystem is assumed to be both finite-dimensional and *neutrally stable* as in [56] and, in this case, we can appeal to powerful center manifold methods to obtain some nontrivial insights and results. We emphasize the fact that these local techniques are not simply an appeal to linearization. Even in the lumped nonlinear case the problem of output regulation for the linearization does not solve the output regulation problem for the nonlinear problem.

4.2   Nonlinear State Feedback Regulation Problem

Most of the material in this section derives from the work in [11], where the authors extended the geometric theory of regulation, as developed by C.I. Byrnes and A. Isidori (see for example [55, 56]), in the special case of bounded input operators, as considered in Chapter 1, and for (possibly unbounded) output operators, that are bounded in an appropriate sense. We proceed by considering a very general situation for a nonlinear system in the form

$$\frac{dz}{dt}(t) = f(z, d, u),$$

$$y(t) = c(z(t)), \qquad \qquad \text{(Measured output)}$$

$$z(0) = z_0,$$

where z is the state of the system in an infinite-dimensional Hilbert state space $\mathcal{Z}$, u is the control input, y is the measured output, $z_0 \in \mathcal{Z}$ is the initial state of the system and d is a disturbance.

Remark 4.1. In what follows we will assume that the origin of $\mathcal{Z}$ is an equilibrium for the uncontrolled system $z_t = f(z, 0, 0)$. Concerning this assumption we remind the reader that our theory, which at some level is based on the existence of an attractive invariant manifold (for example a center manifold), is a local theory that requires solving the regulation problem in the neighborhood of some fixed equilibrium point z_0 (e.g., $f(z_0, 0, 0) = 0$). If this point is not the origin, then we can make a simple change of variables by setting $\widetilde{z} = (z - z_0)$ and define a new nonlinear function $\widetilde{f}(\widetilde{z}, 0, 0) = f(\widetilde{z} + z_0, 0, 0)$ and consider the system $\widetilde{z}_t = \widetilde{f}(\widetilde{z}, 0, 0)$ where the nonlinear function $\widetilde{f}$ satisfies $\widetilde{f}(0, 0, 0) = 0$.

In addition we assume that there exists a *neutrally stable* [56] finite-dimensional exogenous system

$$\frac{dw}{dt} = s(w),$$
$$w(0) = w_0 \in \mathcal{W},$$

(here we assume that $\mathcal{W}$ is a finite-dimensional Hilbert vector space) that generates both the reference signal y_r and the disturbance d. Namely, we assume

$$y_r(t) = q(w(t)) \qquad q : \mathcal{W} \mapsto \mathcal{Y},$$
$$d(t) = p(w(t)) \qquad p : \mathcal{W} \mapsto \mathcal{Z}.$$

As we have already discussed in Remark 4.1 we assume that the origin in $\mathcal{Z} \times \mathcal{W} \times \mathcal{U}$ space, $\mathbf{O} = (0, 0, 0)$, is an equilibrium point, i.e., $f(\mathbf{O}) = 0$, and that f can be decomposed, about this equilibrium, as a sum of its linear and nonlinear terms, namely

$$f(z, w, u) = \left.\frac{\partial f}{\partial z}\right|_{\mathbf{O}} z + \left.\frac{\partial f}{\partial w}\right|_{\mathbf{O}} w + \left.\frac{\partial f}{\partial u}\right|_{\mathbf{O}} u + F(z, w, u),$$

where the nonlinear term $F(z, w, u)$ satisfies $F(\mathbf{O}) = 0$, $F_z(\mathbf{O}) = 0$, $F_u(\mathbf{O}) = 0$ and $F_w(\mathbf{O}) = 0$.

Next we define the linear operators

$$A = \left.\frac{\partial f}{\partial z}\right|_{\mathbf{O}} \in \mathcal{L}(\mathcal{Z}), \quad B_{\text{in}} = \left.\frac{\partial f}{\partial u}\right|_{\mathbf{O}} \in \mathcal{L}(\mathcal{U}, \mathcal{Z}).$$

We also have

$$p(w) = \left.\frac{\partial p}{\partial w}\right|_{\mathbf{O}} + \widehat{p}(w), \quad \widehat{p}(0) = 0, \quad \frac{\partial \widehat{p}}{\partial w}(0) = 0,$$

$$q(w) = \left.\frac{\partial q}{\partial w}\right|_{\mathbf{O}} + \widehat{q}(w), \quad \widehat{q}(0) = 0, \quad \frac{\partial \widehat{q}}{\partial w}(0) = 0.$$

We set

$$B_{\text{d}} = \left.\frac{\partial p}{\partial w}\right|_{\mathbf{O}} \in \mathcal{L}(\mathcal{D}, \mathcal{Z}), \quad Q = \left.\frac{\partial q}{\partial w}\right|_{\mathbf{O}} \in \mathcal{L}(\mathcal{W}, \mathcal{Y}), \text{ and } P = \left.\frac{\partial p}{\partial w}\right|_{\mathbf{O}} \in \mathcal{L}(\mathcal{W}, \mathcal{D}).$$

For the measured output we have

$$c(z) = \frac{\partial c}{\partial z}\bigg|_{\mathcal{O}} + \widehat{c}(z), \quad \widehat{c}(0) = 0, \quad \frac{\partial \widehat{c}}{\partial z}(0) = 0,$$

and we set $C = \dfrac{\partial c}{\partial z}\bigg|_{\mathcal{O}}$ so that $C \in \mathcal{L}(\mathcal{Z}, \mathcal{Y})$ and

$$y(t) = Cz(t) + \widehat{c}(z(t)).$$

Finally we also have

$$s(w) = Sw + \widehat{s}(w) \quad \widehat{s}(0) = 0, \quad \frac{\partial \widehat{s}}{\partial w}(0) = 0.$$

Here $S \in \mathcal{L}(\mathcal{W})$ (so in a fixed basis S is given as an $N_{\mathcal{W}} \times N_{\mathcal{W}}$ matrix).

Problem 4.1 (Nonlinear State Feedback Regulator Problem). *The objective of output regulation is to find a control law $u = \gamma(w)$ such that the closed-loop trajectories exist and remain bounded for all time, the error $e(t) = y(t) - y_r(t)$ exists for all time and*

$$e(t) = c(z(t)) - y_r(t) \xrightarrow{t \to +\infty} 0.$$

We consider a special class of problems in which we impose the following assumptions. First, recall that an operator is *accretive* if the numerical range lies in the right half plane, i.e., Re $(\Theta(A)) = $ Re $\langle A\varphi, \varphi \rangle \geq 0$ for $\varphi \in D(A) \subset Z$. An accretive operator A is *m-accretive* if for all Re $(\lambda) < 0$ (in addition to accretive) we have that the resolvent satisfies

$$(\lambda I - A)^{-1} \in \mathcal{L}(Z), \quad \|(\lambda I - A)^{-1}\| \leq \frac{1}{|\text{Re }(\lambda)|}.$$

Recall that $\mathcal{L}(Z)$ denotes the space of bounded linear maps on $\mathcal{Z}$. An operator is called quasi-accretive (or quasi-m-accretive) if $(\alpha I - A)$ is accretive (or quasi-accretive) for some scalar α. This is equivalent to the condition $\Theta(A)$ is contained in a half plane Re $(\lambda) \geq$ const.

A quasi-accretive operator A is called *sectorial* if the numerical range is not only contained in a half space Re $(\lambda) \geq$ const but also is contained in a sector

$$|\arg(\lambda - \gamma)| \leq \theta < \pi/2.$$

Here γ is called the vertex and θ is the semiangle. A is called *m-sectorial* if it is sectorial and A is quasi-m-accretive. For a detailed discussion of accretive operators see [47].

Finally, following Pazy [72], we say that A is *dissipative* provided $(-A)$ is (maximal) accretive.

In order to simplify the exposition in this book we will impose the following assumptions.

Assumption 4.1. *We assume the input and measured output operators are linear functions of the state of the control system, and that reference signal and disturbance are linear functions of the state of the exosystem. We also assume that $F(z, w, u) = F(z)$ so that, in particular, the uncontrolled problem is autonomous. Thus we assume $\widehat{q} = 0$, $\widehat{p} = 0$, $\widehat{q} = 0$, $\widehat{c} = 0$, $\widehat{s} = 0$ and we have*

$$c(z) = Cz, \quad q(w) = Qw, \quad s(w) = Sw.$$

Assumption 4.2.

1. *$(-A)$ is a sectorial operator with compact resolvent. (See Section 2.2.1.)*

2. *The analytic semigroup $T(t) = e^{At}$ is exponentially stable.*

3. *$B_{in} \in \mathcal{L}(\mathcal{U}, \mathcal{Z})$ and $B_d \in \mathcal{L}(\mathcal{D}, \mathcal{Z})$ are bounded.*

4. *$C \in \mathcal{L}(Z_\alpha, Y)$ for some $\alpha > 0$, i.e., there is a constant C_α so that*

$$\|C\varphi\|_Y \leq C_\alpha \|\varphi\|_\alpha.$$

5. *We assume the spectrum of S is on the imaginary axis and has no non-trivial Jordan blocks.*

Remark 4.2. In some of our examples, and quite often in practice, the operator A is self-adjoint. We also note that, since we assume $(-A)$ is sectorial, we work with the semigroup $\exp(At)$ rather than $\exp(-At)$ as is done in Henry [47]. Indeed, the various definitions of accretive, dissipative and sectorial operator vary somewhat from one reference to another. In particular a sectorial operator in Engel & Nagel [38] is the negative of a sectorial operator in T. Kato [57].

4.2.1 A Local State Feedback Result

Under the assumptions made in the previous section we obtain the abstract nonlinear system

$$z_t = Az + F(z) + B_{in}u + B_d d, \tag{4.1}$$

$$w_t = Sw, \tag{4.2}$$

$$z(0) = z_0, \quad w(0) = w_0, \tag{4.3}$$

$$y = Cz, \quad y_r = Qw, \quad d = Pw. \tag{4.4}$$

In the present setting the local nonlinear state feedback regulator problem for (4.1)–(4.4) is to find a control $u = \gamma(w)$ such that the error satisfies

$$e(t) := y(t) - y_r(t) \to 0, \text{ as } t \to \infty, \tag{4.5}$$

for sufficiently small initial conditions z_0, w_0, and reference and disturbance signals y_r and d.

Theorem 4.1. *Under Assumptions 4.1 and 4.2, the local state feedback regulator problem for* (4.1)–(4.5) *is solvable if, and only if, there exist mappings* $\pi : \mathcal{W} \to D(A) \subset \mathcal{Z}$ *and* $\gamma : \mathcal{W} \to \mathcal{U}$ *satisfying the "regulator equations,"*

$$\frac{\partial \pi(w)}{\partial w} S w = A\pi(w) + F(\pi(w)) + B_{in}\gamma(w) + B_d P w, \tag{4.6}$$

$$C\pi(w) = Qw, \tag{4.7}$$

for all w *in some neighborhood of the origin in* $\mathcal{W}$. *In this case a feedback law solving the state feedback regulator problem is given by*

$$u(t) = \gamma(w)(t). \tag{4.8}$$

In spite of the inherent technical difficulties that arise in infinite-dimensions, the proof of Theorem 4.1 can be obtained using an argument similar to that given in [56]. Indeed, under the assumptions on A, B_{in}, B_{d} and C, we can appeal to a version of the Center Manifold Theorem to aid in the proof.

Remark 4.3 (Comments on the proof). We give essentially the proof of Theorem 4.1 found in [11]. First, we note that to be able to adapt the necessary results from the Center Manifold Theory, it is useful to formulate the problem in the composite state space $\mathcal{X} = \mathcal{Z} \times \mathcal{W}$. Namely, with $u = \gamma(w) = \Gamma w + \widetilde{\gamma}(w)$ in (4.8) we have

$$\frac{dX}{dt} = \mathcal{A}X + \mathcal{F}(X), \ X(0) = X_0, \tag{4.9}$$

$$X = \begin{pmatrix} z \\ w \end{pmatrix}, \quad \mathcal{F}(X) = \begin{pmatrix} F(z) + \widehat{\gamma}(w) \\ 0 \end{pmatrix},$$

$$\mathcal{A} = \begin{pmatrix} A & (B_{\mathrm{in}}\Gamma + B_{\mathrm{d}}P) \\ 0 & S \end{pmatrix},$$

$$e = Cz - Qw.$$

In addition to Assumption 4.2 we first assume γ to be a fixed (but otherwise arbitrary) continuously differentiable function. We also assume that $\mathcal{F}$ is continuously differentiable in the Hilbert scale of spaces $\mathcal{X}^\alpha$, generated by $\mathcal{A}$ in $\mathcal{X}$, for some $\alpha > 0$, and satisfies

$$\mathcal{F}(0) = 0, \quad \mathcal{F}_X(0) = 0$$

where, due to our assumptions on A and S, we have that the operator $(-\mathcal{A})$ is a sectorial operator, and therefore $\mathcal{A}$ generates an analytic semigroup $\mathcal{T}(t)$. Furthermore, we have

$$\sigma(\mathcal{A}) = \sigma(A) \cup \sigma(S), \quad \sigma(A) \cap \sigma(S) = \emptyset,$$

and for some $\beta > 0$,

$$\sigma(A) \subset \mathbb{C}_{-\beta}^- = \{\zeta : \operatorname{Re}(\zeta) \le -\beta\}, \text{ and } \sigma(S) \subset \mathbb{C}_0 = i\mathbb{R},$$

and the spectrum of S consists of a finite number of eigenvalues and there is a complete set of associated eigenfunctions.

According to D. Henry's book, [47, Theorem 6.2.1], for every γ (as above), the space $\mathfrak{X} = \mathfrak{X}_1 \otimes \mathfrak{X}_2$ can be decomposed into a direct sum of $\mathcal{A}$ invariant subspaces $\mathfrak{X}_j$ for $j = 1, 2$, i.e., $\mathcal{A}\mathfrak{X}_j \subset \mathfrak{X}_j$. Indeed, the subspace $\mathfrak{X}_2 = \mathcal{P}_2(\mathfrak{X}) \cong \mathcal{W}$ is the eigenspace spanned by the finitely many eigenvalues of $\mathcal{A}_2 = \mathcal{A}\big|_{\mathfrak{X}_2}$ corresponding to S acting in $\mathcal{W}$. From [55, 56] and since the exosystem is assumed to be neutrally stable we recall, once again, that $\sigma(S) = \sigma(\mathcal{A}) \cap \mathbb{C}_0$ consists of finitely many eigenvalues of finite multiplicity on the imaginary axis each having geometric and algebraic multiplicity equal. The subspace $\mathfrak{X}_1$ is defined by $\mathfrak{X}_1 = \mathcal{P}_1\mathfrak{X} = (I - \mathcal{P}_2)\mathfrak{X}$ and the operator $\mathcal{A}_1 = \mathcal{A}\big|_{\mathfrak{X}_1}$ corresponding to $\mathcal{A}$ has an infinite spectrum bounded away from the imaginary axis in the left half plane.

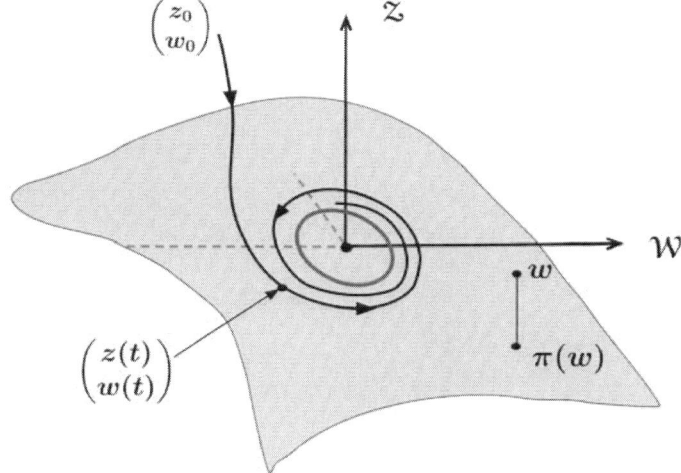

Fig. 4.1: Attractive Center Manifold.

Under the imposed assumptions, the key point of the theorem follows from [47, Theorem 6.2.1], which guarantees the existence of a C^1 local invariant manifold Σ (see Fig. 4.1) defined in a neighborhood U of the origin in $\mathfrak{X}^\alpha$ and a mapping $\pi : \mathfrak{X}_2 \to \mathfrak{X}$ so that, after identifying $\mathcal{W}$ and $\mathfrak{X}_2$, we have

$$\Sigma = \left\{ \begin{pmatrix} \pi(w) \\ w \end{pmatrix} : w \in U \subset \mathcal{W} \right\}$$

where U is a small neighborhood of the origin of $\mathcal{W}$. Also $\mathcal{A}_1$ generates an exponentially stable analytic semigroup $\mathfrak{T}_1$ in $\mathcal{X}_1$.

Invariance of Σ implies that for initial data $X(0) = (\pi(w_0), w_0)^\top \in \Sigma$ the solution to (4.9) satisfies $z(t) = \pi(w(t))$. In particular, on Σ we have

$$\begin{pmatrix} z \\ w \end{pmatrix} = \begin{pmatrix} \pi(w) \\ w \end{pmatrix},$$

and from the equations of motion we have

$$\frac{dz}{dt} = A\pi(w) + F(\pi(w)) + B_{\text{in}}\gamma(w) + B_{\text{d}}Pw,$$

and

$$\frac{dz}{dt} = \frac{d\pi(w)}{dt} = \frac{\partial\pi(w)}{\partial w}\frac{dw}{dt} = \frac{\partial\pi(w)}{\partial w}Sw.$$

Therefore, invariance of Σ is equivalent to

$$\frac{\partial\pi(w)}{\partial w}Sw = A\pi(w) + F(\pi(w)) + B_{\text{in}}\gamma(w) + B_{\text{d}}Pw, \qquad (4.10)$$

and (4.10) is exactly the first regulator equation given in (4.6). Furthermore, for initial data $(z_0, w_0)^\top \in U$, i.e., for sufficiently small initial data, the attractivity of the center manifold implies that the solution $(z(t), w(t))^\top$ satisfies

$$\|z(t) - \pi(w(t))\|_\alpha \leq Ke^{-\alpha_0 t}\|z_0 - \pi(w_0)\|_\alpha, \qquad (4.11)$$

where K depends on α. The estimate (4.11) is precisely the statement that the center manifold is locally exponentially attractive.

In the general theory of dynamical systems, center manifolds are not unique. However, as it was shown in [56] using an ω-limit argument, the neutral stability of the exosystem implies uniqueness of the center manifold Σ for any choice of γ.

By the same ω-limit argument and the attractivity of the center manifold, it follows that if $u = \gamma(w)$ solves the output regulation problem, then the center manifold is error zeroing, i.e., the error $e(t)$ is identically zero for all time on that unique center manifold Σ. This means that

$$e = C\pi(w) - Qw = 0,$$

must hold in a neighborhood of the origin. In particular, the solvability of the output regulation problem implies the solvability of the regulator equations (4.6) and (4.7).

Conversely, if the regulator equations hold then $z = \pi(w)$ is an invariant manifold which, by reversing the arguments above, must be the center manifold Σ for the closed-loop system. Since Σ is error zeroing and exponentially attractive, and c is continuous from $\mathcal{X}_\alpha$ to $\mathcal{Y}$ we have

$$\|e(t)\|_\mathcal{Y} = \|Cz(t) - Qw(t)\|_\mathcal{Y}$$

$$= \|[Cz(t) - C\pi(w(t))] + [C\pi(w(t)) - Qw(t)]\|_\mathcal{Y}$$

$$= \|Cz(t) - C\pi(w(t))\|_\mathcal{Y} \leq C_\alpha \|z(t) - \pi(w(t))\|_\alpha$$

$$\leq C_\alpha K e^{-\alpha_0 t} \|z_0 - \pi(w_0)\|_\alpha \ \to 0 \quad \text{as} \ t \to \infty,$$

which implies that the regulator problem is solvable. This concludes the proof of Theorem 4.1.

Remark 4.4. The assumption that the control input operator is bounded in the state space is a major restriction that excludes many important and interesting applications. The extension of Theorem 4.1 to include unbounded input operators B_{in} is not a simple task. The problems in this case are well known even in the linear case discussed in Chapter 2. For nonlinear systems these technical difficulties are even more challenging. At the time of writing of this book there are a number of research groups working to produce extensions of Theorem 4.1 with one or another set of assumptions concerning the properties of the nonlinear system. In this book we will take a different approach to the case of unbounded inputs. Rather than attempt to prove necessary and sufficient conditions for solvability as in Theorem 4.1 we provide assumptions on the nonlinear system that will guarantee solvability of the local state feedback regulator problem. This may seem to be a serious shortcoming not to face such difficulty squarely but we remind that this monograph was never intended to be a theoretical reference source but rather a practical guide to solving problems of output regulation for distributed parameter systems. At this point in time, to our knowledge, the theoretical machinery has simply not yet been organized in a way that gives a clean and complete theory. By some time, in the not so distant future, we are sure that this gap will be filled.

Example 4.1. We present a simple one-dimensional example of Theorem 4.1 by way of a set-point control problem for a controlled Chafee-Infante reaction diffusion equation. In particular we consider a system defined on a unit spatial interval $0 \leq x \leq 1$ for $t \geq 0$ given by

$$z_t(x,t) = z_{xx}(x,t) + (z(x,t) - z(x,t)^3) + B_{\text{in}}u,$$
$$0 = -z_x(0,t) + k_0 z(0,t),$$
$$0 = z_x(1,t) + k_1 z(1,t) = 0,$$
$$z(x,0) = \varphi(x),$$
$$y(t) = (Cz)(t) = z(x_1,t), \ 0 < x_1 < 1,$$
$$y_r(t) = M.$$

The operator A in this case is

$$A = \frac{d^2}{dx^2}, \ D(A) = \{\varphi \in H^2(0,1) : -\varphi'(0) + k_0\varphi(0) = 0, \ \varphi'(1) + k_1\varphi(1) = 0\}.$$

The operator A is self-adjoint in $\mathcal{Z} = L^2(0,1)$ and generates an exponentially stable semigroup in $\mathcal{Z}$. Even more is true, $(-A)$ also satisfies all the conditions in Assumption 4.2. Namely, $(-A)$ is a sectorial operator and it generates an exponentially stable semigroup in the infinite scale of Hilbert spaces $\mathcal{H}^\alpha = D((-A)^{\alpha/2})$.

Remark 4.5. In the above description we have not strictly followed our assumptions from Section 4.2 concerning the nonlinear term $F(z)$. Technically we should define $Az = z_{xx} + z$ and then set $F(z) = -z^3$. But, sticking with the situation most often encountered in the literature for the Chafee-Infante equation, we will set $Az = z_{xx}$ and $F(z) = z - z^3$. This convention does not pose any technical difficulties in solution of the regulator problem.

In this example we consider a bounded input operator B_{in}. Let us denote by I_0 the interval $[(x_0 - \nu_0), (x_0 + \nu_0)]$ where $0 < x_0 < 1$ and ν_0 is small enough so that $I_0 \subset (0, 1)$. Then we define

$$(B_{\text{in}}u)(x, t) = \frac{1}{2\nu_0}\mathbf{1}_{I_0}(x)u(t).$$

Here $\mathbf{1}_{I_0}(x)$ denotes the indicator function of the set I_0, i.e.,

$$\mathbf{1}_{I_0}(x) = \begin{cases} 1, & x \in I_0, \\ 0, & \text{otherwise.} \end{cases}$$

Therefore, the input operator is a bounded operator which for small ν_0 provides a bounded approximation to the Dirac delta function supported at x_0. Notice that the output operator, point evaluation at x_1, is unbounded but it is bounded on $\mathcal{H}^\alpha$ (see Section 2.2.1). So all the required conditions are met.

In the case of set-point control we have a one-dimensional exosystem, so that $\mathcal{W} = \mathbb{R}$, $w_t = 0$ with $w(0) = M$, and we have $w(t) = M$, for all $t \geq 0$. Therefore

$$y_r(t) = q(w(t)) = M \quad \text{or} \quad q(w) = w.$$

For this problem we seek mappings

$$\pi : \mathcal{W} \to D(A) \subset \mathcal{Z}, \quad \gamma : \mathcal{W} \to \mathcal{Y} = \mathbb{R},$$

satisfying the regulator equations (4.6) and (4.7), which in this case become

$$0 = A\pi(w) + \pi(w) - \pi(w)^3 + B_{\text{in}}\gamma(w),$$
$$\pi(w)(x_1) = w.$$

In [10], an efficient numerical algorithm for solving these equations, based on an interpretation of the regulator equations as a fixed point problem, was presented and used to solve for an approximate control using Newton iteration. Here we solve the problem using the methodology developed in [2] and described in the next section of this monograph.

As a specific numerical example we take $M = .75$, $x_0 = .75$, $\nu_0 = .02$, $x_1 = .25$, $k_0 = 1$, $k_1 = 2$ and $\varphi(x) = .5\cos(\pi x)$. The desired control $u = \gamma(M)$ is a nonlinear function of M but is independent of x and t. And for this specific example we get

$$u = \gamma(.75) = 41.80740.$$

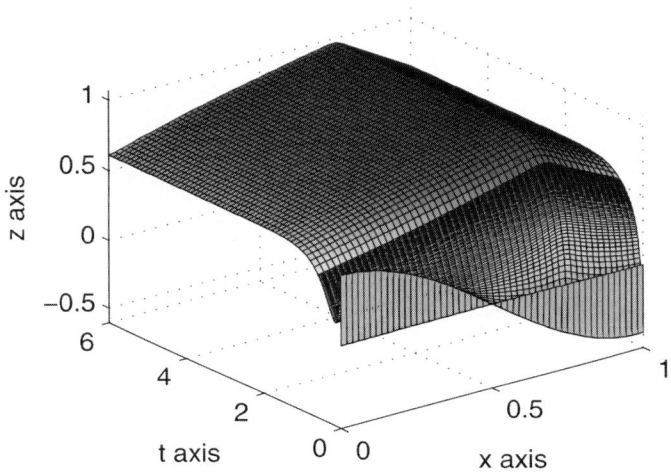

Fig. 4.2: Surface $z(x,t)$.

In Fig. 4.2 we have plotted the solution surface for the closed loop system. In Fig. 4.3 we have plotted both the measured output y for the closed loop system and the desired reference trajectory $y_r = M$. Finally in Fig. 4.4 we have plotted the error $e(t) = y(t) - y_r(t)$. The figures clearly suggest the desired result that the error approaches (quite rapidly) zero as t tends to infinity.

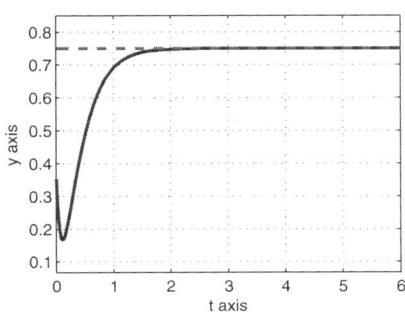

Fig. 4.3: Plot of $y(t)$, $y_r(t)$.

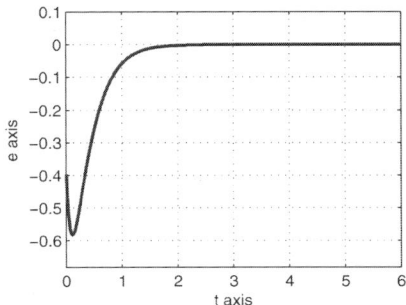

Fig. 4.4: Error $e(t)$.

4.3   Set-Point Regulation for Nonlinear Systems

In this section we describe the approach alluded to in Remark 4.4 in which we will provide a set of assumptions that allow us to use geometric thinking to solve a broad class of regulation problems modeled by nonlinear distributed parameter systems. In this section we focus on set-point regulation problems (i.e., when the reference signals are time independent) since this case most easily and more closely resembles the problems solved up to this point using the regulator equations approach. Certainly, hidden behind the main assumptions are the regulator equations. We first assume the existence of a set of control gain parameters for which a certain nonlinear elliptic boundary value problem also has a unique classical solution. This nonlinear elliptic boundary value problem corresponds to the first regulator equation. Then there is a second assumption concerning the attractivity, of solutions to the closed loop system, to the solution of the nonlinear elliptic boundary value problem described above. The control law is obtained from the gain parameters described in the first assumption. Since we are not attempting to state and prove necessary and sufficient conditions for solvability (as in Theorem 4.1) we proceed to show only that if the two assumptions are satisfied then we can solve the problem of regulation for the set-point control problem. For more general time-dependent reference signals and disturbances it is not so easy and we will turn to this situation later in the book. Part of the problem is the necessity of having sufficient knowledge of the center manifold to be able to find an initial condition that starts out on this unknown invariant (center) manifold. The problem of finding such initial conditions on an unknown invariant manifold has been the topic of many research articles in which numerous approximations schemes have been devised, even in the case of finite-dimensional nonlinear systems. To our knowledge there are no definitive results for general problems of this type at this time.

As we will show, the geometric approach provides a simple and extremely flexible design strategy for solving a wide class of "set-point" regulation problems for nonlinear parabolic boundary control systems. We emphasize that in this section the signals to be tracked and disturbances to be rejected are time independent. The theoretical underpinnings of our approach is still the regulator equations from the geometric theory of regulation applicable in the neighborhood of an equilibrium. In most of our examples Theorem 4.1 is not directly applicable since we consider unbounded boundary control but the method works equally well for bounded or unbounded input and output operators and even finite-dimensional nonlinear control systems. Along the way we provide some discussion to demonstrate how the method can be altered to provide many alternative control mechanisms. In particular, we show how the method can be adapted to solve tracking and disturbance rejection for piecewise constant time-dependent signals.

As we have already mentioned several times the main objective of this work is as a guide to practical application of the geometric methodology and not the detailed mathematical investigation of the partial differential equation models that appear in our application examples. In our opinion such a diversion would seriously detract from the main point of the work which is to exhibit the utility of the design methodology and its numerical implementation. So, for example, we do not go into any details concerning Hilbert space formulations for weak solutions in generalized Sobolev spaces and the elliptic estimates which are typically used to guarantee existence and prove regularity of solutions.

The main reference for this section is the work [2] which is primarily concerned with set-point tracking and disturbance rejection for nonlinear parabolic control systems in the form

$$z_t(x,t) = Az(x,t) + F(z(x,t)) + \sum_{j=1}^{n_\mathrm{d}} B_\mathrm{d}^j(x)d_j(t) + \sum_{j=1}^{n_\mathrm{in}} B_\mathrm{in}^j(x)u_j(t), \quad (4.12)$$

$$z(x,0) = z_0(x), \quad z_0 \in \mathcal{Z} = L^2(\Omega), \quad (4.13)$$

$$y_i(t) = (C_i z)(x,t), \quad i = 1, \ldots, n_\mathrm{c}, \quad (4.14)$$

with $x \in \Omega$, an open bounded subset of $\mathbb{R}^n$ with piecewise C^2 boundary, and $t \geq 0$. Here $z(x,t)$ is the state variable and it can be either a scalar or a vector. The terms

$$B_\mathrm{d}^j(x)d_j(t), \quad j = 1, \ldots, n_\mathrm{d}, \quad (4.15)$$

$$B_\mathrm{in}^j(x)u_j(t), \quad j = 1, \ldots, n_\mathrm{in}, \quad (4.16)$$

represent disturbances and control inputs, respectively. Note that in Equations (4.15)–(4.16) each term is a multiplicative operator between a space dependent function and the corresponding input function which is considered to be time-dependent only. The expressions $B_\mathrm{d}^j(x)$ and $B_\mathrm{in}^j(x)$ are assumed to be known functions, and may also be unbounded (e.g., for example, in the case of boundary control they typically are given by delta distributions supported on a portion of the boundary). In general B_d^j refers to a *disturbance input* operator and B_in^j refers to a *control input* operator.

The state operator A is assumed to be a linear differential operator in an infinite-dimensional Hilbert state space $\mathcal{Z} = L^2(\Omega)$. It is assumed that the operator A defined on a dense domain $D(A)$ generates an exponentially stable C_0 semigroup in $\mathcal{Z}$. In our intended applications the operators C_i in Equation (4.14) are typically point evaluation or weighted integrals of the solution $z(x,t)$ in some parts of the domain Ω or its boundary. Therefore these operators are generally densely defined and not usually bounded in the state space $\mathcal{Z}$. The most common situation is that $(-A)$ is an accretive operator that generates a Hilbert scale of spaces $\mathcal{H}^\alpha$ for $\alpha \in \mathbb{R}$ and the domain of C_i, denoted by $D(C_i)$, is contained in $\mathcal{H}^{\alpha_0}$ for some $\alpha_0 > 0$. So we assume that

$C_i : D(C_i) \to \mathbb{R}$ for each i (see, e.g., [38, 47, 57] for a discussion of sectorial operators).

Remark 4.6. We assume that the piecewise C^2 boundary of Ω, denoted by $\partial\Omega$, is represented by the union of $(n-1)$-dimensional connected hypersurfaces S_j, which are subsets of $\partial\Omega$ and whose interiors are pairwise disjoint.

Here the nonlinear function F is a smooth function with $F(0) = 0$ so that the uncontrolled control system has the origin in $\mathcal{Z}$ as an exponentially stable equilibrium.

Remark 4.7. Just as we have already seen in the linear case in Chapters 1 and 2, stability of the origin for the uncontrolled nonlinear problem, i.e., the problem with all $u_j = 0$ and $d_j = 0$, is a fundamentally important part of the theoretical development (see [55]) of the geometric approach to regulation based on center manifold theory. Here by stability we mean that for all sufficiently small initial data $z_0 \in \mathcal{Z}$, say $\|z_0\| \leq \delta$, there exists positive constants M and α (depending on δ) so the solution of

$$z_t(x,t) = Az(x,t) + F(z(x,t)),$$
$$z(x,0) = z_0(x),$$

satisfies

$$\|z(\cdot,t)\| \leq Me^{-\alpha t} \quad \text{for all } t \geq 0.$$

Once again, as in Chapters 1 and 2, if the operator A does not generate a stable semigroup then we must first introduce a feedback mechanism that stabilizes the control system. The problem of finding such a feedback law is the stabilization problem and is not the same as the regulator problem considered in this book.

As we learned in Chapter 2, many distributed parameter control results are stated in what is often referred to as standard system form. Introducing a few new notations allows us to write the control system (4.12)–(4.14) in the standard system theoretic state space form. To this end, let us define

$$D = \begin{bmatrix} d_1 \\ d_2 \\ \vdots \\ d_{n_d} \end{bmatrix}, \quad U = \begin{bmatrix} u_1 \\ u_2 \\ \vdots \\ u_{n_{in}} \end{bmatrix}, \quad Y = \begin{bmatrix} y_1 \\ y_2 \\ \vdots \\ y_{n_c} \end{bmatrix}, \quad Y_r = \begin{bmatrix} y_{r,1} \\ y_{r,2} \\ \vdots \\ y_{r,n_c} \end{bmatrix}.$$

With this notation we can rewrite our control system in standard form as

$$z_t = Az + F(z) + B_d D + B_{in} U,$$

$$Y = Cz,$$

where we have written the input, disturbance and output terms in matrix form as

$$B_\mathrm{d} D = \sum_{j=1}^{n_\mathrm{d}} B_\mathrm{d}^j(x)\, d_j(t), \quad B_\mathrm{in} U = \sum_{j=1}^{n_\mathrm{in}} B_\mathrm{in}^j(x)\, u_j(t), \quad Y = Cz = \begin{bmatrix} C_1(z) \\ C_2(z) \\ \vdots \\ C_{n_c}(z) \end{bmatrix},$$

and where B_d and B_in are disturbance input and control input operators respectively, and C denotes the output operator.

Again referring to Chapter 2 we recall that for distributed parameter systems governed by partial differential equations it very often happens that the control inputs and the disturbances influence the system through unbounded operators in the Hilbert state space, for example, through a boundary control. We remind the reader that we will follow the development in [2] in order to deal with this situation. We note that when we write the system (4.12)–(4.14) the boundary conditions corresponding to controls or disturbances that enter through boundary conditions on the hypersurfaces S_j or at points or on hypersurfaces inside the domain are included in the input operators B_d^j and B_in^j, for some set of indices. Let the sequences B_d^j and B_in^j be ordered so that the first n_d^b and n_in^b elements correspond to boundary operators $\mathcal{B}_\mathrm{d}^j$ and $\mathcal{B}_\mathrm{in}^j$ defined on the hypersurfaces $S_\mathrm{d}j$ and $S_\mathrm{in}j$ respectively. Then the control system (4.12) can be written in the equivalent form

$$z_t(x,t) = A_0 z(x,t) + F(z(x,t)) + \sum_{j=n_\mathrm{d}^b+1}^{n_\mathrm{d}} B_\mathrm{d}^j(x) d_j(t) \tag{4.17}$$

$$+ \sum_{j=n_\mathrm{in}^b+1}^{n_\mathrm{in}} B_\mathrm{in}^j(x) u_j(t),$$

$$(\mathcal{B}_\mathrm{d}^j z)(x,t) = \mathcal{B}_\mathrm{d}^j(x)\, d_j(t), \quad x \in S_\mathrm{d}j, \quad j = 1,\ldots,n_\mathrm{d}^b, \tag{4.18}$$

$$(\mathcal{B}_\mathrm{in}^j z)(x,t) = \mathcal{B}_\mathrm{in}^j(x)\, u_j(t), \quad x \in S_\mathrm{in}j, \quad j = 1,\ldots,n_\mathrm{in}^b. \tag{4.19}$$

where (4.18), (4.19) correspond to the boundary disturbance and control input terms which replace the homogeneous boundary conditions hidden in the definition of $D(A)$.

In this case the functions $\mathcal{B}_\mathrm{d}^j(x)$ in (4.18) and $\mathcal{B}_\mathrm{in}^j(x)$ in (4.19) are not the same as the distributional functions B_d^j and B_in^j given in (4.15) and (4.16). Indeed, the functions $\mathcal{B}_\mathrm{d}^j(x)$ and $\mathcal{B}_\mathrm{in}^j(x)$ are typically smooth functions, not distributions. The reformulation of problem (4.12)–(4.16) into the form (4.17)–(4.19) is discussed in [6, 7, 78, 84]. In particular, it is well known that under suitable assumptions there exist operators B_d^j and B_in^j so that system (4.17)–(4.19) can be written in the form (4.12) (cf. [6, 7, 78, 84]). Here the operator A_0 is typically a linear elliptic partial differential operator engendered with only a partial set of boundary conditions. We denote the dense domain of A_0 in $\mathcal{Z}$ by $D(A_0)$.

The operator A is the same linear elliptic partial differential operator A_0 with domain, denoted by $D(A)$, given by

$$D(A) = \{\varphi \, : \, \mathcal{B}_d^j \varphi = 0 \ (j = 1, \dots, n_d^b),$$
$$\mathcal{B}_{in}^j \varphi = 0 \ (j = 1, \dots, n_{in}^b)\} \cap D(A_0).$$

The structure of the boundary operators depends on the structure of the operator A and other physical properties of the particular problem. Generally speaking the boundary operators $\mathcal{B}_d^j$ and $\mathcal{B}_{in}^j$ can represent any of the classical boundary conditions, including Dirichlet, Neumann and Robin, etc. The explicit examples in this work provide a clear description of the general types of boundary conditions that can be handled with this methodology.

Remark 4.8. The output operators C_i are often given as point evaluation or as the weighted average of the state $z(x, t)$ on a hypersurface $\mathcal{S}_i$ inside or on the boundary of Ω, i.e.,

$$y_i(t) = C_i z = \frac{1}{|\mathcal{S}_i|} \int_{\mathcal{S}_i} z(x, t) \, d\sigma_x,$$

where by $d\sigma_x$ we denote the natural hypersurface measure on $\mathcal{S}_i$. For example it could be that $\mathcal{S}_i$ is one of the boundary patches $\mathcal{S}_j$ (see Remark 4.6). We note that the operators C_i are well defined in our setting since in a typical parabolic problem the state $z(\cdot, t)$ for $t > 0$ is contained in $C^\infty(\overline{\Omega})$ so that the trace on the boundary of Ω is a continuous function.

4.3.1   Statement of the Set-Point Control Problem

As we have noted, for nonlinear distributed parameter systems with unbounded control there is not yet a version of Theorem 4.1 that would provide necessary and sufficient conditions for solvability of a regulation problem in terms of the regulator equations. Nevertheless, in this section of the book we discuss a general strategy capable of delivering control laws that solve a wide variety of set-point regulation problems, i.e., tracking/disturbance rejection problems for *time-independent* reference signals $y_{r_i} \in \mathbb{R}$, $i = 1, \dots, n_c$, and disturbances, $d_j(t) = d_j \in \mathbb{R}$, $i = j, \dots, n_d$.

In Section 4.4 we discuss a methodology for solving tracking and disturbance rejection problems for signals that are time-dependent but *piecewise constant* over specified time intervals. We show that for these very special time-dependent signals the corresponding tracking problems can be solved with a slight modification of the results of this section.

Problem 4.2. *Our design objective is to find a set of time-independent controls $u_j(t) = \gamma_j$, $j = 1, \dots, n_{in}$, for the system (4.12)–(4.14) so that the error* *defined by*

$$e(t) = \|Y(t) - Y_r\|_\infty = \sup_{1 \le i \le n_c} |y_i(t) - y_{r_i}|, \tag{4.20}$$

satisfies

$$e(t) \xrightarrow{t \to \infty} 0. \tag{4.21}$$

while the state of the closed loop control system remains bounded for all time.

The methodology for solving Problem 4.2 is based on the following two main assumptions:

Assumption 4.3. *There exist constants γ_j, $j = 1, \ldots, n_{in}$, and a classical solution $\overline{z}(x) \in \mathcal{Z}$ of the nonlinear elliptic boundary value problem (4.22) satisfying the constraints given in (4.23):*

$$0 = A\overline{z}(x) + F(\overline{z}(x)) + \sum_{j=0}^{n_d} B_d^j(x)d_j + \sum_{j=0}^{n_{in}} B_{in}^j(x)\gamma_j, \tag{4.22}$$

$$C_i\overline{z} = y_{r_i}, \quad i = 1, \ldots, n_c. \tag{4.23}$$

Assumption 4.4. *For sufficiently close initial data*

$$\|z_0(x) - \overline{z}(x)\| < \delta,$$

the solution $z(x, t)$ of the system (4.12)–(4.14) with controls $u_j = \gamma_j$, i.e.,

$$z_t(x, t) = Az(x, t) + F(z(x, t)) + \sum_{j=0}^{n_d} B_d^j(x)d_i + \sum_{j=0}^{n_{in}} B_{in}^j(x)\gamma_j,$$

$$z(x, 0) = z_0(x),$$

satisfies

$$\lim_{t \to \infty} |C_i z(\cdot, t) - C_i\overline{z}(\cdot)| = 0, \quad i = 1, \ldots, n_{in}. \tag{4.24}$$

The above assumptions clearly entail the main ingredients of the sufficiency part of Theorem 4.1 which says that under the above assumptions the regulator problem is solvable. Indeed the first assumption is simply the statement that the regulator equations are solvable and the second assumption corresponds to the local exponential attractivity of the center invariant manifold. The condition in (4.24) implies the asymptotic error condition (4.21), i.e., under Assumptions 4.3 and 4.4, it is obvious that

$$y_i(t) = C_i z \xrightarrow{t \to \infty} C_i\overline{z} = y_{r_i}, \quad i = 1, \ldots, n_{in}.$$

Thus the controls solving our set-point control problem are $u_j = \gamma_j$, $j = 1, \ldots, n_{in}$, which are obtained, along with $\overline{z}$, by solving the system (4.22)–(4.23).

It is further assumed that the disturbances $d_j(t)$ and signals to be tracked $y_{r_i}(t)$ are generated as outputs of a neutrally stable, finite-dimensional exogenous system. The case of set-point control, where all the d_j and y_{r_i} are time independent, satisfies this requirement. In particular the exosystem in this case is given by

$$w_t = Sw, \quad w(0) = w_0, \tag{4.25}$$

where $w = [w_1, w_2, \cdots, w_{n_c+n_d}]^\mathsf{T} \in \mathcal{W} = \mathbb{R}^{n_c+n_d}$, S is the $(n_c+n_d) \times (n_c+n_d)$ zero matrix, and

$$w_0 = [y_{r_1}, \cdots, y_{r_{n_c}}, d_1, \cdots, d_{n_d}]^\mathsf{T}. \tag{4.26}$$

Clearly the solution to the initial value problem (4.25) is the constant vector $w(t) = w_0$ for all times.

In the geometric theory we seek controls u_j as a feedback of the state of the exosystem, i.e., $u_j = \gamma_j(w)$, which we will often denote simply by γ_j. So in particular in this case we seek controls that are time independent. In matrix form we have

$$U = \begin{bmatrix} u_1 \\ \vdots \\ u_{n_{in}} \end{bmatrix} = \begin{bmatrix} \gamma_1 \\ \vdots \\ \gamma_{n_{in}} \end{bmatrix}.$$

By assumption A is the generator of an exponentially stable semigroup (see Remark 4.7 and the discussion proceeding it) and therefore its spectrum lies in the strict left half complex plane. In this case the closed loop system, consisting of (4.12)–(4.14) coupled with (4.25) and with controls $u_j = \gamma_j(w)$, is given in the state space $\mathcal{Z} \times \mathcal{W}$ as

$$z_t = Az + F(z) + \sum_{j=0}^{n_d} B_d^j d_j(w) + \sum_{j=0}^{n_{in}} B_{in}^j \gamma_j(w), \tag{4.27}$$

$$w_t = Sw, \tag{4.28}$$

$$z(x, 0) = \varphi(x), \quad w(0) = w_0.$$

The linearization of this problem has spectrum consisting of the spectrum of A together with the spectrum of S. So there are $n_d + n_c$ eigenvalues at zero (on the imaginary axis) and the remainder of the spectrum is in the left half complex plane. In the terminology of dynamical systems, the problem has an infinite-dimensional stable manifold and a $(n_d + n_c)$-dimensional center manifold. Further, solutions beginning in a sufficiently small neighborhood of the origin in $\mathcal{Z} \times \mathcal{W}$ converge exponentially to a solution on the $(n_d + n_c)$-dimensional center manifold. In the set-point case this solution is a point on the center manifold corresponding to the single vector w_0.

In order to obtain the controls γ_j, in the terminology introduced by C.I. Byrnes and A. Isidori [56, 55], we seek an "error zeroing" center manifold. Thus we seek an invariant manifold for the dynamics of the closed loop system on which the error, defined in (4.20), is identically zero. We note that it is possible that such an invariant manifold may not exist. But if it does then we can solve the corresponding regulator problem as follows. We seek a mapping $\bar{z}(w) : \mathcal{W} \to \mathcal{Z}$ that expresses the invariance of the z dynamics of the closed loop system. At least locally, i.e., in a neighborhood $\mathcal{X}_0$ of the origin in $\mathcal{X} = \mathcal{Z} \times \mathcal{W}$, the center manifold Σ is given as the graph of a function in $\mathcal{Z} \times \mathcal{W}$ space, i.e.,

$$\Sigma = \left\{ \begin{pmatrix} \bar{z}(w) \\ w \end{pmatrix} : w \in W_0 \right\}$$

for some neighborhood W_0 of the origin in $\mathbb{R}^3$.

As for invariance, first we note that, by the chain rule, (4.28), and since $S = 0$,

$$\bar{z}_t = \frac{\partial \bar{z}}{\partial w} w_t = 0,$$

so we obtain from (4.27)

$$0 = A\bar{z}(w) + F(\bar{z}(w)) + \sum_{j=0}^{n_{\mathrm{d}}} B_{\mathrm{d}}^j d_j(w) + \sum_{j=0}^{n_{\mathrm{in}}} B_{\mathrm{in}}^j \gamma_j(w), \qquad (4.29)$$

which is precisely (4.22). The requirement that the invariant manifold be error zeroing means, in addition, that $\bar{z}$ must satisfy

$$C_i \bar{z}(w) - w_i = 0, \quad i = 1, .., n_{\mathrm{c}}, \qquad (4.30)$$

for all w in a neighborhood of the origin. Recall that by our choice of initial conditions in (4.26), we have $w_i = y_{r_i}$. We conclude that equations (4.29) and (4.30) are precisely equations (4.22), (4.23) and we see that the Assumptions 4.3 and 4.4 are simply the requirements of the existence of an error zeroing attractive invariant manifold.

4.3.2 Solution Strategy

Turning now to the problem of finding gain parameters γ_j and the solution $\bar{z}$ of (4.29) satisfying (4.30) (i.e., the solutions of (4.22)–(4.23)) we proceed in a way almost identical to the approach that evolved in Chapter 2. Indeed, we use the same notation as we did in solving the examples presented in Chapters 2 and 3. First we solve the n_{in} linear boundary value problems given by

$$0 = AX_j(x) + B_{\mathrm{in}}^j(x), \quad j = 1, \ldots, n_{\mathrm{in}} \text{ and } x \in \Omega. \qquad (4.31)$$

Notice that as long as the coefficients in these elliptic boundary value problems are sufficiently smooth the solution X_j will also be smooth by elliptic regularity so that $X_i \in D(C_j)$.

With this we can assemble the matrix $G_{n_{\mathrm{c}} \times n_{\mathrm{in}}}$, whose entries are

$$g_{ij} = C_i X_j, \quad i = 1, \ldots, n_{\mathrm{c}}, \ j = 1, \ldots, n_{\mathrm{c}}. \qquad (4.32)$$

Each component X_j belongs to $\mathcal{Z}$, and it is the response of the linear operator A to the input B_{in}^j, namely

$$X_j = -A^{-1}B_{\text{in}}^j. \tag{4.33}$$

We rewrite Equation (4.29) as

$$\overline{z}(x) = -A^{-1}\left(F(\overline{z}(x)) + \sum_{j=0}^{n_{\text{d}}} B_{\text{d}}^j(x)d_j \right) + \sum_{j=0}^{n_{\text{in}}} \left(-A^{-1}B_{\text{in}}^j(x)\ \gamma_j \right), \tag{4.34}$$

and let $\widetilde{z}(x)$ be the solution of

$$0 = A\widetilde{z} + F(\overline{z}(x)) + \sum_{j=0}^{n_{\text{d}}} B_{\text{d}}^j(x)d_j. \tag{4.35}$$

Here, $\widetilde{z}$ is the response of the linear operator A to the sum of the nonlinear term $F(\overline{z}(x))$ and all the disturbances B_{d}^j, namely

$$\widetilde{z} = -A^{-1}\left(F(\overline{z}) + \sum_{j=0}^{n_{\text{d}}} B_{\text{d}}^j(x)d_j \right). \tag{4.36}$$

Substituting Equations (4.33) and (4.36) in Equation (4.34) yields

$$\overline{z} = \widetilde{z} + \sum_{j=0}^{n_{\text{in}}} X_j\gamma_j.$$

Applying the operators C_i to each side of the above equation, and substituting Equations (4.30) and (4.32) it follows that

$$y_{r_i} = C_i\overline{z} = C_i\widetilde{z} + \sum_{j=0}^{n_{\text{in}}} g_{ij}\gamma_j, \quad i = 1,\ldots,n_{\text{c}}.$$

These equations can be written in matrix form as

$$G\Gamma = Y_r - \widetilde{Y}, \tag{4.37}$$

where

$$\Gamma = [\gamma_1, \gamma_2, \cdots, \gamma_{n_{\text{in}}}]^\mathsf{T}, \ Y_r = [y_{r_1}, y_{r_2}, \cdots, y_{r_{n_{\text{c}}}}]^\mathsf{T}, \ \widetilde{Y} = [C_1\widetilde{z}, C_2\widetilde{z}, \cdots, C_{n_{\text{c}}}\widetilde{z}]^\mathsf{T}.$$

The matrix G is, in general, a rectangular matrix, and we choose Γ to be the minimal solution of Equation (4.37). Because of Assumption 4.3, system (4.37) is assumed to be consistent, and the minimal solution is either the unique solution or, in case of multiple solutions, the solution having the least Euclidean norm.

Each equation in (4.31) is decoupled from the others, thus can be solved individually. In contrast, Equations (4.34), (4.35) and (4.37) are fully coupled and should be solved together.

4.3.3   Numerical Example

Just as we did in Chapter 2 we now present a relatively simple one-dimensional example of tracking and disturbance rejection for a nonlinear parabolic boundary control system. We leave to Chapter 5 the more complicated boundary control systems that will emphasize how easily these more challenging problems can be handled. Certainly the methodology described in Section 4.3.1 can be applied to linear or nonlinear problems with bounded or unbounded observation and control.

Example 4.2 (Burgers' Equation). In our first example we consider a boundary controlled viscous Burgers' equation

$$z_t(x,t) = \nu z_{xx}(x,t) - z(x,t)z_x(x,t), \ 0 \le x \le 1, \tag{4.38}$$

with initial condition

$$z(x,0) = \varphi(x).$$

Here ν is a kinematic viscosity and is considered constant on the interval.

The equation (4.38) is supplemented with a non-homogeneous constant Dirichlet boundary condition at $x = 0$,

$$z(0,t) = d,$$

which we treat as a disturbance.

In addition we have a pair of measured outputs given by point evaluation at the points $x = 0.25$ and $x = 0.75$, respectively

$$y_1(t) = C_1(z) = z(0.25,t),$$
$$y_2(t) = C_2(z) = z(0.75,t).$$

And, finally, we are given a pair of constant reference signals $y_{r_1}, y_{r_2} \in \mathbb{R}$ to be tracked.

Our objective is to find two constant control inputs $u_j = \gamma_j$ with $j = 1,2$ so that the measured outputs y_j of the closed loop system track the reference signals y_{r_j} while rejecting the disturbance d. The first control u_1 enters as a point source in the domain at the point $x = 0.5$, and the second control u_2 enters through a Neumann boundary condition at the right end of the interval. In particular we have the following conditions

$$[z(x,t)]_{x=0.5} = 0, \tag{4.39}$$

$$[\nu z_x(x,t)]_{x=0.5} = u_1, \tag{4.40}$$

$$\nu z_x(1,t) = u_2, \tag{4.41}$$

where the notation $[\varphi]_{x=x_0}$ denotes the jump at x_0 defined by

$$[\varphi]_{x=x_0} = \varphi(x_0^+) - \varphi(x_0^-).$$

Here we have used the notation $x_0^{\pm}$ for the limit from the right $(+)$ and the limit from the left $(-)$ at x_0.

The above description of the problem is illustrated in the Fig. 4.5.

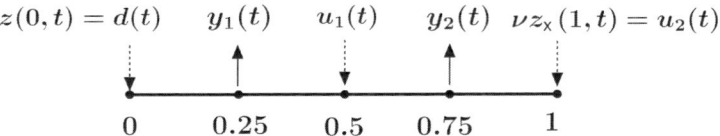

$$z(0,t) = d(t) \quad y_1(t) \quad u_1(t) \quad y_2(t) \quad \nu z_{\mathrm{x}}(1,t) = u_2(t)$$

$$0 \qquad 0.25 \qquad 0.5 \qquad 0.75 \qquad 1$$

Fig. 4.5: Burgers' Example Domain.

Let us define $A_0 = \nu d^2/dx^2$ in $L^2(0,1)$ with domain $D(A_0) = H^2(0,1)$. With this, the problem formulated as in (4.12)–(4.14) can be written using the boundary operators

$$\mathcal{B}_{\mathrm{d}}^1 z = z(0,t) = d, \tag{4.42}$$

$$\mathcal{B}_{\mathrm{in}}^1 z = \begin{cases} [z(x,t)]_{x=0.5} = 0, \\ [\nu z_x(x,t)]_{x=0.5} = u_1, \end{cases} \tag{4.43}$$

$$\mathcal{B}_{\mathrm{in}}^2 z = \nu z_x(1,t) = u_2. \tag{4.44}$$

The same problem can be formulated in state space form as in Equations (4.17)–(4.19) by introducing the equivalent distributional terms

$$B_{\mathrm{d}}^1 d = \frac{d\delta_0}{dx}\, d, \tag{4.45}$$

$$B_{\mathrm{in}}^1 u_1 = -\delta_{0.5}\, u_1, \tag{4.46}$$

$$B_{\mathrm{in}}^2 u_2 = \delta_1\, u_2, \tag{4.47}$$

where δ_{x_0} is the Dirac delta distribution supported at $x = x_0$.

For this example the nonlinear term in (4.27) is $F(z) = -z\, z_x$.

Remark 4.9. It should be noticed that the boundary operators in (4.42) and (4.44) produce the classical Dirichlet and Neumann boundary conditions, and the operator (4.42) imposes a jump in the first derivative of the solution. The operators (4.45)–(4.47) are their equivalent distributional counterpart. When solving a partial differential equation with finite element methods it is generally easier to deal with the Dirichlet boundary operator (4.42) rather than (4.45); in case of Neumann boundary conditions (4.44) and (4.47) result in the same forcing term; to force the jump in the solution first derivative it is straightforward to use (4.46) rather then (4.43). In the rest of this work we will consider the operators that are more convenient, keeping in mind that there always exist these alternative distributional counterparts.

From here the problem is solved numerically by proceeding exactly as described in Section 4.3.2. First we must solve for X_j, $j = 1, 2$, but rather than solving the equations as written in (4.31) we solve

$$\nu \frac{d^2 X_1}{dx^2} = \delta_{0.5}, \quad \text{for } 0 < x < 1, \tag{4.48}$$

$$X_1(0) = 0, \quad \nu \frac{dX_1}{dx}(1) = 0,$$

for X_1, and

$$\nu \frac{d^2 X_2}{dx^2} = 0, \quad \text{for } 0 < x < 1,$$

$$X_2(0) = 0, \quad \nu \frac{dX_2}{dx}(1) = 1,$$

for X_2. These still produce the desired functions

$$X_1 = (-A)^{-1} B_{in}^1 \quad \text{and} \quad X_2 = (-A)^{-1} B_{in}^2,$$

and the entries of the 2×2 matrix

$$G = \begin{pmatrix} C_1 X_1 & C_1 X_2 \\ C_2 X_1 & C_2 X_2 \end{pmatrix}.$$

With this, we now solve the following steady state coupled system for the unknown values γ_1, γ_2, $\overline{z}$ and $\widetilde{z}$.

$$0 = \nu \frac{d^2 \overline{z}}{dx^2} - \overline{z} \frac{d\overline{z}}{dx} - \delta_{0.5} \gamma_1, \quad \text{for } 0 < x < 1,$$

$$\overline{z}(0) = d, \quad \nu \frac{d\overline{z}}{dx}(1) = \gamma_2,$$

$$0 = \nu \frac{d^2 \widetilde{z}}{dx^2} - \overline{z} \frac{d\overline{z}}{dx}, \quad \text{for } 0 < x < 1,$$

$$\widetilde{z}(0) = d, \quad \nu \frac{d\widetilde{z}}{dx}(1) = 0,$$

$$\Gamma = \begin{bmatrix} \gamma_1 \\ \gamma_2 \end{bmatrix} = G^{-1} \begin{bmatrix} -C_1 \widetilde{z} + y_{r_1} \\ -C_2 \widetilde{z} + y_{r_2} \end{bmatrix}. \tag{4.49}$$

Finally, we use the inputs γ_1 and γ_2 in (4.39)–(4.41) to obtain the desired closed loop system

$$z_t = \nu z_{xx} - z z_x - \delta_{0.5} \gamma_1, \quad \text{for } 0 < x < 1 \text{ and } 0 < t \leq T, \tag{4.50}$$

$$z(0,t) = d, \quad \nu z_x(1,t) = \gamma_2, \tag{4.51}$$

$$z(x,0) = \varphi(x). \tag{4.52}$$

For our numerical simulation we have chosen the parameters

$$\nu = 0.2,\ d = 0.35,\ y_{r_1} = 0.5\ \text{and}\ y_{r_2} = 0.75.$$

After following step-by-step the procedure in Equations (4.48)–(4.49), and evaluating the input parameters, γ_1 and γ_2, we solve the closed loop system (4.50)–(4.52) on the time interval $0 < t \leq T$ with $T = 10$ with initial data $\varphi(x) = 0$. In Figs. 4.6 and 4.7, we display the signals to be tracked y_{r_1} and y_{r_2} and the corresponding outputs C_1 and C_2 for the closed loop solution $z(x,t)$. As expected after a short transient the two graphs overlap.

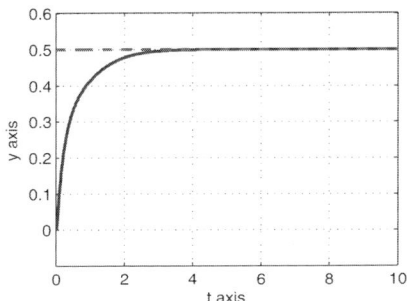
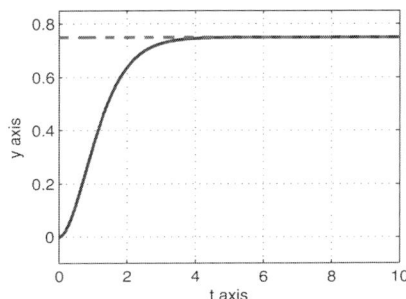

Fig. 4.6: $y_{r_1}(t)$ and $C_1(z(t))$ for $0 \leq t \leq 10$.

Fig. 4.7: $y_{r_2}(t)$ and $C_2(z(t))$ for $0 \leq t \leq 10$.

In Figs. 4.8 and 4.9 the errors $e_1(t) = y_{r_1} - C_1(z)$ and $e_2(t) = y_{r_1} - C_1(z)$ are given for all times.

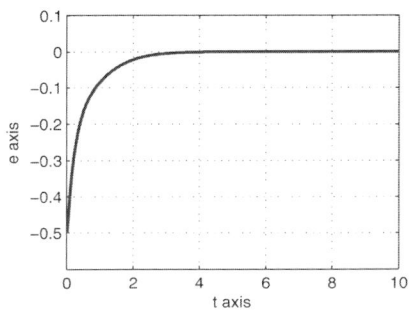
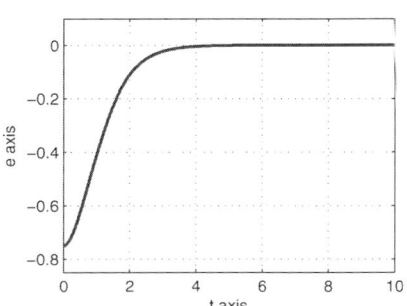

Fig. 4.8: $e_1 = y_{r_1} - C_1(z)$ for $0 \leq t \leq 10$.

Fig. 4.9: $e_2 = y_{r_2} - C_2(z)$ for $0 \leq t \leq 10$.

In Fig. 4.10 we display the solution profile $z(x,t)$ both in space and time. We note that at $x = 0$ the numerical solution satisfies $z(0,t) = 0.35$ for all time.

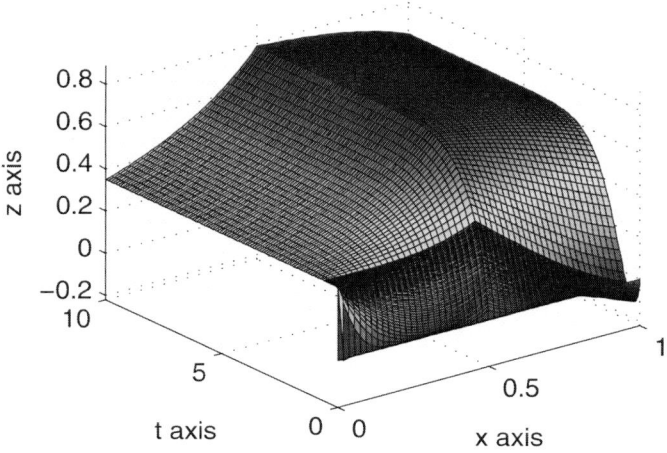

Fig. 4.10: Solution $z(x,t)$ for $0 \leq x \leq 1$ and $0 \leq t \leq 10$.

4.4 Tracking/Rejection of Piecewise Constant Signals

In this section we show that the set-point methodology described in Section 4.3.1 can be easily modified to track time-dependent disturbances and reference signals consisting of piecewise constant time-dependent signals, when the inertial terms are not negligible. However some numerical instabilities can occur and a certain adjustment, involving an approximation, in the algorithm is required. We note that a more general and detailed discussion for the time-dependent signals, without the approximation, is the focus of the next section.

In this section we consider piecewise constant, in time, reference and disturbance signals. We once again consider a system in the form

$$z_t(x,t) = Az(x,t) + F(z(x,t)) + \sum_{j=0}^{n_{\mathrm{d}}} B_{\mathrm{d}}^j(x)\, d_j(t) + \sum_{j=0}^{n_{\mathrm{in}}} B_{\mathrm{in}}^j(x)\, u_j(t), \quad (4.53)$$

$$z(x,0) = z_0(x), \qquad z_0 \in \mathcal{Z} = L^2(\Omega), \tag{4.54}$$

$$y_i(t) = (C_i z)(t), \quad i = 1, \ldots, n_{\mathrm{c}}, \tag{4.55}$$

where now we seek time-dependent control gains $u_j(t) = \gamma_j(t)$, in order to track and reject piecewise constant time-dependent signals $y_{r_i}(t)$ and $d_j(t)$. As a first approximation since $y_{r_i}(t)$ and $d_j(t)$ are assumed to be piecewise constant in time, we could split the time interval into sub-intervals

$0 = t_0, t_1, \ldots, t_N$, such that both $y_{r_i}(t)$ and $d_j(t)$ are constant on each sub-interval, i.e.,

$$y_{r_i}(t) = y_{r_{i,k}} \text{ and } d_j(t) = d_{j,k}, \text{ for } t \in [t_k, t_{k+1}).$$

Then, according to Section 4.3.1, for the kth sub-interval, we could find a set of control gains $\gamma_{j,k}$, and build piecewise constant time-dependent controls

$$u_j(t) = \gamma_{j,k} , \text{ for } t \in [t_k, t_{k+1}).$$

Using this procedure, we find that any time $y_{r_i}(t)$ or $d_j(t)$ change, a transient with

$$y_{r_i}(t) \neq y_i(t) = (C_i z)(x)$$

occurs. As long as the inertial terms remain small, the error is quickly reabsorbed. However, if the inertial terms are not negligible then we need to take into account their effect in the control design strategy.

In the latter case, in order to determine the controls $\gamma_j(t)$ we proceed by following a procedure similar to the one described in Section 4.3.2 with the exception that we consider the auxiliary state variables $\bar{z}$ and $\tilde{z}$ now to be time-dependent.

We use as initial data for $\bar{z}(x, 0)$ the steady state solutions $\overline{Z}(x)$ needed to solve the set-point control problem

$$Z_t(x, t) = AZ(x, t) + F(Z(x, t)) + \sum_{j=0}^{n_d} B_d^j(x)\, d_{j,0} + \sum_{j=0}^{n_{in}} B_{in}^j(x)\, \gamma_{j,0}, \quad (4.56)$$

$$Z(x, 0) = z_0(x), \tag{4.57}$$

$$y_i(t) = (C_i Z)(t), \quad i = 1, \ldots, n_c, \tag{4.58}$$

where the disturbances and the signals to be tracked are

$$d_{j,0} = d_j(0) , \quad y_{r_{i,0}} = y_{r_i}(0),$$

respectively. Thus, they are time *independent*. Notice that to find $\overline{Z}(x)$, we follow the strategy described in Section 4.3.2, and determine the solutions X_j, the matrix G and the controllers $u_{j,0} = \gamma_{j,0}$, however, there is no need to solve the closed loop system (4.56)–(4.58).

Next we seek time-dependent control gains $u_j(t) = \gamma_j(t)$ such that $\bar{z}(x, t)$ satisfy

$$\bar{z}_t(x, t) = A\bar{z}(x, t) + F(\bar{z}(x, t)) + \sum_{j=0}^{n_d} B_d^j(x)\, d_j(t) + \sum_{j=0}^{n_{in}} B_{in}^j(x)\, \gamma_j(t), \quad (4.59)$$

$$\bar{z}(x, 0) = \overline{Z}(x), \tag{4.60}$$

$$y_{r_i}(t) = (C_i \bar{z})(t), \quad i = 1, \ldots, n_c, \tag{4.61}$$

for all time. Notice that, because of the initial condition (4.60), Equation (4.61) is automatically satisfied for $t = 0$.

We rewrite Equation (4.59) as

$$\overline{z}(x,t) = -A^{-1}\left(-\overline{z}_t(x,t) + F(\overline{z}(x,t)) + \sum_{j=0}^{n_{\mathrm{d}}} B_{\mathrm{d}}^j(x)\, d_j(t)\right) \tag{4.62}$$

$$+ \sum_{j=0}^{n_{\mathrm{in}}}\left(-A^{-1}B_{\mathrm{in}}^j(x)\,\gamma_j(t)\right).$$

Let $\widetilde{z}(x,t)$ be the solution of

$$\overline{z}_t(x,t) = A\widetilde{z}(x,t) + F(\overline{z}(x,t)) + \sum_{j=0}^{n_{\mathrm{d}}} B_{\mathrm{d}}^j(x)d_j(t). \tag{4.63}$$

Here, $\widetilde{z}$ is the response of the linear operator A to the sum of the nonlinear term $F(\overline{z}(x))$, the inertial term $-\overline{z}_t(x,t)$ and all the disturbances $B_{\mathrm{d}}^j d_j$, namely

$$\widetilde{z}(x,t) = -A^{-1}\left(F(\overline{z}(x,t)) - \overline{z}_t(x,t) + \sum_{j=0}^{n_{\mathrm{d}}} B_{\mathrm{d}}^j(x)d_j(t)\right).$$

Substituting Equations (4.33) and (4.63) in Equation (4.62) yields

$$\overline{z}(x,t) = \widetilde{z}(x,t) + \sum_{j=0}^{n_{\mathrm{in}}} X_j\gamma_j(t).$$

Applying the operator C_i to each side of the above equation, and substituting Equations (4.61) and (4.32) it follows that

$$y_{r_i}(t) = (C_i\overline{z})(t) = (C_i\widetilde{z})(t) + \sum_{j=0}^{n_{\mathrm{in}}} g_{ij}\gamma_j(t), \quad i = 1,\dots,n_{\mathrm{c}}.$$

The above equation in matrix form is equivalent to

$$G\Gamma(t) = Y_r(t) - \widetilde{Y}(t), \tag{4.64}$$

where

$$\Gamma(t) = [\gamma_1(t), \gamma_2(t), \cdots, \gamma_{n_{\mathrm{in}}}(t)]^\mathsf{T},$$
$$Y_r(t) = [y_{r_1}(t), y_{r_2}(t), \cdots, y_{r_{n_{\mathrm{c}}}}(t)]^\mathsf{T},$$

and

$$\widetilde{Y}(t) = [(C_1\widetilde{z})(t), (C_2\widetilde{z})(t), \cdots, (C_{n_{\mathrm{c}}}\widetilde{z})(t)]^\mathsf{T}.$$

The initial boundary value problem (4.59)–(4.61), Equation (4.63), and the coupling condition (4.64) should be solved together for all time. Once the time-dependent control gains $u_j(t) = \gamma_j(t)$ are known, they can be used in the closed loop system (4.53)–(4.55).

Remark 4.10. The system of equations (4.59)–(4.61), (4.63) represents a singular DAE and we have observed that introducing the inertial term in Equations (4.59) and (4.63) leads to numerical instabilities. Further, since there is no inertial term involving $\widetilde{z}_t(x,t)$ we see that although $\widetilde{z}$ is time-dependent the solution of (4.63) does not require any initial data for $\widetilde{z}$. In order to deal with the numerical instability we have found that slightly reducing the inertial effects of $\overline{z}_t(x,t)$ in (4.63) produces a stable numerical system. Although this leads to an approximate solution for $\overline{z}(x,t)$, and therefore to

$$y_{r_i}(t) \simeq (C_i \overline{z})(t), \quad i = 1, \ldots, n_c,$$

in most of the cases this error is very small, and is immediately reabsorbed. This idea is explained in much more detail in the next section devoted to tracking and rejecting more general time-dependent signals. We also explain an iterative scheme that allows us to systematically improve the approximation.

Thus we consider replacing the coefficient of $\overline{z}_t(x,t)$ in (4.63) by $(1 - \beta)$, for small $\beta > 0$, in the system (4.59)–(4.61), (4.63) to obtain

$$\overline{z}_t(x,t) = A\overline{z}(x,t) + F(\overline{z}(x,t)) + \sum_{j=0}^{n_d} B_d^j(x)\, d_j(t) + \sum_{j=0}^{n_{in}} B_{in}^j(x)\, \gamma_j(t),$$

$$(1 - \beta)\overline{z}_t(x,t) = A\widetilde{z}(x,t) + F(\overline{z}(x,t)) + \sum_{j=0}^{n_d} B_d^j(x) d_j(t),$$

$$\overline{z}(x,0) = \overline{Z}(x),$$

$$y_{r_i}(t) = (C_i \overline{z})(t), \quad i = 1, \ldots, n_c.$$

In our numerical simulation in the next section we have set $\beta = .05$ which seems to work quite well.

A detailed discussion and analysis of the errors for the so-called β-iteration procedure for time-dependent signals will discussed in a forthcoming paper [3].

Example 4.3 (Chafee-Infante). In this section, we present an example of tracking and disturbance rejection for piecewise constant reference signals for a one-dimensional Chafee-Infante equation. This example is similar to the Burgers' example presented in Example 4.2.

Consider the following control problem with state variable $z = z(x,t)$

$$z_t = z_{xx} - \lambda(z^3 - z) + \delta_{0.5} u_1(t), \quad 0 \leq x \leq 1, \tag{4.65}$$

with initial data

$$z(x,0) = \varphi(x),$$

boundary conditions

$$z(0,t) = d(t),$$
$$z_x(1,t) + \alpha z(1,t) = u_2(t),$$

and measured outputs given by

$$y_1(t) = C_1(z) = z(0.25, t), \quad y_2(t) = C_2(z) = z(0.75, t). \tag{4.66}$$

The disturbance $d(t)$ is a piecewise constant time-dependent function, and the flux boundary condition on the right side of the domain is a mixed Robin boundary condition. Also the reference signals to be tracked, $y_{r_1}(t)$ and $y_{r_2}(t)$, are now piecewise constant time-dependent functions.

Our objective is to find two control inputs $u_j(t)$ with $j = 1, 2$ so that the measured outputs $y_j(t)$ track the reference signals $y_{r_j}(t)$ while rejecting the disturbance $d(t)$.

The physical description of the problem is illustrated in Fig. 4.11.

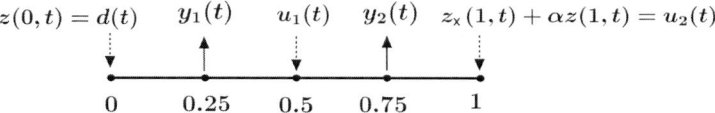

Fig. 4.11: Chaffee-Infante Example Domain.

The control problem is solved numerically by proceeding exactly as described in the first part of Section 4.4. First we must solve for X_j $(j = 1, 2)$, in steady-state problems

$$\frac{d^2 X_1}{dx^2} = \delta_{0.5}, \quad \text{for } 0 < x < 1$$

$$X_1(0) = 0, \quad \frac{dX_1}{dx}(1) + \alpha X_1(1) = 0,$$

for X_1, and

$$\frac{d^2 X_2}{dx^2} = 0, \quad \text{for } 0 < x < 1,$$

$$X_2(0) = 0, \quad \frac{dX_1}{dx}(1) + \alpha X_1(1) = 1,$$

for X_2. These produce the entries of the 2×2 matrix

$$G = \begin{pmatrix} C_1 X_1 & C_1 X_2 \\ C_2 X_1 & C_2 X_2 \end{pmatrix}.$$

Then, we solve the steady state coupled system

$$0 = \frac{d^2 \overline{Z}}{dx^2} - \lambda(\overline{Z}^3 - \overline{Z}) - \delta_{0.5}\gamma_{1,0},$$

$$\overline{Z}(0) = d_0, \quad \frac{d\overline{Z}}{dx}(1) + \alpha \overline{Z}(1) = \gamma_{2,0},$$

$$0 = \frac{d^2\widetilde{Z}}{dx^2} - \lambda(\overline{Z}^3 - \overline{Z}),$$

$$\widetilde{Z}(0) = d_0, \quad \frac{d\widetilde{Z}}{dx}(1) + \alpha\widetilde{Z}(1) = 0,$$

$$\Gamma = \begin{bmatrix} \gamma_{1,0} \\ \gamma_{2,0} \end{bmatrix} = G^{-1} \begin{bmatrix} -C_1\widetilde{Z} + y_{r_{1,0}} \\ -C_2\widetilde{Z} + y_{r_{2,0}} \end{bmatrix},$$

where $d_0 = d(0)$, $y_{r_{1,0}} = y_{r_1}(0)$ and $y_{r_{2,0}} = y_{r_2}(0)$. This system provides the initial condition $\overline{Z}$ for solving the time-dependent problem. With this, we now solve the initial boundary value problem

$$\overline{z}_t = \frac{d^2\overline{z}}{dx^2} - \lambda(\overline{z}^3 - \overline{z}) - \delta_{0.5}\gamma_1(t),$$

$$\overline{z}(x,0) = \overline{Z}(x),$$

$$\overline{z}(0,t) = d(t), \quad \frac{d\overline{z}}{dx}(1,t) + \alpha\overline{z}(1,t) = \gamma_2(t),$$

$$0.9\,\widetilde{z}_t = \frac{d^2\widetilde{z}}{dx^2} - \lambda(\overline{z}^3 - \overline{z}),$$

$$\widetilde{z}(0,t) = d(t), \quad \frac{d\widetilde{z}}{dx}(1,t) + \alpha\widetilde{z}(1,t) = 0,$$

$$\Gamma = \begin{bmatrix} \gamma_1(t) \\ \gamma_2(t) \end{bmatrix} = G^{-1} \begin{bmatrix} -(C_1\widetilde{z})(t) + y_{r_1}(t) \\ -(C_2\widetilde{Z})(t) + y_{r_2}(t) \end{bmatrix}.$$

Finally, we set $u_1(t) = \gamma_1(t)$ and $u_2(t) = \gamma_2(t)$ and solve the closed loop system (4.65)–(4.66).

For this example we choose the parameters $\lambda = 10$ and $\alpha = 10$. Moreover, we set

$$d(t) = 0.75 + 0.25\,U(t-50) - 0.25\,U(t-100), \tag{4.67}$$

$$y_{r_1}(t) = 1 - 0.25\,U(t-25) + 0.5\,U(t-75) - 0.25\,U(t-125), \tag{4.68}$$

$$y_{r_2}(t) = 0.5 - 0.25\,U(t-40) + 0.5\,U(t-90) - 0.25\,U(t-140), \tag{4.69}$$

with U the Heaviside function defined by

$$U(x - x_0) = \begin{cases} 0 & \text{for } x < x_0, \\ 1 & \text{for } x \geq x_0. \end{cases}$$

For our numerical simulation we solve on the time interval $0 < t \leq T$ with $T = 150$ and initial data $\varphi(x) = \cos(x)$. The numerical solution of the weak formulation of the above systems is obtained using the predefined PDE Coefficient tool of the COMSOL Multiphysics package. The geometry is discretized with 128 quadratic Lagrange elements. No stabilization method is used for the advection term.

In Figs. 4.12 and 4.13, we display the reference signals $y_{r_1}(t)$ and $y_{r_2}(t)$ and the quantities $(C_1 z)(t)$ and $(C_2 z)(t)$ for all the time T. After a short transient $C_1 z$ converges to y_{r_1} and $C_2 z$ converges to y_{r_2}. Any time y_{r_1}, y_{r_2} or d change, there are small spikes in the graphs of $C_1 z$ and $C_2 z$ that are immediately reabsorbed. These spikes are due to the inertial approximation and are more evident in the plot of the errors. In Fig. 4.14 we display the time-dependent disturbance $d(t)$ for all time.

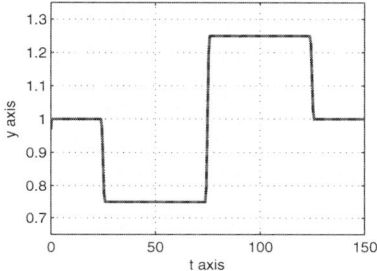
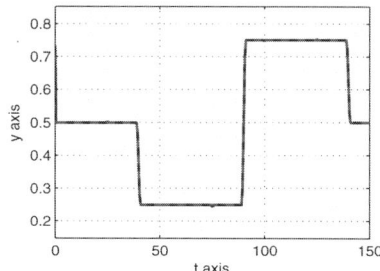

Fig. 4.12: $y_{r_1}(t)$, $C_1(z(t))$.　　　　**Fig. 4.13**: $y_{r_2}(t)$, $C_2(z(t))$.

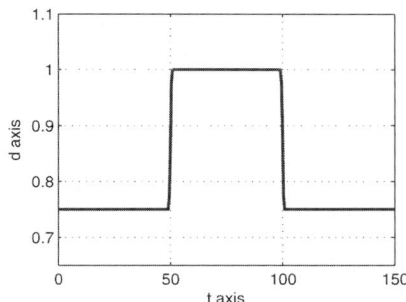

Fig. 4.14: $d(t)$.

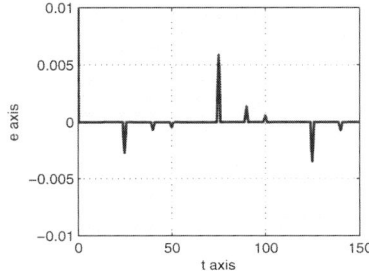
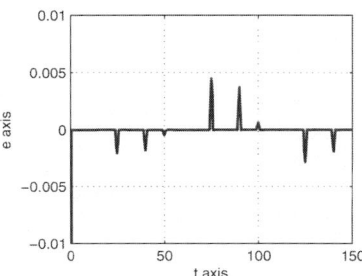

Fig. 4.15: $e_1(t)$.　　　　**Fig. 4.16**: $e_2(t)$.

In Fig. 4.15, we display the error $e_1(t)$ and in Fig. 4.16, we display the error e_2 for all time t. Near a time in which y_{r_1}, y_{r_2} or d change, there is a short transient. After that, both graphs return to essentially zero until new changes occur. In Fig. 4.17 we have plotted the solution surface $z(x, t)$ for $0 \le x \le 1$ and $0 \le t \le 150$.

1. At $x = 0$ (on the left side of the figure) we obtain the curve $z(0, t)$ which approximates $d(t)$ in (4.67).

2. At $x = 0.25$ the solution $z(0.25, t)$ quickly approximates $y_{r_1}(t)$ in (4.68).

3. At $x = 0.75$ the solution $z(0.75, t)$ quickly approximates $y_{r_2}(t)$ in (4.69).

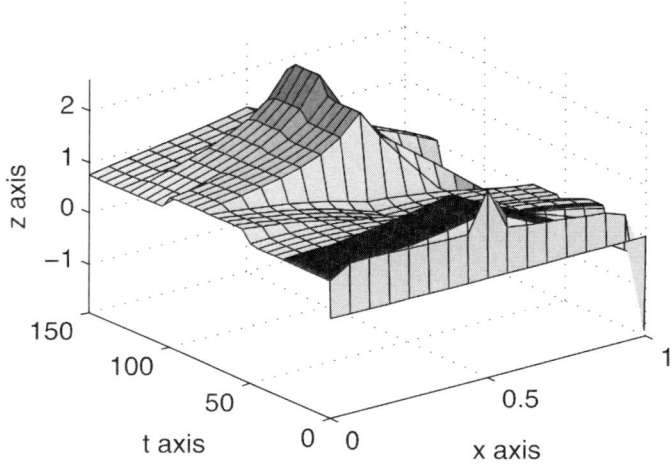

Fig. 4.17: Solution Surface $z(x, t)$ for $0 \le x \le 1$ and $0 \le t \le 150$.

4.5   Nonlinear Regulation for Time-Dependent Signals

In the previous section we presented a technique that was primarily applicable for set-point control and the methodology was based on the geometric theory of output regulation. In this section we show that a generalization, motivated by the same theory, can be derived to design feedback laws for solving regulation problems involving very general time-dependent reference and disturbance signals. In this work the usual assumptions used in the geometric theory do not strictly apply, but nevertheless, we show that some analogs of the regulator equations from the geometric theory can be derived and used to obtain accurate approximations of the control inputs. We also note that this methodology is not a simple extension of the set-point tracking case. In particular the time-dependent infinite-dimensional controller involves the solution of a singular DAE. An important part of this work is the development of an iterative regularization scheme that provides a sequence of time-dependent control laws providing successively more accurate approximate solutions of the regulation problem.

In this section we demonstrate the method for a multi-input multi-output regulation problem involving a tracking/disturbance rejection problem for a nonlinear distributed parameter system governed by a one-dimensional viscous Burgers' equation. More complex examples are given in Chapter 5.

The problem of designing control laws to achieve regulation of complex nonlinear systems is a difficult task and has received considerable attention in the engineering literature. This is particularly true in the case of nonlinear systems governed by partial differential equations and especially for the so-called boundary control case where the sensors and actuators produce unbounded operators in the Hilbert state space. As discussed in the previous section, in the recent papers [1, 2] the authors have developed an approach for obtaining *time-independent* control laws in the case of set-point control, where the signals to be tracked and disturbances to be rejected are *time-independent*, i.e., the case of set-point control.

Also in the previous section it was shown that a slight generalization of the method could produce time-dependent feedback laws capable of tracking and rejecting *piecewise constant* reference signals and disturbances. In this case the resulting DAE is time-dependent and the geometric regulator theory, based on center manifold analysis, is not directly applicable since the reference signals and disturbances cannot be generated by a neutrally stable finite-dimensional exosystem. Indeed to generate such time periodic signals would require an infinite-dimensional signal generator. Furthermore a piecewise constant signal is not smooth and therefore also violates the conditions required by the center manifold theorem. Nevertheless, as we have demonstrated in Example 4.3, in practical numerical simulations we are able to handle these cases.

For more general time-dependent reference signals and disturbances the methodology adopted in [2] doesn't work. Indeed, it only works in the case of piecewise constant signals, since these are a sequence of set-point problems for which the method developed in [2] was designed. For general time-dependent reference signals and disturbances a more complicated design strategy is required. The theoretical aspects of this topic are the main focus of a more lengthy discussion, which is the subject of our current research efforts (TTU preprint [3]). In this section we describe the general methodology and apply it to two regulation examples for the one-dimensional viscous Burgers' equation.

4.5.1   The Regulation Problem

Consider a nonlinear parabolic control system acting in a bounded domain $\Omega \subset \mathbb{R}^n$ and $t \geq 0$.

$$z_t(x,t) = Az(x,t) + F(z(x,t)) + \sum_{j=1}^{n_d} B_d^j(x)d_j(t)$$

$$+ \sum_{j=1}^{n_{in}} B_{in}^j(x)u_j(t), \tag{4.70}$$

$$z(x,0) = z_0(x), \quad z_0 \in \mathcal{Z} = L^2(\Omega), \tag{4.71}$$

$$y_i(t) = (C_i z)(x,t), \quad i = 1, \ldots, n_c. \tag{4.72}$$

Here $z(x,t)$ is the state variable at position $x \in \Omega$ at time t. The terms

$$B_d^j(x)d_j(t), \quad j = 1, \ldots, n_d, \tag{4.73}$$

$$B_{in}^j(x)u_j(t), \quad j = 1, \ldots, n_{in}, \tag{4.74}$$

represent disturbances and control inputs, respectively. Note that in Equations (4.73)–(4.74) each term is a multiplicative operator between a spatial dependent function and the corresponding input function which is considered to be time-dependent only. The expressions $B_d^j(x)$ and $B_{in}^j(x)$ are assumed to be known functions, and may also be unbounded (e.g., for example, they could be given by delta functions supported on the boundary) or at points on the interior of the spatial interval. In general B_d^j refers to a *disturbance input* operator and B_{in}^j refers to a *control input* operator.

The state operator A is assumed to be a linear differential operator (plus appropriate homogeneous boundary conditions) in the infinite-dimensional Hilbert state space $\mathcal{Z} = L^2(\Omega)$. It is assumed that the operator A defined on a dense domain $\mathcal{D}(A)$ generates an exponentially stable C_0 semigroup in $\mathcal{Z}$. In our intended applications the operators C_i in Equation (4.72) are typically point evaluation or a weighted integral of the solution $z(x,t)$ over some subregion of Ω. Therefore these operators can be bounded or unbounded in the state space $\mathcal{Z}$. The most common situation is that A is a sectorial operator [47] that generates an analytic semigroup and $(-A)$ generates a Hilbert scale of spaces $\mathcal{Z}_\alpha$ for $\alpha \in \mathbb{R}$. In the case that an operator C_i is unbounded we will assume that the domain of C_i, denoted by $\mathcal{D}(C_i)$, is contained in some $\mathcal{Z}_{\alpha_0}$ for some $\alpha_0 > 0$, and $C_i : \mathcal{D}(C_i) \to \mathcal{Y}$ (the output space) for each i (see, e.g., [47, 57]). We also assume that all the operators B_d^j and B_{in}^j are A-bounded with arbitrarily small bound (see [57, 38]).

Here the nonlinear function F is assumed to be a smooth function with $F(0) = 0$, so that the uncontrolled control system has the origin in $\mathcal{Z}$ as an equilibrium. We also assume that this equilibrium for the uncontrolled system is exponentially stable.

For our tracking/disturbance rejection control problems we are given *time-dependent* reference signals $y_{r_i}(t) \in \mathbb{R}$, $i = 1, \ldots, n_c$, and disturbances, $d_j(t) \in \mathbb{R}$, $i = j, \ldots, n_d$, defined for all $t \geq 0$. More specific conditions on smoothness assumptions and the long-time behavior of these functions will be given below.

Let us define

$$Y_r(t) = [y_{r_1}(t), y_{r_2}(t), \cdots, y_{r_{n_c}}(t)]^\mathsf{T}, \tag{4.75}$$

and

$$Y(t) = [(C_1 z)(t), (C_2 z)(t), \cdots, (C_{n_c} z)(t)]^\mathsf{T}.$$

The regulation problem can be stated as follows.

Problem 4.3. *Our design objective is to find a set of time-dependent controls $u_j(t) = \gamma_j(t)$, $j = 1, \ldots, n_{in}$, for the system (4.70)–(4.72) such that the error defined by*

$$e(t) = \|Y(t) - Y_r\|_\infty = \sup_{1 \leq i \leq n_c} |y_i(t) - y_{r_i}|,$$

satisfies

$$e(t) \xrightarrow{t \to \infty} 0,$$

while the state of the closed loop control system remains bounded for all time.

The methodology for solving Problem 4.3 is based on two main assumptions:

Assumption 4.5. *There exists (a) an initial condition $\varphi(x)$, (b) time-dependent functions $\gamma_j(t)$, $j = 1, \ldots, n_{in}$, and (c) a classical solution $\bar{z}(x, t)$ of the nonlinear initial boundary value problem (4.76)–(4.77) satisfying the constraints given in (4.78) :*

$$\bar{z}_t = A\bar{z} + F(\bar{z}) + \sum_{j=0}^{n_d} B_d^j(x)d_j(t) + \sum_{j=0}^{n_{in}} B_{in}^j(x)\gamma_j(t), \tag{4.76}$$

$$\bar{z}(x, 0) = \varphi(x), \tag{4.77}$$

$$C_i \bar{z}(x, t) = y_{r_i}(t), \quad i = 1, \ldots, n_c. \tag{4.78}$$

Assumption 4.6. *For sufficiently close initial data*

$$\|z_0(x) - \varphi(x)\| < \delta,$$

the solution $z(x, t)$ of the system (4.70)–(4.72) with controls $u_j = \gamma_j$, i.e.,

$$z_t(x, t) = Az(x, t) + F(z(x, t)) + \sum_{j=0}^{n_d} B_d^j(x)d_j(t) + \sum_{j=0}^{n_{in}} B_{in}^j(x)\gamma_j(t),$$

$$z(x, 0) = z_0(x),$$

satisfies

$$\lim_{t \to \infty} \left| C_i z(x, t) - C_i \bar{z}(x, t) \right| = 0, \quad i = 1, \ldots, n_{in}.$$

Remark 4.11. In the traditional geometric theory of regulation the disturbances and reference signals are generated by a neutrally stable exosystem

$$w_t = Sw, \quad w(0) = w_0, \tag{4.79}$$

with

$$y_{r_j} = Q_j w, \ j = 1, \cdots, n_c, \text{ and } d_j = P_j w, \ j = 1, \cdots, n_d.$$

In (4.79), assuming a fixed basis for the ($K < \infty$)-dimensional state space $\mathcal{W}$, S is represented by a $K \times K$ matrix with all eigenvalues on the imaginary axis. In this case, for the composite system consisting of (4.70)–(4.72) coupled with (4.79) and controls $u_j = \gamma_j(w(t)) \equiv \gamma_j(t)$, in the state space $\mathcal{X} = \mathcal{Z} \times \mathcal{W}$,

$$z_t = Az + F(z) + \sum_{j=0}^{n_d} B_d^j d_j(t) + \sum_{j=0}^{n_{in}} B_{in}^j \gamma_j(t), \tag{4.80}$$

$$w_t = Sw, \tag{4.81}$$

$$z(x, 0) = z_0(x), \quad w(0) = w_0, \tag{4.82}$$

the center manifold theorem [31, 47], provides the existence of an attractive K-dimensional center manifold Σ which is given, at least locally, by the graph of a function. Indeed, Σ can be written as the graph of the map $w \mapsto \bar{z}(w)$ in the composite $\mathcal{X}$ space as

$$\Sigma = \left\{ \begin{pmatrix} \bar{z}(w) \\ w \end{pmatrix} \in \mathcal{X} : w \in W_0 \right\},$$

for some neighborhood W_0 of the origin of $\mathbb{R}^K$.

With the center manifold map defined by $\bar{z}(x, w) = \bar{z}$, we observe that the chain rule gives (4.81)

$$\bar{z}_t = \frac{\partial \bar{z}}{\partial w} w_t = \frac{\partial \bar{z}}{\partial w} Sw.$$

So from (4.80) under the assumption of invariance (suppressing dependence on w) we obtain

$$\frac{\partial \bar{z}}{\partial w} S = A\bar{z} + F(\bar{z}) + \sum_{j=0}^{n_d} B_d^j d_j + \sum_{j=0}^{n_{in}} B_{in}^j \gamma_j. \tag{4.83}$$

In addition, the requirement that the invariant manifold be error zeroing implies that $\bar{z}$ must satisfy the additional constraints

$$C_i \bar{z}(w) - Q_i w = 0, \quad i = 1, .., n_c, \tag{4.84}$$

for all w in a neighborhood of the origin. The equations (4.83)–(4.84) when taken together are the so-called Regulator or FBI Equations [56, 55, 11, 10]. In general it is extremely difficult to solve these equations explicitly, and even to approximate solutions numerically.

Returning to our Assumptions 4.5 and 4.6, the equation (4.76) expresses the invariance of Σ assuming that we have an initial condition on Σ. In other words, we would need to have

$$w(0) = w_0 \quad \text{and} \quad \varphi(x) = \overline{z}(x, w_0).$$

Unfortunately, finding such initial data is about as difficult as solving the regulator equations. But we have found that for special initial conditions in combination with a certain regularization scheme, we are able to obtain accurate approximations of the desired dynamic controls $\gamma_j(t)$.

4.5.2 The Controller Design

To determine the desired controls $\gamma_j(t)$ we proceed in the following way. First we solve the n_{in} linear boundary value problems given by

$$0 = AX_j(x) + B_{\text{in}}^j(x), \quad j = 1, \ldots, n_{\text{in}} \text{ and } x \in \Omega.$$

Notice that as long as the coefficients in these elliptic boundary value problems are sufficiently smooth, and the boundary of the domain Ω is sufficiently regular, the solution X_j will also be smooth, by elliptic regularity, so that $X_i \in \mathcal{D}(C_j)$.

With this we can assemble the matrix $G_{n_c \times n_{\text{in}}}$, whose entries are

$$g_{ij} = C_i X_j, \quad i = 1, \ldots, n_c, \ j = 1, \ldots, n_c. \tag{4.85}$$

Each component X_j belongs to a Hilbert scale space $\mathcal{Z}_\alpha \subset \mathcal{D}(C_i)$ for some $\alpha > 0$, and it is the response of the linear operator A to the input B_{in}^j, namely

$$X_j = -A^{-1} B_{\text{in}}^j. \tag{4.86}$$

Next, from Assumption 4.5 we know that there exist control gains $u_j(t) = \gamma_j(t)$ and initial data $\overline{z}(x, t) = \varphi(x)$ such that $\overline{z}(x, t)$ satisfies (4.76) and (4.78) for all time.

We rewrite Equation (4.76) as

$$\overline{z}(x, t) = -A^{-1}\left(-\overline{z}_t(x, t) + F(\overline{z}(x, t)) \right) \tag{4.87}$$

$$+ \sum_{j=0}^{n_d} B_d^j(x) \, d_j(t) \right) + \sum_{j=0}^{n_{\text{in}}} \left(-A^{-1} B_{\text{in}}^j(x) \, \gamma_j(t) \right).$$

Let $\widetilde{z}(x,t)$ be the solution of

$$\overline{z}_t(x,t) = A\widetilde{z}(x,t) + F(\overline{z}(x,t)) + \sum_{j=0}^{n_d} B_d^j(x)d_j(t). \tag{4.88}$$

Here, $\widetilde{z}$ is the response of the linear operator A to the sum of the nonlinear term $F(\overline{z}(x))$, the inertial term $-\overline{z}_t(x,t)$ and all the disturbances $B_d^j d_j(t)$, namely

$$\widetilde{z}(x,t) = -A^{-1}\left(F(\overline{z}(x,t)) - \overline{z}_t(x,t) + \sum_{j=0}^{n_d} B_d^j(x)d_j(t) \right).$$

Substituting Equations (4.86) and (4.88) in Equation (4.87) yields

$$\overline{z}(x,t) = \widetilde{z}(x,t) + \sum_{j=0}^{n_{in}} X_j\gamma_j(t).$$

Applying the operators C_i to each side of the above equation, and substituting Equations (4.78) and (4.85) it follows that for each $i = 1,\ldots,n_c$,

$$y_{r_i}(t) = (C_i\overline{z})(t) = (C_i\widetilde{z})(t) + \sum_{j=0}^{n_{in}} g_{ij}\gamma_j(t).$$

The above equation in matrix form is equivalent to

$$G\Gamma(t) = Y_r(t) - \widetilde{Y}(t), \tag{4.89}$$

where

$$\Gamma(t) = [\gamma_1(t), \gamma_2(t), \cdots, \gamma_{n_{in}}(t)]^\mathsf{T},$$

$$\widetilde{Y}(t) = [(C_1\widetilde{z})(t), (C_2\widetilde{z})(t), \cdots, (C_{n_c}\widetilde{z})(t)]^\mathsf{T},$$

and $Y_r(t)$ was defined in (4.75). The initial boundary value problem (4.76)–(4.77), Equation (4.88), and the coupling condition (4.89) should be solved together for all time. Once the controllers $u_j(t) = \gamma_j(t)$ are known, they can be used in the closed loop system (4.80)–(4.82). Note that since there is no inertial term involving $\widetilde{z}_t(x,t)$ we see that although $\widetilde{z}$ is time-dependent the solution of (4.88) does not require any initial data for $\widetilde{z}$.

1. The system of equations (4.76)–(4.88) and (4.89) represents a singular DAE, due to the inertial term in Equation (4.88), which leads to numerical instabilities. In order to deal with the numerical instability we have found that slightly reducing the inertial effects of $\overline{z}_t(x,t)$ in (4.88) produces a stable numerical system.

2. Further, the initial data $\bar{z}(x, 0) = \varphi(x)$ in Assumption 4.5 is not known. We choose as an approximate initial data for this unknown function the solution $\bar{\varphi}(x)$ of the following steady-state DAE:

$$0 = A\bar{\varphi}(x) + F(\bar{\varphi}) + \sum_{j=0}^{n_{\mathrm{d}}} B_{\mathrm{d}}^j(x) \, d_j(0) + \sum_{j=0}^{n_{\mathrm{in}}} B_{\mathrm{in}}^j(x) \, v_j,$$

$$y_{r_i}(0) = C_i\bar{\varphi}, \quad i = 1, \ldots, n_{\mathrm{c}}. \tag{4.90}$$

Note that, because of the constraint (4.90), Equation (4.78) is automatically satisfied for $t = 0$.

3. Replacing the coefficient of $\bar{z}_t(x, t)$ by $(1 - \beta)$, for small $\beta > 0$, and the initial data by its approximation $\bar{\varphi}$, the system (4.76), (4.77), (4.88) and (4.89) becomes

$$\bar{z}_t(x, t) = A\bar{z}(x, t) + F(\bar{z}(x, t)) + \sum_{j=0}^{n_{\mathrm{d}}} B_{\mathrm{d}}^j(x) \, d_j(t) \tag{4.91}$$

$$+ \sum_{j=0}^{n_{\mathrm{in}}} B_{\mathrm{in}}^j(x) \, \gamma_j(t),$$

$$(1 - \beta)\bar{z}_t(x, t) = A\bar{z}(x, t) + F(\bar{z}(x, t)) + \sum_{j=0}^{n_{\mathrm{d}}} B_{\mathrm{d}}^j(x) d_j(t), \tag{4.92}$$

$$\bar{z}(x, 0) = \bar{\varphi}(x), \tag{4.93}$$

$$G\Gamma(t) = Y_r(t) - \tilde{Y}(t), \tag{4.94}$$

although these lead to an approximate solution for $\bar{z}(x, t)$, and therefore to

$$y_{r_i}(t) \simeq (C_i\bar{z})(t), \quad i = 1, \ldots, n_{\mathrm{c}},$$

in most cases the error is small, i.e., we achieve approximate rather than exact regulation. A detailed derivation of the errors for time-dependent signals is the subject of a lengthy forthcoming paper [3].

4.5.3   An Iterative Scheme

In order to improve the accuracy of the solution obtained in the previous section we introduce an iterative scheme to obtain updated values for the state variable $\bar{z}$ and the control gains γ. In particular we seek the iterative values in the form

$$\bar{z}_n = \sum_{i=1}^{n} \bar{z}^i, \qquad \gamma_n = \sum_{i=1}^{n} \gamma^i,$$

where we start at $n = 1$ and continue to add new terms aiming to reduce the error obtained at the previous step. Let the error at the ith iteration be

$$E_i(t) = Y_r(t) - Y_i(t), \quad \text{with}$$
$$Y_i(t) = [(C_1\overline{z}_i)(t), (C_2\overline{z}_i)(t), \cdots, (C_{n_c}\overline{z}_i)(t)].$$

For $i = 1$ we calculate $\overline{z}^1$ and γ^1 that satisfy the DAE (4.91)–(4.94). Then for $1 < i \leq n$, we solve the DAE

$$\overline{z}_t^i(x,t) = A\overline{z}^i(x,t) + F(\overline{z}_i(x,t)) - F(\overline{z}_{i-1}(x,t)) + \sum_{j=0}^{n_{\text{in}}} B_{\text{in}}^j(x)\,\gamma_j^i(t),$$

$$(1 - \beta)\overline{z}_t^i(x,t) = A\widetilde{z}^i(x,t) + F(\overline{z}_i(x,t)) - F(\overline{z}_{i-1}(x,t)), \quad (4\,95)$$

$$\overline{z}^i(x,0) = 0, \qquad G\gamma^i(t) = E_{i-1} - \widetilde{Y}_i,$$

for $\overline{z}^i$ and γ^i. Note that every time a new i^{th} step is added, the new control input γ^i is built to reject the error obtained at the previous step. Again the constant $(1-\beta)$ in front of the inertial term in Equation (4.95) has been added to handle the numerical instability of the DAE.

We have observed that using the control Γ_n obtained on the nth iteration produces an error E_n satisfying

$$\|E_n(t)\|_\infty \leq H_n(t) + \left(\frac{\beta}{|\lambda_{\max A}|}\right)^n M_n, \quad (4.96)$$

where $H_n(t)$ satisfies

$$H_n(t) \leq M_0 \left(\frac{t}{\beta}\right)^n e^{-\alpha_0 t}, \text{ for } \alpha_0, M_0 > 0. \quad (4.97)$$

$\lambda_{\max A}$ is the largest eigenvalue of the operator A (which is the smallest eigenvalue in absolute value) and M_n is a constant that depends on

$$\|Y_r^{(i)}\|_\infty = \max_{1 \leq i \leq (n+1)} \left(\sup_{0 \leq t < \infty} \|Y_r^{(i)}(t)\|_\infty\right).$$

Roughly speaking we can say that the approximate initial data $\overline{\varphi}$ is responsible for the error $H_n(t)$, which is reabsorbed exponentially in time, while the inertial approximation $(1 - \beta)$ generates the error $(\beta/|\lambda_{\max A}|)^n$. Note that this last error converges geometrically to zero with n, only if M_n, which depends on all the first $(n + 1)$ derivatives of $Y_r(t)$, doesn't grow faster than $(|\lambda_{\max A}|/\beta)^n$.

Remark 4.12. For a linear system, i.e., $F(z) = 0$, with bounded input and output operators, i.e., the functions $B_d^j(x)$ and $B_{\text{in}}^j(x)$ are in $\mathcal{Z}$ and the output operators C_i are bounded in $\mathcal{Z}$, we can establish the estimate (4.96). We conjecture that the same result holds in the nonlinear case, and even in case of unbounded input and output operators. Our optimism is based on consistent observations for a significant number of numerical examples. The proof of (4.96), for linear SISO harmonic reference signals is lengthy and somewhat involved [67]. A proof of (4.96) for very general bounded smooth reference and disturbance signals and with A-bounded output operator C can be found in the PhD thesis [68] and in the paper [3].

Example 4.4 (Burgers' Equation). In boundary control form we consider the special case of a viscous Burgers' equation

$$z_t(x,t) = c\, z_{xx}(x,t) - z(x,t)z_x(x,t),\ 0 \le x \le 1, \tag{4.98}$$
$$z(0,t) = d(t) = M_2 + A_2 \sin(\alpha_2 t), \tag{4.99}$$
$$c\, z_x(1,t) = u(t), \tag{4.100}$$

with initial condition

$$z(x,0) = \varphi(x). \tag{4.101}$$

Here c is a kinematic viscosity and is considered constant on the interval.

In addition we have a measured output given by point evaluation at the point $x = x_0 \in (0,1)$

$$y(t) = C(z) = z(x_0,t),$$

and we are given a reference signal $y_r(t) = M_1 + A_1 \sin(\alpha_1 t)$ to be tracked.

We rewrite equation system (4.98)–(4.101) in the standard systems form (4.70)–(4.72) with appropriate distributional operators B_d and B_in (see Example 2.1)

$$z_t = Az + F(z) + B_\mathrm{d}d(t) + B_\mathrm{in}u(t), \tag{4.102}$$
$$z(x,0) = \varphi(x). \tag{4.103}$$

First we seek an initial condition $\overline{\varphi}(x)$, that satisfies the set-point regulation problem

$$0 = A\overline{\varphi} + F(\overline{\varphi}) + B_\mathrm{d}M_2 + B_\mathrm{in}\gamma_0,$$
$$C(\overline{\varphi}) = y_r(0) = M_1.$$

The solution $(\overline{\varphi}(x),\ \gamma_0)$ of the above problem is found using the strategy developed in Section 4.3.2 for nonlinear set-point control.

Then we seek the solution of the system

$$\overline{z}_t = A\overline{z} + F(\overline{z}) + B_\mathrm{d}d(t) + B_\mathrm{in}\gamma(t), \tag{4.104}$$
$$\overline{z}(x,0) = \overline{\varphi}(x),$$
$$C\overline{z} = y_r(t). \tag{4.105}$$

As already pointed out, the system (4.104)–(4.105) corresponds to a dynamic version of the usual regulator equations. We seek a sequence of approximate solutions obtained by using the iterative scheme previously described.

The analysis proceeds by setting

$$X = -A^{-1}B_{\text{in}},$$
$$G = CX,$$
$$\overline{z}_0 = \overline{\varphi},$$

and solving the steady state DAE

$$0 = A\overline{z}_0 + F(\overline{z}_0) + B_{\text{d}}M_2 + B_{\text{in}}\Gamma_0,$$
$$0 = A\tilde{z}_0 + F(\overline{z}_0) + B_{\text{d}}M_2,$$
$$\gamma_0 = G^{-1}(M_1 - C(\tilde{z}_0)).$$

Then set

$$\overline{Z}_i = \begin{cases} 0, & i = 0, \\ z_i, & i = 1, \cdots n, \end{cases}$$

where $\overline{z}_i = \sum_{j=1}^{i} \overline{z}^j,$

$$e_i(t) = \begin{cases} y_r, & i = 0, \\ y_r - C(\overline{Z}_i) & i = 1, \cdots, n, \end{cases}$$

$$\overline{z}^{i,0} = \begin{cases} \overline{z}_0, & i = 1, \\ 0, & i = 2, \cdots n, \end{cases}$$

$$d^i(t) = \begin{cases} d(t) = M_2 + A_2 \sin(\alpha_2 t), & i = 1, \\ 0, & i = 2, \cdots n, \end{cases}$$

and solve the iterative, coupled, time-dependent DAE

$$\overline{z}_t^i = A\overline{z}^i + F(\overline{Z}_i) - F(\overline{Z}_{i-1}) + B_{\text{d}}d^i(t) + B_{\text{in}}\gamma^i(t),$$

$$(1 - \beta)\overline{z}_t^i = A\tilde{z}^i + F(\overline{Z}_i) - F(\overline{Z}_{i-1}) + B_{\text{d}}d^i(t),$$

$$\overline{z}^i(x, 0) = \overline{z}^{i,0},$$

$$\gamma^i(t) = G^{-1}(e_{i-1} - C(\tilde{z}^i)),$$

for $i = 1, \cdots, n$. Finally, set

$$u(t) = \sum_{i=1}^{n} \gamma^n, \qquad (4.106)$$

and solve the closed loop system (4.102)–(4.103).

In our specific numerical simulation we have chosen

$$c = 1, \qquad x_0 = 0.5, \qquad \varphi(x) = M_2,$$
$$M_1 = 0.5, \qquad A_1 = 0.25, \qquad \alpha_1 = 2\pi,$$
$$M_2 = 0.75, \qquad A_2 = 1, \qquad \alpha_2 = \pi,$$
$$\beta = 0.05, \qquad n = 4.$$

Remark 4.13. The signal to be tracked $y_r(t)$ and the disturbance to be rejected $d(t)$ are periodic functions. On the invariant manifold we expect the control $u(t)$ also to be a periodic function of time. However, if in the closed loop system we use the formula (4.106) for all time we are going to include in $u(t)$ also the contributions in the gains $\gamma^i(t)$ needed to suppress the error $H_n(t)$ (see formula (4.96)). We already pointed out that the approximate initial data $\overline{\varphi}$ is responsible for this error and that the same is reabsorbed exponentially in time. However, initially the error has an oscillatory behavior that can lead to numerical instabilities. Also, in Equation (4.97), the norm bound of $H_n(t)$ is proportional to the term $\left(\frac{t}{\beta}\right)^n$. Thus it increases both with n and with $\frac{1}{\beta}$. This implies that every time we introduce a new iteration we increase the error bound of H_n, and that a large β helps in keeping $H_n(t)$ small. On the other hand a large β makes the second (and most important) part of the error in (4.96) large, so it is not recommended. We overcome these difficulties by adopting the following two workaround strategies.

1. We only solve the closed loop system after the error $H_n(t)$ has been reabsorbed and the gains $\gamma^i(t)$ have become nearly periodic. In the specific simulation we have solved the iterative DAE for $t \in (-10, 6)$, and the closed loop system only for $t \in (0, 6)$.

2. When solving the iterative DAE in the first moments we use a large β and then smoothly allow it to change to the desired small value. In the specific simulation we have set

$$\beta = 0.5 - 0.045 \times \text{flc2hs}(t + 6, 3),$$

where flc2hs$(t - t_0, \Delta T)$ is a smooth approximation of the Heaviside step function, with continuous second derivative, centered at $t = t_0$, and with a smooth transient in the interval $(t_0 - \Delta T, t_0 + \Delta T)$. This is an already built-in function in the Comsol Multiphysics package.

In Fig. 4.18, we display the reference signal $y_r(t)$ and the output $C(z)(t)$ for $t \in (0, 6)$. Notice that after a short transient $C(z)$ converges to y_r. In Fig. 4.19, we display the disturbance signal $d(t)$ for $t \in (0, 6)$.

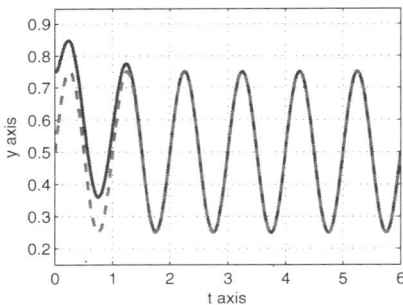

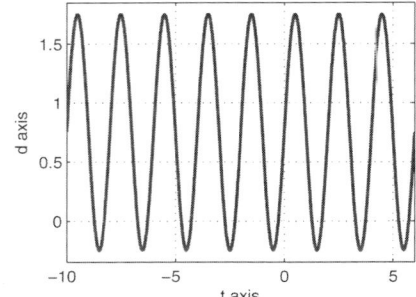

Fig. 4.18: $y_r(t)$, $(Cz)(t)$, $0 \le t \le 6$.   **Fig. 4.19**: $d(t)$ for $0 \le t \le 6$.

In Figs. 4.20, 4.21, and 4.22 we display the errors $e_1(t)$, $e_2(t)$, and $e_4(t)$ for $t \in (-10, 6)$ obtained after the first, the second and the fourth iteration, respectively. As already pointed out initially some oscillations in the errors arise, however they are reabsorbed in a short time. Also notice how the asymptotic l_∞ norm of the error decreases as the number of iterations increases.

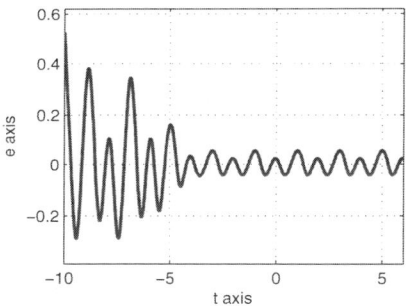

Fig. 4.20: $e_1(t)$ for $-10 \le t \le 6$.   **Fig. 4.21**: $e_2(t)$ for $-10 \le t \le 6$.

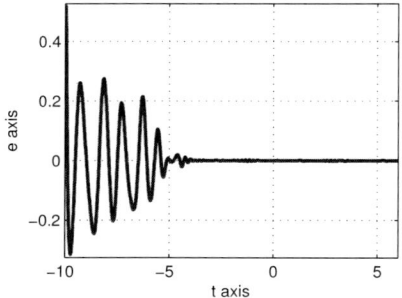

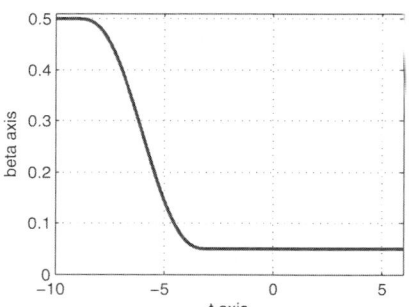

Fig. 4.22: $e_4(t)$ for $-10 \le t \le 6$.   **Fig. 4.23**: $\beta(t)$ for $-10 \le t \le 6$.

In Fig. 4.23 we have displayed the time-dependent $\beta(t)$ needed to reduce the amplitude of the error initial oscillations in the iterative procedure.

In Fig. 4.24 we have plotted the solution surface $z(x, t)$ for $0 \leq x \leq 1$ and $0 \leq t \leq 6$.

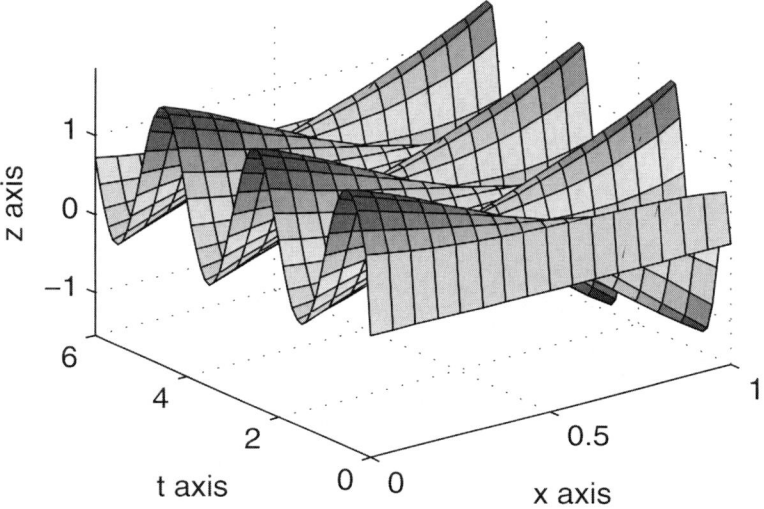

Fig. 4.24: Solution Surface $z(x, t)$ for $0 \leq x \leq 1$ and $0 \leq t \leq 6$.

4.6 Fourier Series Methods for Nonlinear Regulation

In this section we consider an extension of the results of Section 3.8 to nonlinear systems. In particular we will consider a regulator problem for a general abstract nonlinear system in the form

$$z_t(t) = Az(t) + f(z) + B_{\mathrm{d}}d + B_{\mathrm{in}}u, \quad t > 0,$$
$$z(0) = z_0,$$
$$y(t) = (Cz)(t),$$

where, as usual, we assume that $f(\cdot)$ is a smooth nonlinear function satisfying $f(0) = 0$ so that the origin in the Hilbert state space $\mathcal{Z}$ is an equilibrium which is assumed to be asymptotically stable for the uncontrolled problem.

Here, once again, the objective is to find a vector of controls u so that the vector of outputs $y(t)$ tracks a vector of periodic reference signals $y_r(t)$, while rejecting a vector of periodic disturbances d, as t tends to infinity. Here we assume there are n_{in} inputs, n_{out} outputs and n_{d} disturbances. In the most common case we will have $n_{\text{in}} = n_{\text{out}}$. So we have

$$B_{\text{in}} u = B_{\text{in}}^1 u_1 + B_{\text{in}}^2 u_2 + \cdots B_{\text{in}}^{n_{\text{in}}} u_{n_{\text{in}}},$$

$$B_{\text{d}} d = B_{\text{d}}^1 u_1 + B_{\text{d}}^2 u_2 + \cdots B_{\text{d}}^{n_{\text{d}}} u_{n_{\text{d}}},$$

$$Cz = \left[C_1 z,\ C_2 z, \cdots, C_{n_{\text{out}}} z \right].$$

We seek control laws

$$u_j(t) = \gamma_j(w(t)),$$

so that the errors

$$e_j(t) = y_j(t) - y_{r,j}(t) \xrightarrow{t \to \infty} 0.$$

In the nonlinear case it is not possible to decouple the problem in order to compute the controls $u_j(t)$ by solving a sequence of simpler problems, as was done in Section 3.8, since the superposition principle doesn't hold.

For nonlinear problems it is useful to assume that there is a single global period P so that all the reference and disturbance signals could be expanded in terms of Fourier series expansions using the single period P. This becomes an issue due to the necessity of evaluating the nonlinear term F at a truncated Fourier series approximation of the state variable obtained when solving the "regulator equations".

Thus in contrast to our assumptions in Chapter 3, in this section we assume that there exists a global period $T = 2P$, with $P > 0$, such that all signals to be tracked and disturbances to be rejected have a period which divides evenly into P. Thus the signals to be tracked $y_{r,j}(t)$ for $j = 1, \cdots, n_c$ and the disturbances d_j for $j = 1, \cdots, n_d$ are assumed to be periodic with non-minimum period P. That is to say, for each periodic signal $y_{r,j}$ there exists a minimum period $2P_{r,j}$ such that $K_j^r P_{r,j} = P$, for some positive integer K_j^r. And again, for each of the disturbances $d_j(t)$ there exists a minimum period $2P_{d,j}$ such that $K_j^d P_{d,j} = P$, for some positive integer K_j^d.

In what follows we are interested in representing all the reference and disturbance signals in terms of a Fourier series using the non-minimum period P. While it may be obvious to some, we take a small amount of space to derive the relation between the coefficients of the Fourier series with fundamental period $2P_j$ for a function $f_j(t)$, and the coefficients of the Fourier series with fundamental period $2P$ for the same function $f_j(t)$, with the assumption that $K_j P_j = P$ for some integer K_j. To this end let

$$f_j(t) = \frac{a_0}{2} + \sum_{n=1}^{\infty} \left\{ a_n \cos\left(\frac{n\pi t}{P} \right) + b_n \sin\left(\frac{n\pi t}{P} \right) \right\},$$

and

$$f_j(t) = \frac{a_0^j}{2} + \sum_{k=1}^{\infty} \left\{ a_k^j \cos\left(\frac{k\pi t}{P_j} \right) + b_k^j \sin\left(\frac{k\pi t}{P_j} \right) \right\},$$

be the two distinct Fourier series representations with period $2P$ and $2P_j$, respectively.

For a_0 we have

$$a_0 = \frac{1}{P} \int_0^{2P} f_j(t)\, dt = \frac{1}{K_j P_j} \int_0^{2K_j P_j} f_j(t)\, dt$$

$$= \frac{K_j}{K_j P_j} \int_0^{2P_j} f_j(t)\, dt = \frac{1}{P_j} \int_0^{2P_j} f_j(t)\, dt = a_0^j.$$

Fix $n \geq 1$

$$a_n = \frac{1}{P} \int_0^{2P} f_j(t) \cos\left(\frac{n\pi t}{P}\right) dt$$

$$= \frac{1}{P} \int_0^{2P} \left[\frac{a_0^j}{2} + \sum_{k=1}^{\infty} \left\{ a_k^j \cos\left(\frac{k\pi t}{P_j}\right) + b_k^j \sin\left(\frac{k\pi t}{P_j}\right) \right\} \right] \cos\left(\frac{n\pi t}{P}\right) dt$$

$$= \frac{1}{P} \int_0^{2P} \left[\frac{a_0^j}{2} + \sum_{k=1}^{\infty} \left\{ a_k^j \cos\left(\frac{kK_j\pi t}{P}\right) + b_k^j \sin\left(\frac{kK_j\pi t}{P}\right) \right\} \right] \cos\left(\frac{n\pi t}{P}\right) dt$$

$$= \begin{cases} 0, & \text{if } n \neq kK_j \text{ for some } k, \\ a_k^j, & \text{if } n = kK_j \text{ for some } k. \end{cases}$$

Similarly for b_n we have

$$b_n = \begin{cases} 0, & \text{if } n \neq kK_j \text{ for some } k, \\ b_k^j, & \text{if } n = kK_j \text{ for some } k. \end{cases}$$

Now every integer n can be written in one of K_j forms as

$$n = kK_j + \ell \text{ for some } \ell = 0, 1, \cdots, (K_j - 1).$$

So we see that

$$f_j(t) = \frac{a_0}{2} + \sum_{n=1}^{\infty} \left\{ a_n \cos\left(\frac{n\pi t}{P}\right) + b_n \sin\left(\frac{n\pi t}{P}\right) \right\}$$

$$= \frac{a_0^j}{2} + \sum_{k=1}^{\infty} \sum_{\ell=0}^{K_j-1} \left\{ a_{kK_j+\ell} \cos\left(\frac{(kK_j + \ell)\pi t}{P}\right) + b_{kK_j+\ell} \sin\left(\frac{(kK_j + \ell)\pi t}{P}\right) \right\}$$

$$= \frac{a_0^j}{2} + \sum_{k=1}^{\infty} \left\{ a_{kK_j} \cos\left(\frac{(kK_j)\pi t}{P}\right) + b_{kK_j} \sin\left(\frac{(kK_j)\pi t}{P}\right) \right\}$$

$$= \frac{a_0^j}{2} + \sum_{k=1}^{\infty} \left\{ a_k^j \cos\left(\frac{k\pi t}{P_j}\right) + b_k^j \sin\left(\frac{k\pi t}{P_j}\right) \right\}.$$

As we have already mentioned, in contrast to Section 3.8, for nonlinear problems we cannot appeal to the superposition principle. Therefore we cannot solve the problem by solving a sequence of simpler problems in which there is a single reference signal or disturbance for each simple problem. Rather, we are forced to consider a single global Fourier expansion for all the unknowns to be solved for in the problem. Notice then, in the case of several signals to be tracked and disturbances to be rejected, that the truncation of the single Fourier series with period $2P$ could involve many terms in order to ensure an accurate answer.

In order to demonstrate our methodology without getting unnecessarily bogged down in notation let us consider a SISO system in which we have only one signal to be tracked and one disturbance to be rejected. In this case we will assume that the reference signal y_r is a periodic function with period $2P^r$, with the following Fourier series representation:

$$ y_r(t) = \frac{a_0^r}{2} + \sum_{k=1}^{\infty} \left\{ a_k^r \cos\left(\frac{k\pi t}{P^r}\right) + b_k^r \sin\left(\frac{k\pi t}{P^r}\right) \right\}. $$

We also assume that the disturbance is periodic with period $2P^r$, with Fourier representation

$$ d(t) = \frac{a_0^d}{2} + \sum_{k=1}^{\infty} \left\{ a_k^d \cos\left(\frac{k\pi t}{P^d}\right) + b_k^d \sin\left(\frac{k\pi t}{P^d}\right) \right\}. $$

In order to simplify the exposition we will impose the following two assumptions:

Assumption 4.7.

1. *There exists a period P common to the reference signal y_r and the disturbance d;*

2. *The signals $y_r(t)$ and $d(t)$ are in the class $C^{(p)}[-P, P]$, for some $p \geq 1$.*

With these assumptions the coefficients in the series satisfy

$$ \sum_{k=1}^{\infty} |k^p a_k^\ell| < \infty \quad \text{and} \quad \sum_{k=1}^{\infty} |k^p b_k^\ell| < \infty, \text{ for } \ell = r, d, $$

and in particular this implies that

$$ k^p a_k^\ell \xrightarrow{k\to\infty} 0 \quad \text{and} \quad k^p b_k^\ell \xrightarrow{k\to\infty} 0, \text{ for } \ell = r, d. $$

It follows that given any desired level of tracking accuracy, we can choose a truncation level N for the infinite Fourier series that will produce a finite-dimensional exosystem, and at the same time, achieve the desired approximation of the exact control law.

Thus, we truncate the infinite series (if it is indeed infinite) at the value N to obtain

$$y_r(t) \simeq y_r^N(t) = \frac{a_0^r}{2} + \sum_{k=1}^{N} \left\{ a_k^r \cos\left(\frac{k\pi t}{P}\right) + b_k^r \sin\left(\frac{k\pi t}{P}\right) \right\},$$

and

$$d(t) \simeq d^N(t) = \frac{a_0^d}{2} + \sum_{k=1}^{N} \left\{ a_k^d \cos\left(\frac{k\pi t}{P}\right) + b_k^d \sin\left(\frac{k\pi t}{P}\right) \right\}.$$

We consider the control problem in standard state space form as

$$z_t = Az + f(z) + B_d d + B_{\text{in}} u,$$
$$z(0) = z_0,$$
$$y = Cz,$$

where A is an (unbounded) linear operator in a Hilbert state space $\mathcal{Z}$. As usual we assume that A is sectorial and has a compact resolvent $(sI - A)^{-1}$ for $s \in \rho(A)$, and that A generates an exponentially stable (analytic) semigroup in the Hilbert scale of spaces $\mathcal{H}^\alpha = D((-A)^{\alpha/2})$, $\alpha > 0$, the domain of $(-A)^{\alpha/2}$. f is a smooth function satisfying $f(0) = 0$, $f'(0) = 0$.

We notice that by our assumptions the signal to be tracked y_r and the disturbance d are generated by an infinite-dimensional block diagonal exosystem

$$w_t = Sw, \quad w(0) = w_0 \in \mathcal{W},$$
$$y_r = Qw, \quad d = Pw,$$

where we denote

$$\alpha_k = \frac{k\pi}{P}, \quad S_k = \begin{bmatrix} 0 & \alpha_k \\ -\alpha_k & 0 \end{bmatrix}, \quad k = 0, 1, \cdots, N,$$

and

$$S = \text{diag}\begin{bmatrix} S_0, & S_1, & S_2, & \cdots \end{bmatrix}.$$

We seek a control $u(t) = \gamma(w(t))$ obtained by solving the "regulator equations" for $\pi(w)$ and $\gamma(w)$ as we have throughout the text:

$$\frac{\partial \pi(w)}{\partial w} Sw = \frac{d\pi(w)}{dt} = A\pi(w) + f(\pi(w)) + B_d Pw + B_{\text{in}} \gamma(w),$$
$$C\pi(w) = Qw,$$

for all w in a neighborhood of zero in $\mathcal{W}$. If the regulator equations are solvable then a feedback law solving the state feedback regulator problem is given by

$$u(t) = \gamma(w)(t).$$

Note that a trajectory of the exosystem $w(t)$ is periodic with period $2P$ and therefore $\pi(w(t))$ and $\gamma(w(t))$ are likewise periodic with the same period. Let us suppress the x dependence of the coefficients of the Fourier series and denote

$$\pi(t) = \frac{a_0^\pi}{2} + \sum_{k=1}^{\infty} \{a_k^\pi \cos(\alpha_k t) + b_k^\pi \sin(\alpha_k t)\},$$

where a_0^π, a_k^π and b_k^π are functions of x. Similarly we seek γ in the form

$$\gamma(t) = \frac{a_0^\gamma}{2} + \sum_{k=1}^{\infty} \{a_k^\gamma \cos(\alpha_k t) + b_k^\gamma \sin(\alpha_k t)\}.$$

Here we have

$$a_0^\pi = \frac{1}{P} \int_{-P}^{P} \pi(t)\,dt,$$

$$a_k^\pi = \frac{1}{P} \int_{-P}^{P} \cos(\alpha_k t)\pi(t)\,dt,$$

$$b_k^\pi = \frac{1}{P} \int_{-P}^{P} \sin(\alpha_k t)\pi(t)\,dt,$$

and

$$a_0^\gamma = \frac{1}{P} \int_{-P}^{P} \gamma(t)\,dt,$$

$$a_k^\gamma = \frac{1}{P} \int_{-P}^{P} \cos(\alpha_k t)\gamma(t)\,dt,$$

$$b_k^\gamma = \frac{1}{P} \int_{-P}^{P} \sin(\alpha_k t)\gamma(t)\,dt.$$

Obviously we will not be able to solve for $\pi(t)$ and $\gamma(t)$ explicitly so we apply a Galerkin methodology to find approximate solutions for π and γ by truncating the infinite sums and by requiring the Galerkin residuals to be zero. Namely, for a fixed N, we set

$$\pi^N(t) = \frac{a_0^\pi}{2} + \sum_{k=1}^{N} \{a_k^\pi \cos(\alpha_k t) + b_k^\pi \sin(\alpha_k t)\},$$

and

$$\gamma^N(t) = \frac{a_0^\gamma}{2} + \sum_{k=1}^{N} \{a_k^\gamma \cos(\alpha_k t) + b_k^\gamma \sin(\alpha_k t)\},$$

and we seek functions a_0^π, a_j^π, b_j^π, and coefficients α_0^γ, a_j^γ and b_j^γ for $j = 1, \cdots, N$ from a system of equations obtained by taking the inner product of equation (4.107) with the test functions 1, $\sin(\alpha_n t)$ and $\cos(\alpha_n t)$, respectively, for $n = 1, \cdots, N$ and requiring equality to hold.

Namely, we first substitute all the approximate Fourier series into the first regulator equation

$$\sum_{k=1}^{N} (\alpha_k^{\pi}) \left\{ -a_k^{\pi} \sin(\alpha_k t) + b_k^{\pi} \cos(\alpha_k t) \right\} = A\frac{a_0^{\pi}}{2}$$

$$+ \sum_{k=1}^{N} \left\{ Aa_k^{\pi} \cos(\alpha_k t) + Ab_k^{\pi} \sin(\alpha_k t) \right\} + f(\pi^N(t))$$

$$+ B_{\mathrm{d}} \left(\frac{a_0^d}{2} + \sum_{k=1}^{N^d} \left\{ a_k^d \cos(\alpha_k t) + b_k^d \sin(\alpha_k t) \right\} \right)$$

$$+ B_{\mathrm{in}} \left(\frac{a_0^{\gamma}}{2} + \sum_{k=1}^{N} \left\{ a_k^{\gamma} \cos(\alpha_k t) + b_k^{\gamma} \sin(\alpha_k t) \right\} \right), \qquad (4.107)$$

and into the second regulator equation

$$C\pi(w(t)) = Qw(t) = y_r.$$

This last equation implies

$$C\frac{a_0^{\pi}}{2} + \sum_{k=1}^{N} \left\{ Ca_k^{\pi} \cos(\alpha_k t) + Cb_k^{\pi} \sin(\alpha_k t) \right\}$$

$$= \frac{a_0^r}{2} + \sum_{k=1}^{N^r} \left\{ a_k^r \cos(\alpha_k t) + b_k^r \sin(\alpha_k t) \right\},$$

which, in turn, implies

$$Ca_k^{\pi} = a_k^r, \quad k = 0, \cdots, N \quad Cb_k^{\pi} = b_k^r, \quad k = 1, \cdots, N. \qquad (4.108)$$

Next we obtain a system of coupled elliptic boundary value problems by multiplying (4.107) by the test functions 1, $\sin(\alpha_n t)$ and $\cos(\alpha_n t)$, respectively, for $n = 1, \cdots, N$ and integrating over the $[-P, P]$ time interval. Here we apply the obvious orthogonality results to obtain the following equations.

1. Multiplying by 1 and integrating from $-P$ to P we obtain

$$0 = Aa_0^{\pi} + 2 \int_{-P}^{P} f(\pi^N(t)) \, dt + B_{\mathrm{d}} a_0^d + B_{\mathrm{in}} a_0^{\gamma}.$$

2. Multiplying by $\cos(\alpha_n t)$ and again integrating from $-P$ to P we obtain

$$\alpha_n b_n^{\pi} = Aa_n^{\pi} + \int_{-P}^{P} f(\pi^N(t)) \cos(\alpha_n t) \, dt + B_{\mathrm{d}} a_n^d + B_{\mathrm{in}} a_n^{\gamma}.$$

3. Finally, multiplying by $\sin(\alpha_n t)$ and again integrating between $-P$ and P we obtain

$$-\alpha_n a_n^\pi = Ab_n^\pi + \int_{-P}^P f(\pi^N(t)) \sin(\alpha_n t)\, dt + B_{\mathrm{d}} b_n^d + B_{\mathrm{in}} b_n^\gamma.$$

We now introduce a notation to represent the nonlinear terms that will be eventually simplified once a particular nonlinear function f is given in a particular example. In general, we define

$$f_0 = 2 \int_{-P}^P f(\pi^N(t))\, dt, \qquad (4.109)$$

$$f_n^c = \int_{-P}^P f(\pi^N(t)) \cos(\alpha_n t)\, dt, \qquad (4.110)$$

$$f_n^s = \int_{-P}^P f(\pi^N(t)) \sin(\alpha_n t)\, dt. \qquad (4.111)$$

With this we can define the following vector expressions of length $2N+1$:

$$\Pi = \begin{bmatrix} a_0^\pi \\ a_1^\pi \\ b_1^\pi \\ \vdots \\ a_N^\pi \\ b_N^\pi \end{bmatrix}, \; \mathcal{S}\Pi = \begin{bmatrix} 0 \\ \alpha_1 b_1^\pi \\ -\alpha_1 a_1^\pi \\ \vdots \\ \alpha_N b_N^\pi \\ -\alpha_N a_N^\pi \end{bmatrix}, \; \mathcal{D} = \begin{bmatrix} a_0^d \\ a_1^d \\ b_1^d \\ \vdots \\ a_N^d \\ b_N^d \end{bmatrix}, \; \mathcal{F}(\Pi) = \begin{bmatrix} f_0 \\ f_1^c \\ f_1^s \\ \vdots \\ f_N^c \\ f_N^s \end{bmatrix}, \; \mathcal{G} = \begin{bmatrix} a_0^\gamma \\ a_1^\gamma \\ b_1^\gamma \\ [1ex] a_N^\gamma \\ b_N^\gamma \end{bmatrix}.$$

Then we can write the first regulator equation as the following system in vector form:

$$\mathcal{S}\Pi = A\Pi + \mathcal{F}(\Pi) + B_{\mathrm{d}}\mathcal{D} + B_{\mathrm{in}}\mathcal{G}.$$

This system is supplemented with boundary conditions and constraints due to the second regulator equation in (4.108). Let us denote the vector $\mathcal{Y}_r$ by

$$\mathcal{Y}_r = \begin{bmatrix} a_0^r, & a_1^r, & b_1^r, & \cdots, & a_N^r, & b_N^r \end{bmatrix}^\mathsf{T}.$$

Then we proceed by solving the system

$$0 = A\Pi - \mathcal{S}\Pi + \mathcal{F}(\Pi) + B_{\mathrm{d}}\mathcal{D} + B_{\mathrm{in}}\mathcal{G},$$
$$0 = A\widetilde{\Pi} - \mathcal{S}\Pi + \mathcal{F}(\Pi) + B_{\mathrm{d}}\mathcal{D},$$
$$\mathcal{G} = G^{-1}(\mathcal{Y}_r - C(\widetilde{\Pi})),$$

where, as usual, $G = C(-A^{-1})B_{\mathrm{in}}$.

Example 4.5 (Burgers' Example). We consider exactly the same regulator problem discussed in Example 4.4 for the nonlinear viscous Burgers' equation

$$z_t(x,t) = c\, z_{xx}(x,t) - z(x,t) z_x(x,t),\ 0 \le x \le 1,$$
$$z(0,t) = d(t) = M_2 + A_2 \sin(\alpha_2 t),$$
$$c\, z_x(1,t) = u(t)$$

with initial condition
$$z(x,0) = \varphi(x),$$

measured output given by point evaluation at the point $x = x_0 \in (0,1)$

$$y(t) = C(z) = z(x_0, t),$$

and reference signal $y_r(t) = M_1 + A_1 \sin(\alpha_1 t)$. We also choose the same parameters as in Example 4.4

$$
\begin{array}{lll}
c = 1, & x_0 = 0.5, & \varphi(x) = M_2, \\
M_1 = 0.5, & A_1 = 0.25, & \alpha_1 = 2\pi, \\
M_2 = 0.75, & A_2 = 1, & \alpha_2 = \pi, \\
\beta = 0.05, & n = 4.
\end{array}
$$

Notice that given the particular structure of the signals y_r and d we can choose the common period to be
$$P = \frac{2\pi}{\alpha_2} = 2\,.$$

With this, we fix $N = 5$ and we proceed explicitly as we described above producing the following algebro-differential equation system

$$0 = A\Pi - S\Pi + \mathcal{F}(\Pi) + B_{\mathrm{d}}\mathcal{D} + B_{\mathrm{in}}\mathcal{G}\,, \tag{4.112}$$

$$0 = A\widetilde{\Pi} - S\Pi + \mathcal{F}(\Pi) + B_{\mathrm{d}}\mathcal{D}\,, \tag{4.113}$$

$$\mathcal{G} = G^{-1}(\mathcal{Y}_r - C(\widetilde{\Pi}))\,, \tag{4.114}$$

where
$$\mathcal{D} = \left[2M_2,\ 0,\ A_2,\ 0,\ 0,\ 0,\ 0,\ 0,\ 0,\ 0,\ 0\right]^{\mathsf{T}},$$
$$\mathcal{Y}_r = \left[2M_1,\ 0,\ 0,\ 0,\ A_1,\ 0,\ 0,\ 0,\ 0,\ 0,\ 0\right]^{\mathsf{T}},$$

and
$$\mathcal{F}(\Pi) = -\frac{1}{2}\frac{d}{dx}\left[a_0^f,\ a_1^f\, b_1^f,\ \cdots,\ a_5^f\, b_5^f\right]^{\mathsf{T}}.$$

In the last expression the coefficients are given by

$$a_0^f = \frac{1}{2}(a_0^\pi)^2 + \sum_{i=1}^{N}\left[(a_i^\pi)^2 + (b_i^\pi)^2\right], \tag{4.115}$$

and for $n = 1, \cdots, N$

$$
a_n^f = a_0^\pi a_n^\pi + \frac{1}{2} \sum_{i=1}^{N} \sum_{j=1}^{N} \left[a_i^\pi a_j^\pi \left(\delta_{i,j+n} + \delta_{j,i+n} + \delta_{n,i+j} \right) \right.
$$
$$
\left. + b_i^\pi b_j^\pi \left(\delta_{i,j+n} + \delta_{j,i+n} - \delta_{n,i+j} \right) \right] , \tag{4.116}
$$
$$
b_n^f = a_0^\pi b_n^\pi + \sum_{i=1}^{N} \sum_{j=1}^{N} \left[a_i^\pi b_j^\pi \left(-\delta_{i,j+n} + \delta_{j,i+n} + \delta_{n,i+j} \right) \right] , \tag{4.117}
$$

which have been obtained integrating equations (4.109)–(4.111) while using the specific structure $(-zz_x)$ of the nonlinear term in the Burger's equation. In Equations (4.115)–(4.117), we used $\delta_{l,m}$ to denote the standard Kronecker delta function. Also notice that these formulas are valid for any N. The equation system, (4.112)–(4.114), is then solved for $t \in (0, 10)$.

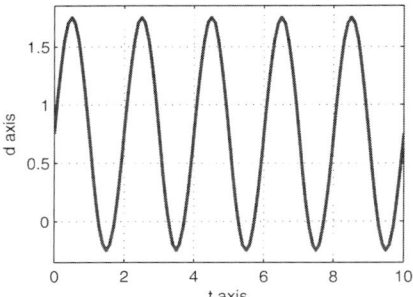

Fig. 4.25: $d(t)$ for $0 \leq t \leq 10$.

Fig. 4.26: $y(t)$ and $C(z)$ for $0 \leq t \leq 10$.

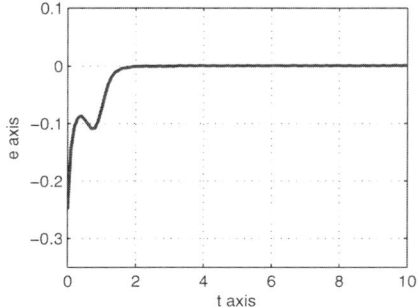
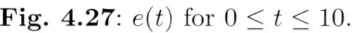

Fig. 4.27: $e(t)$ for $0 \leq t \leq 10$.

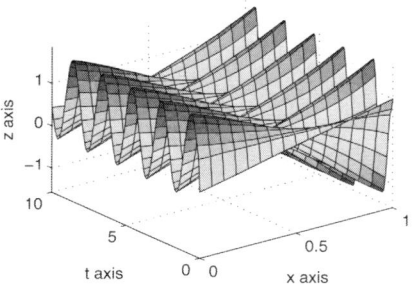

Fig. 4.28: Solution Surface.

In Fig. 4.25, we display the disturbance signal $d(t)$ for $t \in (0, 10)$ and in Fig. 4.26, we display the reference signal $y_r(t)$ and the output $C(z(t))$ for $t \in (0, 10)$. Notice that after a short transient $C(z)$ converges to y_r. In Fig. 4.27 we display the error $e(t) = y_r(t) - C(z)$ for the same time interval. Finally, In Fig. 4.28 we have plotted the solution surface $z(x, t)$, for $0 \leq x \leq 1$ and $0 \leq t \leq 10$.

4.7 Zero Dynamics Design for Nonlinear Systems

In this section we consider the nonlinear analog of the methodology explained in the examples presented in Section 3.9, where we introduced the so-called zero dynamics design methodology and showed that it derives from the geometric constructs associated with system zeros. In Examples 3.2 and 3.3 we showed how the zero dynamics design strategy can be applied for both boundary and interior design regulation problems. In this section we show that this methodology can also be applied to nonlinear systems but, as one might expect for nonlinear systems, the proofs of convergence are somewhat more challenging. In [13] the authors showed that zero dynamics design can be applied to a one-dimensional Kuramoto-Sivashinsky model and in [14] the same type of analysis is carried out for a viscous Burgers' equation. The main results of the papers [13, 14] are presented in Examples 4.6 and 4.7. The method of proof requires establishing the stability of a certain system in a Sobolev space in which the (unbounded) output operator is continuous. This approach can be difficult to apply and must be handled on a case-by-case basis. Essentially what is proved is that the solution to the zero dynamics system is attractive for solutions of the nearby closed loop system provided that the proper control is derived from the zero dynamics system. The main drawback of this methodology is that its applicability is limited to special types of co-located boundary control systems. On the other hand one particularly attractive feature is that the disturbances to be rejected or the signals to be tracked can be very general and not necessarily generated by a neutrally stable finite-dimensional exosystem. Indeed the only requirement is for the signals to be smooth functions in time.

The zero dynamics design methodology presented here has evolved over a series of papers [13, 14, 16, 18, 15, 17] for asymptotic regulation of linear and nonlinear boundary control systems. The main idea behind this methodology derives from a nonequilibrium theory of output regulation as developed in [22, 28] where it was applied successfully for certain classes of finite-dimensional nonlinear systems. The methodology in [22, 28] is particularly useful in the case in which a well defined zero dynamics is available. As it turns out, many co-located boundary control systems provide a setting in which a natural notion of zero dynamics is readily available. This type of construction was exploited in the work [26] in which a nonlinear enhancement of classical root-locus constructs [9] was established by proving convergence of trajectories for the closed loop system (with proportional error feedback) to the trajectories of the associated zero dynamics in the high gain limit. Additionally, semiglobal stabilization was also established in [26] using the convergence of trajectories results and the fact that the system is *exponentially minimum phase*, i.e., has globally exponentially stable zero dynamics. These results in [26] are used extensively in the detailed proofs of stability given in [14]. In there, the authors showed that using a proportional error boundary feedback control law, the resulting closed-loop trajectories tend (in a uniform way) to the trajectories of the open-loop zero dynamics as the gain parameters are increased to infinity. We note that in the case of nonlinear lumped or distributed systems the frequency domain concepts of zeros and poles (as discussed in Section 3.9) do not exist. Nevertheless, as demonstrated in [27, 59, 54, 55], there is a well defined analog of system zeros for nonlinear lumped systems. This state space counterpart of the notion of systems zeros has played an important role in the development of nonlinear feedback design, for example, this concept lies at the heart of the recent solution of the regulator problem for lumped nonlinear systems [56, 55].

4.7.1   Zero Dynamics Inverse Design

In this section we present an extension to nonlinear systems of the methodology introduced in Section 3.9. The feedback design problem considered here is asymptotic signal tracking and disturbance rejection for co-located boundary controlled systems. Our basic problem therefore involves the following main ingredients.

1. *The Control System:* We consider a boundary control system, which we denote by Σ_p, evolving on a Hilbert space $\mathcal{Z}$ according to

$$z_t = A_0 z + F(z) + B_\mathrm{d}d, \ z(0) = z_0, \tag{4.118}$$
$$\mathcal{B}z(t) = u(t), \qquad \text{(Boundary input)} \quad (4.119)$$
$$y(t) = \mathcal{C}z(t), \qquad \text{(Boundary output)} \quad (4.120)$$

   with state variable z, control u and output y. Here A_0 is an unbounded densely defined operator in $\mathcal{Z}$. $\mathcal{B}$ is a boundary control input operator and $\mathcal{C}$ is a co-located boundary operator producing a measured output.

2. *The Signals:* We are given a known signal vector $y_r(\cdot) \in \mathbb{R}^{m_r}$, to be tracked and a disturbance vector $d(\cdot) \in \mathbb{R}^{m_d}$ to be rejected. We recall our standing assumption on exogeneous signals, that y_r and d can be expressed as a linear combination of a certain vector $w = (w_1, \ \cdots, \ w_k)$, i.e.,

$$y_r = Qw \ \text{ and } \ d = Pw,$$

   for some Q and P.

3. *The Measured Variables:* We assume that y (the measured variable for the control system) is available and that w is known.

4. *The Errors:* We consider the tracking error

$$e(t) = y(t) - y_r(t).$$

5. *The Zero Dynamics Controller:* We seek a feedback controller, in terms of the zero dynamics system which we denote by Σ_{zd}, evolving on the Hilbert space $\mathcal{Z}$ according to

$$\xi_t = A_0\xi + F(\xi) + B_\mathrm{d}d, \ \xi(0) = \xi_0, \tag{4.121}$$
$$\mathcal{C}\xi(t) = y_r, \qquad \text{(Boundary input)} \quad \tag{4.122}$$
$$u(t) = \mathcal{B}\xi(t). \qquad \text{(Boundary output)} \quad \tag{4.123}$$

Assumption 4.8. *Our standing assumption on the signals to be tracked or rejected is that w is in the Sobolev space $W^{1,\infty}([0,\infty); \mathbb{R}^r)$. More precisely, we assume*

$$w \in \mathcal{W}_\sigma = \{w : \mathrm{ess\,sup}_{t\in[0,\infty)} \left\{\|w(t)\|, \|w'(t)\|\right\} \leq \sigma\}, \tag{4.124}$$

for some $\sigma < \infty$.

Remark 4.14. The requirement $w \in W^{1,\infty}([0,\infty); \mathbb{R}^r)$ is a technical enhancement of the assumption $w \in L^\infty([0,\infty); \mathbb{R}^r)$, standard in the literature on lumped nonlinear systems [55]. This hypothesis is our preferred choice of one of several alternatives that will guarantee the existence and uniqueness of solutions for a large number of nonlinear distributed parameter systems.

Remark 4.15. For a broad class of nonlinear parabolic systems, the spectrum of the spatial system operator of the linearization of system (4.121)–(4.122) is known to coincide with the zeros of the transfer function for the linearization of the system (4.118)–(4.120), just as in the lumped linear case (thereby partially justifying its name). More importantly, for certain high gain nonlinear systems, it has also been shown that the limits of the closed-loop trajectories converge to trajectories of the zero-dynamics ([19], [26]), thereby providing a nonlinear enhancement of root-locus techniques just as in the lumped linear case.

The ZDI design objective. Our objective is to design a feedback controller such that, in some reasonable topology, for a bounded set $B_\mathrm{p} \subset \mathcal{Z}$ and a bounded set $B_{\mathrm{zd}} \subset \mathcal{Z}$ of initial conditions (z_0, ξ_0):

1. The positive orbit of $B_\mathrm{p} \times B_{\mathrm{zd}}$ is bounded,

2. $\lim\limits_{t\to\infty} \|z(t) - \xi(t)\| = 0$, and

3. $\lim\limits_{t\to\infty} \|y(t) - y_r(t)\| = 0$.

Remark 4.16. This design methodology has been used and rigorously justified in a number of cases. For *linear* distributed parameter systems, we cite [18, 15, 17], and for *nonlinear* systems [13, 14, 16].

The proof derives from the approach discussed in the proof of Proposition 3.1 in Chapter 3. In particular with z in (4.118)–(4.120), ξ in (4.121)–(4.122) and the control u given in (4.123) we define the state variable $\eta = z - \xi$, which solves the problem

$$\eta_t = A\eta + \widetilde{F}_\xi(\eta), \ \eta(0) = \eta_0, \tag{4.125}$$

$$A = A_0 \ \text{with} \ D(A) = D(A_0) \cap \ker \mathcal{B}, \tag{4.126}$$

$$y(t) = \mathcal{B}\eta(t), \tag{4.127}$$

where $\eta_0 \equiv z_0 - \xi_0$ and

$$\widetilde{F}_\xi(\eta) = F(\eta + \xi) - F(\xi). \tag{4.128}$$

Here we impose for A the same conditions we imposed in Assumption 3.1. In addition we assume that A generates an infinite scale of spaces, $\mathcal{H}^\alpha$ ($\alpha \in \mathbb{R}$), as discussed in Section 2.2.1. Then for suitable assumptions on the nonlinear terms $F(\cdot)$ and for sufficiently small initial conditions η_0 and ξ_0 (and therefore also z_0) one can show that the system (4.125)–(4.127) is locally asymptotically stable in some $\mathcal{H}^\alpha$ space for $\alpha > 0$, on which the output operator $\mathcal{C}$ is continuous. In this way we can show that

1. The positive orbit of η is bounded,

2. $\lim\limits_{t\to\infty} \|\eta(t)\| = 0$, and

3. $\lim\limits_{t\to\infty} \|y(t) - y_r(t)\| = \lim\limits_{t\to\infty} \|\mathcal{C}(\eta)(t)\| = 0$.

In the following two examples, we illustrate the above design methodology to achieve asymptotic regulation for a boundary controlled viscous Burgers' system (cf, [14]) in Example 4.6, and for a one-dimensional Kuramoto-Sivashinsky equation (cf. [13]) in Example 4.7.

Example 4.6 (Viscous Burgers' System). Let the control system Σ_p, with state space $\mathcal{Z} = L^2(0,1)$, evolve according to a Robin boundary controlled viscous Burgers' equation on the spatial domain $0 \le x \le 1$ and be given by

$$z_t(x,t) = c z_{xx}(x,t) - z(x,t) z_x(x,t), \ t \ge 0, \tag{4.129}$$

$$\mathcal{B}_0 z = -u_0(t), \quad \mathcal{B}_1 z = u_1(t), \tag{4.130}$$

$$z(x,0) = z_0(x), \tag{4.131}$$

where the boundary control operators are given by

$$\mathcal{B}_0 z = -c z_x(0,t) - k_0 z(0,t), \quad \mathcal{B}_1 z = c z_x(1,t) - k_1 z(1,t).$$

The outputs

$$y_j(t) = (\mathcal{C}_j z)(t) = z(j, t), \quad j = 0, 1, \quad t > 0. \tag{4.132}$$

are assumed to be the measured variables. The functions $u_j(t)$, $j = 0, 1$ represent the control inputs. System (4.129)–(4.132) is a 2×2 MIMO nonlinear distributed parameter system with co-located inputs and outputs. $c > 0$ is a parameter modeling the kinematic viscosity.

Remark 4.17. Precise statements of necessary conditions about solvability and regularity properties of solutions for the problem (4.129)–(4.131) can be found in [14]. It follows from these statements that for sufficiently smooth controls, u_j, $j = 0, 1$ the solution of the above problem is instantly classical for $t > 0$ and, therefore, the point evaluations of $z(x, t)$ in equation (4.132) are well defined.

Given two signals $y_{r,j}(t) \in W^1_{\text{loc}}$, $j = 0, 1$, we define the error function $e(t)$ component-wise by

$$e_j(t) = z(j, t) - w_j(t), \quad j = 0, 1.$$

For this system our objective is to find controls $u_j(t)$ so that for given signals $y_{r,j}(t)$ the errors $e_j(t)$ satisfy

$$e_j(t) \xrightarrow{t \to \infty} 0, \quad j = 0, 1,$$

while the state variable z remains bounded in the Hilbert state space.

As outlined before, the design of our controllers involves the solution of a feedback zero dynamics controller Σ_{zd} that processes the error zeroing constraints, $\mathcal{C}_j \xi = y_{r,j}$, here introduced as boundary controls, and produces the desired controls $u_j = \mathcal{B}_j \xi$ as outputs. These controls are then introduced as inputs into the control system Σ_{p} to force the measured outputs $y_j(t)$ to track the reference signals $y_{r,j}(t)$.

Namely, the zero dynamics controller Σ_{zd} is given by

$$\xi_t(x, t) = c \xi_{xx}(x, t) - \xi(x, t) \xi_x(x, t), \tag{4.133}$$
$$\xi(0, t) = y_{r,0}(t), \tag{4.134}$$
$$\xi(1, t) = y_{r,1}(t), \tag{4.135}$$
$$\xi(x, 0) = \xi_0(x), \tag{4.136}$$

whose output u is defined component-wise by

$$u_j(t) = \mathcal{B}_j z(t) = (-1)^{j+1} c \xi_x(j, t) - k_j y_{r,j}(t), \quad k_j > 0, \quad j = 0, 1. \tag{4.137}$$

We now recall some standard concepts from Lyapunov stability theory.

Definition 4.1.

1. A continuous function $\alpha : [0, s) \to [0, \infty)$ is said to be of class $\mathcal{K}$ if α is strictly increasing and $\alpha(0) = 0$.

2. α in class $\mathcal{K}$ is said to be of class $\mathcal{K}_\infty$ if $s = \infty$ and $\lim_{t \to \infty} \alpha(t) = \infty$.

3. A continuous function $\beta : [0, s) \times [r, \infty) \to [0, \infty)$ is said to belong to class $\mathcal{KL}$, if $\beta(\cdot, \rho)$ is class $\mathcal{K}$ and $\beta(\sigma, \cdot)$ is monotone decreasing in σ with $\lim_{\sigma \to \infty} \beta(\sigma, \cdot) = 0$.

One of the results proven in [14] is that, in the language of lumped nonlinear systems, the zero dynamics system is *input-to-state stable* in both $L^2(0, 1)$ and $H^1(0, 1)$.

Theorem 4.2. *The unforced* $(y_{r,0} = y_{r,1} = 0)$ *zero dynamics system* (4.133)–(4.136) *is globally asymptotically stable on both* $L^2(0, 1)$ *and* $H^1(0, 1)$. *Assume that the conditions* (4.124) *on* $y_{r,j}(t)$, $j = 0, 1$ *are satisfied. Then there exist*

1. *a class* $\mathcal{KL}$ *function* β *and a class* $\mathcal{K}$ *function* γ,

2. *a class* $\mathcal{KL}$ *function* β_1 *and a class* $\mathcal{K}$ *function* γ_1,

such that
$$\|\xi(t)\| \leq \beta(\|\xi_0\|, \sigma) + \gamma(\sigma),$$
$$\|\xi(t)\|_{H^1(0,1)} \leq \beta_1(\|\xi_0\|_{H^1(0,1)}, t) + \gamma_1(\sigma).$$

Moreover, there exist class $\mathcal{K}_\infty$ *functions* R_i, *for* $i = 0, 1$ *such that the solution* ξ *of the zero-dynamics* (4.133)–(4.136) *satisfies the following estimates:*
$$\|\xi(t)\| \leq R_0(\sigma), \quad t \geq T_0, \text{ and}$$
$$\|\xi(t)\|_{H^1(0,1)} \leq R_1(\sigma), \quad t \geq T_0,$$

for a sufficiently large time $T_0 = T_0(\|\xi_0\|, \sigma) > 0$. *In particular,*
$$\varlimsup_{t \to \infty} \|\xi(t)\| \leq R_0(\sigma) \text{ and } \varlimsup_{t \to \infty} \|\xi(t)\|_{H^1(0,1)} \leq R_1(\sigma) \tag{4.138}$$

hold.

Remark 4.18. Equation (4.138) asserts that, for any bounded set Σ of sufficiently small σ, the zero dynamics system is *ultimately bounded* in both L^2 and H^1.

Interconnecting the systems (4.129)–(4.131) and (4.133)–(4.136) coupled through the controls $u_j = \mathcal{B}_j \xi$ (as given in (4.137)) we obtain the closed loop system

$$\begin{aligned}
z_t(x, t) - c z_{xx}(x, t) + z(x, t) z_x(x, t) &= 0, \\
\mathcal{B}_0 z(t) = -c z_x(0, t) - k_0 z(0, t) &= \mathcal{B}_0 \xi(t), \\
\mathcal{B}_1 z(t) = c z_x(1, t) - k_1 z(1, t) &= \mathcal{B}_1 \xi(t), \\
z(x, 0) &= z_0(x),
\end{aligned} \tag{4.139}$$

$$\xi_t(x,t) - c\xi_{xx}(x,t) + \xi(x,t)\xi_x(x,t) = 0,$$
$$\xi(0,t) = y_{r,0}(t),$$
$$\xi(1,t) = y_{r,1}(t),$$
$$\xi(x,0) = \xi_0(x). \tag{4.140}$$

The main result found in [14] concerning the zero dynamics design for the Burgers' system is that it achieves *semiglobal* asymptotic regulation, i.e.,

Theorem 4.3. *There exists σ^* such that if conditions (4.124) on the signals $y_{r,j}(t)$, $j = 0, 1$ are satisfied with $\sigma < \sigma^*$ then, for any $R_0 > 0$, there exists $K(R_0) > 0$ such that if the initial solutions in (4.139) and (4.140) satisfy*

$$z_0, \ \xi_0 \in L^2(0,1), \quad \|z_0\|, \|\xi_0\| \le R_0, \tag{4.141}$$

and the gain parameters satisfy

$$k_0, k_1 \ge K(R_0), \tag{4.142}$$

then

1. *$z(t), \xi(t) \in H^1(0,1)$, for $t > 0$,*

2. *$z(t), \xi(t)$ are ultimately bounded in $H^1(0,1)$, i.e., there exists class $\mathcal{K}_\infty$ functions R, S of σ such that*

$$\varlimsup_{t\to\infty} \|z\|_{H^1(0,1)} \le S(\sigma), \quad \text{and} \quad \varlimsup_{t\to\infty} \|\xi\|_{H^1(0,1)} \le R(\sigma),$$

3. *$\|z(t) - \xi(t)\|_{H^1(0,1)} \xrightarrow{t\to\infty} 0$, and therefore*

4. *$e_j(t) \xrightarrow{t\to\infty} 0, \quad j = 0, 1$.*

In particular, the zero dynamics controller solves the problem of asymptotic regulation.

The proof of Theorem 4.3 consists of checking that the criteria (1–4) for the design methodology are satisfied. A straightforward calculation shows that the difference $\eta = z - \xi$, between the functions $z(x,t)$ and $\xi(x,t)$, satisfies the following initial boundary value problem:

$$0 = \eta_t - c\eta_{xx} + \eta\eta_x + (\eta\xi)_x, \tag{4.143}$$
$$0 = -c\eta_x(0,t) - k_0\eta(0,t), \tag{4.144}$$
$$0 = c\eta_x(1,t) - k_1\eta(1,t), \tag{4.145}$$
$$\eta(x,0) = \eta_0(x) = z_0(x) - \xi_0(x). \tag{4.146}$$

In other words the function $\widetilde{F}_\xi(\cdot)$ defined in (4.128) is given by $\widetilde{F}_\xi(\eta) = -\eta\eta_x - (\eta\xi)_x$. In [14] the authors prove the following result.

Theorem 4.4. *Let η be the solution of the problem* (4.143)–(4.146), *where the coefficient function ξ is the solution of the problem* (4.133)–(4.136). *Assume that conditions* (4.124), (4.141) *and* (4.142) *hold. Then*

$$\|\eta(t)\|_{H^1(0,1)} \xrightarrow{t\to\infty} 0. \tag{4.147}$$

All details of the proofs of the above results can be found in [14]. Here we only note that since (4.147) is basically assertion (3) of Theorem 4.3, the final assertion of the same theorem follows immediately. Indeed, we have

$$|e_j(t)| = |\eta(j,t)| \le \|\eta(t)\|_{C[0,1]} \le M\|\eta(t)\|_{H^1(0,1)}, \ j = 0, 1, \ M > 0.$$

In our specific numerical example we constructed the control laws $u_j(t)$ to track the reference signals

$$y_{r,0}(t) = \sin(t), \qquad y_{r,1}(t) = \sin(2t),$$

i.e., so that the errors $e_j(t) = y_j(t) - y_{r,j}(t)$ go to zero as t goes to infinity.

Here we have set the initial condition in (4.131) to be $z_0(x) = \cos(\pi x)$, and selected the kinematic viscosity as $c = 1$, and the stabilizing gains as $k_0 = 1$ and $k_1 = 2$. For the zero dynamics controller we have taken as initial value in (4.136) $\xi_0(x) = 0$, and we solve for $t \in [0, 10]$.

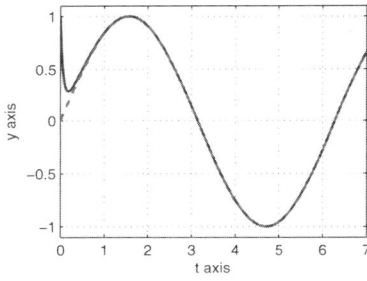

Fig. 4.29: $y_{r,0}(t)$ and $y_0(t)$.

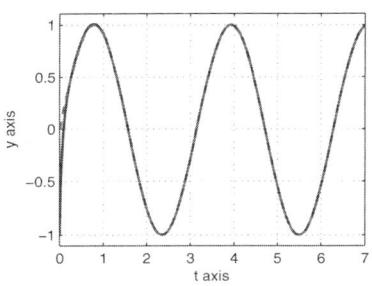

Fig. 4.30: $y_{r,1}(t)$ and $y_1(t)$.

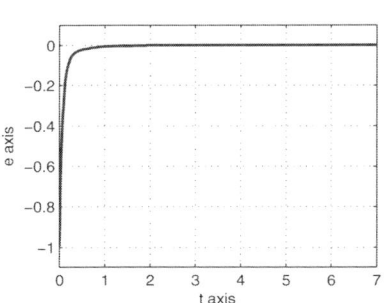

Fig. 4.31: $e_0(t)$.

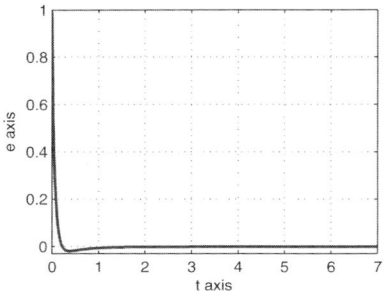

Fig. 4.32: $e_1(t)$.

In Figs. 4.29 and 4.30 we present the reference signals $y_{r,j}$ and the resulting measured outputs $y_j(t) = C_j(z(t))$. In Figs. 4.31 and 4.32 we present the corresponding errors $e_j(t) = y_{r,j}(t) - y_j(t)$. In Fig. 4.33 we depict the numerical solution surface $z(x,t)$ for $0 \leq x \leq 1$ and for $0 \leq t \leq 10$.

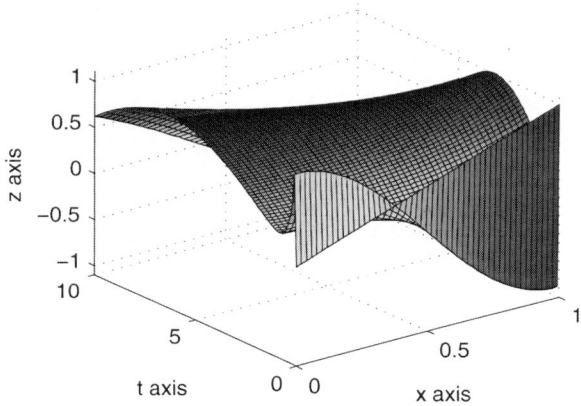

Fig. 4.33: Solution Surface $z(x,t)$.

It is clear from the above figures that the goal of harmonic tracking for the outputs at $x = 0$ and $x = 1$ is achieved.

Example 4.7 (Kuramoto-Sivashinsky equation). The results presented in this example are derived from the papers [13, 12] which are concerned with output regulation for the Kuramoto-Sivashinsky equation using co-located boundary control. The Kuramoto-Sivashinsky equation is given by the fourth-order nonlinear dissipative partial differential equation

$$z_t + u_{xxxx} + \nu z_{xx} + z z_x = 0, \quad 0 < x < 1, \quad t > 0.$$

For details concerning modeling with the Kuramoto-Sivashinsky equation see [13].

Our objective is to design a feedback controller to obtain a closed loop system in which all outputs are driven to track time-dependent reference signals while the state remains bounded. We consider the two special cases, set-point control and harmonic tracking.

The control system is modeled by a boundary controlled Kuramoto-Sivashinsky equation on the spatial interval $0 \leq x \leq 1$. The desired boundary controls $u_j(t)$ enter through boundary operators $\mathcal{B}_j$:

$$z_t(x,t) = -z_{xxxx}(x,t) - \nu z_{xx}(x,t) - z(x,t)z_x(x,t), \tag{4.148}$$

$$\mathcal{B}_0(z)(t) = z_{xxx}(0,t) - k_0 z(0,t) = u_0(t), \tag{4.149}$$

$$\mathcal{B}_1(z)(t) = -z_{xxx}(1,t) - k_1 z(1,t) = u_1(t), \tag{4.150}$$

$$z_x(0, t) = 0, \quad z_x(1, t) = 0,$$
$$z(x, 0) = \varphi(x), \tag{4.151}$$

where k_j are positive constants (called gains) that provide stability and $\nu > 0$ represents the anti-diffusion parameter which we assume satisfies $\nu \leq \frac{1}{2}$.

The co-located measured outputs are taken to be

$$y_j(t) = z(j, t), \quad j = 0, 1.$$

Our objective is to find controls u_0, u_1 so that for given continuous target functions $y_{r,0}(t)$, $y_{r,1}(t)$, the errors defined by

$$e_j(t) = z(j, t) - y_{r,j}(t), \quad j = 0, 1,$$

satisfy

$$e_j(t) \xrightarrow{t \to \infty} 0, \quad j = 0, 1.$$

We make the following assumption concerning the reference signals.

Assumption 4.9. *For the reference signals* $y_{r,j}(t)$, $j = 0, 1$

$$\|y_{r,j}\|_{H^{1,\infty}(0,\infty)} \leq M$$

where $M > 0$ is a constant.

For the system (4.148)–(4.150) we obtain the *Zero Dynamics* by replacing the equations for the boundary conditions with the constraints: $e_j(t) = 0$ ($j = 0, 1$). Thus we obtain the system

$$\xi_t(x, t) = -\xi_{xxxx}(x, t) - \nu \xi_{xx}(x, t) - \xi(x, t)\xi_x(x, t), \tag{4.152}$$
$$\xi(0, t) = y_{r,0}(t), \quad \xi(1, t) = y_{r,1}(t), \tag{4.153}$$
$$\xi_x(0, t) = 0, \quad \xi_x(1, t) = 0, \tag{4.154}$$
$$\xi(x, 0) = \psi(x), \tag{4.155}$$

where $\psi \in L^2(0, 1)$ is arbitrary.

The main goal of this example, just as in the previous example, is to show that the desired controls can be obtained as outputs of the zero dynamics

$$u_j(t) = \mathcal{B}_j(\xi)(t) = (-1)^j \xi_{xxx}(j, t) - k_j \xi(j, t), \quad j = 0, 1, \tag{4.156}$$

where ξ is the solution of the zero dynamics problem (4.152)–(4.155).

The composite closed loop system, consisting of the control system (4.148)–(4.151) and the zero dynamics (here considered as a controller) (4.152)–(4.155) coupled through the boundary controls defined in (4.156), is then given by

$$z_t(x, t) + z_{xxxx}(x, t) + \nu z_{xx}(x, t) + z(x, t)z_x(x, t) = 0, \tag{4.157}$$
$$z_{xxx}(0, t) - k_0 z(0, t) = \xi_{xxx}(0, t) - k_0 y_{r,0}(t),$$

$$- z_{xxx}(1,t) - k_1 z(1,t) = -\xi_{xxx}(1,t) - k_1 y_{r,1}(t),$$
$$z_x(0,t) = 0, \quad z_x(1,t) = 0,$$
$$z(x,0) = \varphi(x), \tag{4.158}$$

$$\xi_t(x,t) + \xi_{xxxx}(x,t) + \nu \xi_{xx}(x,t) + \xi(x,t)\xi_x(x,t) = 0, \tag{4.159}$$
$$\xi(0,t) = y_{r,0}(t), \quad \xi(1,t) = y_{r,1}(t),$$
$$\xi_x(0,t) = 0, \quad \xi_x(1,t) = 0,$$
$$\xi(x,0) = \psi(x). \tag{4.160}$$

The main result of [13, 12] is the following theorem.

Theorem 4.5. *Let M be sufficiently small. For any pair $R_0, R_2 > 0$, there exists $K(R_0, R_2) > 0$, such that the following statement holds for the closed loop system (4.157)–(4.160). Let $\nu \le 1/2$, the initial functions in (4.158) and (4.160) satisfy the conditions*

$$\varphi, \psi \in H^2(0,1),$$

$$(\varphi(0) - \psi(0))^2 + (\varphi(1) - \psi(1))^2 \le R_0^2, \quad \|\partial_{xx}(\varphi - \psi)\|^2 \le R_2^2,$$

and the gain parameters satisfy

$$k_0, k_1 \ge K(R_0, R_2),$$

then, we have

$$\|z(t, \cdot) - \xi(t, \cdot)\| \xrightarrow{t \to \infty} 0, \quad \|z_{xx}(t, \cdot) - \xi_{xx}(t, \cdot)\| \xrightarrow{t \to \infty} 0, \tag{4.161}$$

and therefore

$$e_j(t) \xrightarrow{t \to \infty} 0, \quad j = 0, 1. \tag{4.162}$$

Remark 4.19. Due to unavailability of a priori estimates on the H^1 norm of the solutions of systems (4.157)–(4.158) and (4.159)–(4.160), we do not know whether it is sufficient to assume that the initial functions are in H^1 to guarantee the exponential convergence to zero in (4.161)–(4.162).

Remark 4.20. As was mentioned in the previous example, for Burgers' equation it was proven in [26] that the trajectories of the closed loop system converge to the trajectories of the associated zero dynamics as $k_0, k_1 \to \infty$. The corresponding convergence in the context of the Kuramoto-Sivashinsky equation can also be proven using the same approach.

The proof of Theorem 4.5 in [13, 12] follows exactly the same procedure as the one outlined in the viscous Burgers' example. Namely one considers the difference $\eta = z - \xi$ between the functions $z(x,t)$ and $\xi(x,t)$ and shows that η goes to zero exponentially in $H^2(0,1)$ as $t \to \infty$. Then by Sobolev embedding the difference η (and its first derivative η_x) goes to zero as $t \to \infty$, not only at the end points but also at every point in between, i.e., in $C^1[0,1]$. Furthermore, at the end points

$$\eta(j,t) = z(j,t) - \xi(j,t) = z(j,t) - y_{r,j}(t) = e_j(t) \xrightarrow{t\to\infty} 0.$$

Note that, just as in the Burgers' example, this proof provides a much stronger result than the one stated in Theorem 4.5.

Appealing to (4.157)–(4.160) we see that the function η satisfies the following initial boundary value problem:

$$\eta_t(x,t) + \eta_{xxxx}(x,t) + \nu\eta_{xx}(x,t) + \eta(x,t)\eta_x(x,t) + (\eta\xi)_x(x,t) = 0,$$
$$\eta_{xxx}(0,t) - k_0\eta(0,t) = 0,$$
$$-\eta_{xxx}(1,t) - k_1\eta(1,t) = 0,$$
$$\eta_x(0,t) = 0, \quad \eta_x(1,t) = 0,$$
$$\eta(x,0) = \eta_0(x), = \varphi(x) - \psi(x).$$

4.7.1.1 Set-Point Tracking Example

In our first numerical example we consider a set-point tracking problem. In the case of set-point control we expect to obtain time-independent values for the controls u_j. This is indeed the case as shown in [13]. The reason why this works derives from an analysis of the *steady-state* behavior of the zero dynamics system (4.152)–(4.155). Indeed, the zero dynamics controller has a single global asymptotically stable equilibrium which we denote by $\xi_0(x)$, i.e., the solution of

$$\xi_0'''' + \nu\xi_0''(x) + \xi_0(x)\xi_0'(x) = 0, \tag{4.163}$$
$$\xi_0(0) = M_0, \quad \xi_0(1) = M_1,$$
$$\xi_0'(0) = 0, \quad \xi_0'(1) = 0.$$

Furthermore it was shown in [13] that the convergence of the time-dependent solution of the zero dynamics to the corresponding stationary solution takes place in the following sense.

Proposition 4.1. *For M_0, $M_1 \in \mathbb{R}$ sufficiently small, there exists a unique steady solution $\xi_0 \in H^2(0,1)$ to (4.163). Moreover, ξ_0 is exponentially stable, i.e., for every initial condition $\psi \in L^2(0,1)$, the zero dynamics solution $\xi(x,t)$ satisfies*

$$\|\xi(\cdot,t) - \xi_0(\cdot)\|_{H^2(0,1)} \xrightarrow{t\to\infty} 0.$$

In our example we have taken as an initial condition $\varphi(x) = x - x^2$ and set $y_{r,0} = M_0 = -1$, $y_{r,1} = M_1 = 1$, $\nu = 1/2$. We have also set $k_0 = k_1 = 1$. The static feedback from the stationary solutions (given below in (4.164)) are used for the controls. Thus we consider the problem

$$z_t(x,t) + z_{xxxx}(x,t) + \nu z_{xx}(x,t) + z(x,t)z_x(x,t) = 0,$$

$$z_{xxx}(0,t) - k_0 z(0,t) = u_0,$$
$$-z_{xxx}(1,t) - k_1 z(1,t) = u_1,$$
$$z(x,0) = \varphi(x),$$

and we solve for $t \in [0,10]$.

The controls u_j are obtained by solving the nonlinear BVP (4.163) and setting

$$u_0 = \xi_0'''(0) - k_0 y_{r,0}, \qquad u_1 = -\xi_0'''(1) - k_1 y_{r,1}. \qquad (4.164)$$

In Figs. 4.34 and 4.35 we present the reference signals $y_{r,j} = M_j$ and the resulting measured outputs $y_j(t) = C_j(z(t))$. In Figs. 4.36 and 4.37 we present the corresponding errors $e_j(t) = y_{r,j} - y_j(t)$. In Fig. 4.38 we depict the numerical solution surface $z(x,t)$ for $0 \le x \le 1$ and for $0 \le t \le 10$.

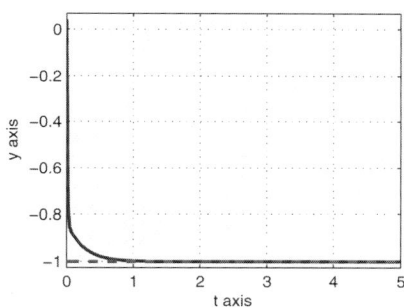

Fig. 4.34: M_0 and $y_0(t)$.

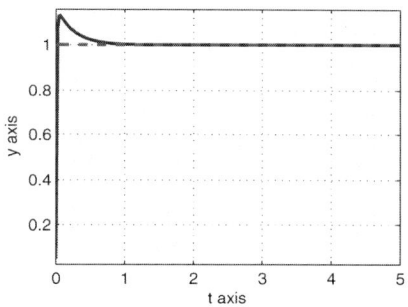

Fig. 4.35: M_1 and $y_1(t)$.

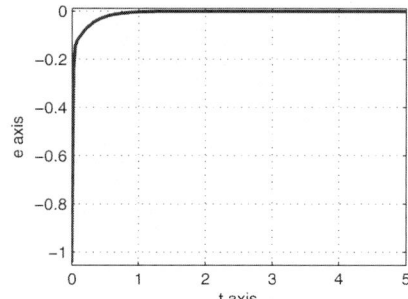

Fig. 4.36: $e_0(t)$.

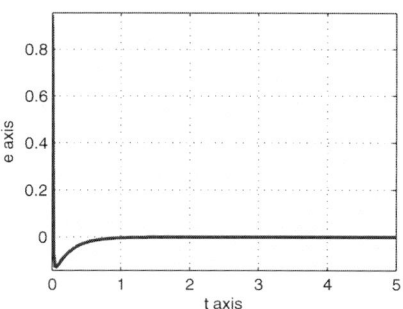

Fig. 4.37: $e_1(t)$.

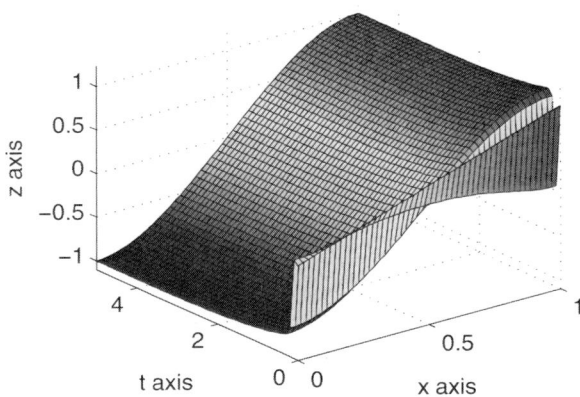

Fig. 4.38: Solution Surface $z(x, t)$.

It is clear from the above figures that the goal of set-point tracking for the outputs at $x = 0$ and $x = 1$ is achieved.

4.7.1.2   Harmonic Tracking Example

In our second numerical example we consider a problem of tracking harmonic signals $y_{r,j}(t) = M_j \sin(\alpha_j t)$ for $j = 0, 1$. In this example we have set $M_0 = 1$, $M_1 = 2$, $\alpha_0 = 2$ and $\alpha_1 = 1$. We have taken as an initial condition $\varphi(x) = \cos(\pi x)$ and set $k_0 = k_1 = 1$, $\nu = 1/2$. Thus we solve the problem

$$z_t(x, t) + z_{xxxx}(x, t) + .5z_{xx}(x, t) + z(x, t)z_x(x, t) = 0,$$
$$z_{xxx}(0, t) - z(0, t) = u_0(t), \quad z_{xxx}(1, t) + z(1, t) = u_1(t),$$
$$z_x(0, t) = 0, \quad z_x(1, t) = 0,$$
$$z(x, 0) = \cos(\pi x),$$

for $t \in [0, 10]$, where the controls u_j are obtained from the zero dynamics

$$\xi_t(x, t) = -\xi_{xxxx}(x, t) - .5\xi_{xx}(x, t) - \xi(x, t)\xi_x(x, t),$$
$$\xi(0, t) = M_0 \sin(\alpha_0 t), \quad \xi(1, t) = M_1 \sin(\alpha_1 t),$$
$$\xi_x(0, t) = 0, \quad \xi_x(1, t) = 0,$$
$$\xi(x, 0) = 0,$$

as

$$u_0(t) = \xi_{xxx}(0, t) - y_{r,0}(t), \quad u_1(t) = -\xi_{xxx}(1, t) - y_{r,1}(t).$$

In Figs. 4.39 and 4.40 we present the reference signals $y_{r,j}(t)$ and the resulting measured outputs $y_j(t) = C_j(z(t))$. In Figs. 4.41 and 4.42 we present the corresponding errors $e_j(t) = y_{r,j} - y_j(t)$. In Fig. 4.43 we depict the numerical solution surface $z(x, t)$ for $0 \le x \le 1$ and for $0 \le t \le 10$.

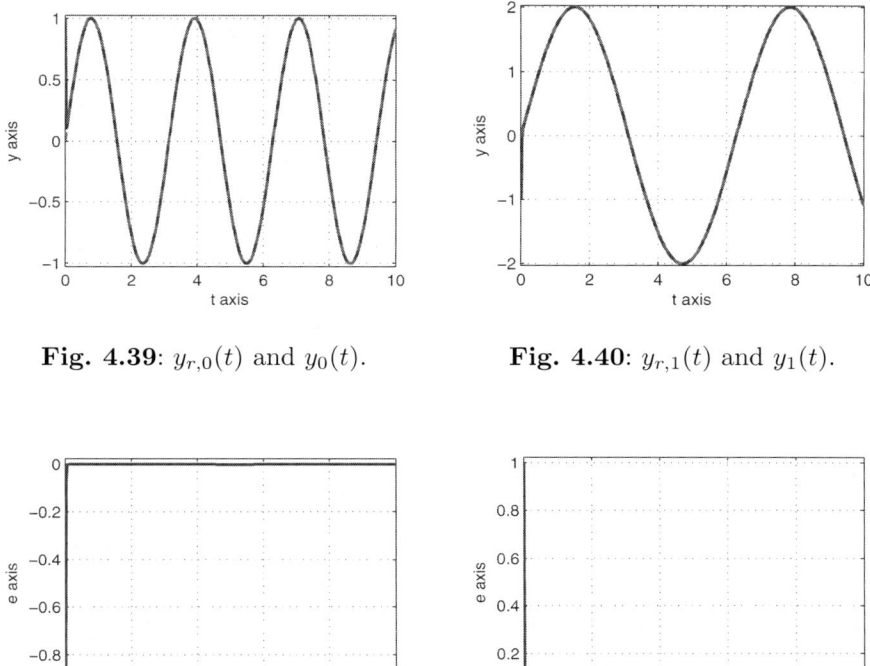

Fig. 4.39: $y_{r,0}(t)$ and $y_0(t)$.

Fig. 4.40: $y_{r,1}(t)$ and $y_1(t)$.

Fig. 4.41: $e_0(t)$.

Fig. 4.42: $e_1(t)$.

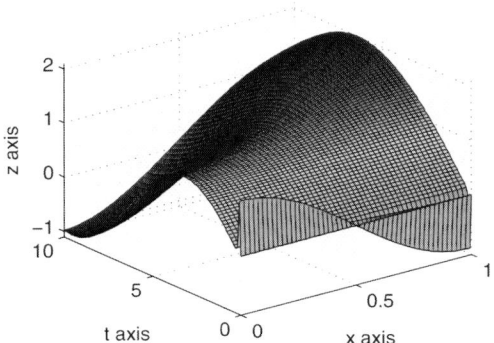

Fig. 4.43: Solution Surface $z(x,t)$.

Once again, it is clear from the above figures that the goal of harmonic tracking for the outputs at $x = 0$ and $x = 1$ is achieved.

Chapter 5

Nonlinear Examples

5.1   Introduction

In order to better familiarize the reader with the different algorithms described in Chapter 4 we present several more specialized examples. These examples are somewhat more complex than the relatively simple motivational examples given in Chapter 4. Furthermore, each of these examples introduces new features and we provide a detailed description of how to implement them. All these examples have been implemented, and solved, using the finite element package COMSOL Multiphysics [65]. This choice is again motivated by its flexibility for solving coupled nonlinear multiphysics problems. In the example in Section 5.2 we apply the methodology introduced in Section 4.3 to design control laws for set-point tracking and disturbance rejection of an incompressible "Navier-Stokes Flow in a 2D Forked Channel". This is the nonlinear version of the example previously presented in Section 3.5 for the linearized Stokes operator. Section 5.3 deals with the tracking and disturbance rejection of a "Non-Isothermal Navier-Stokes Flow in a 2D Box". This problem is related to the example in Section 3.6 in Chapter 3 except that now the Boussinesq approximation of the buoyancy force is taken into account in the momentum equation, resulting in a fully nonlinear coupled system of equations for velocity, pressure and temperature. In Section 5.3.1 we solve "The Set-Point Case" as described in Section 4.3. In Section 5.3.2 we solve "The Piecewise Constant Signal Case" as described in Section 4.4. Finally, in Section 5.3.3 we solve "The Harmonic Case" using the β-iteration as described in Section 4.5. In the example presented in Section 5.4 we describe strategies for the design of control laws for a nonlinear "2D Chafee-Infante with Time-Dependent Regulation". An approximation of the control laws are found using the β-iteration algorithm as described in Section 4.5. In the example in Section 5.5 we describe the design of control laws for the "Regulation of 2D Burgers Using Fourier Series". An approximation of the control laws is found using the Fourier Series algorithm as described in Section 4.6. The example in Section 5.6 deals with the harmonic tracking of a nonlinear "Back-Step Navier-Stokes Flow". In order to find approximate control laws we adopt a mixed method that takes into account the features of both the Fourier Series and β-iteration algorithms. In Section 5.7 we introduce several examples of "Nonlinear Regulation Using Zero Dynamics Design" as described in Section 4.7. In these examples the input and output operators are co-located and are allowed to be infinite-dimensional in both time and space. Three different examples demonstrate the features of the method. Section 5.7.1 deals with the control of a "Vibrating Nonlinear Beam", Section 5.7.2 deals with a "2D Chafee-Infante Equation MIMO" control problem, while Section 5.7.3 deals with a "2D Burgers Equation MIMO" control problem.

5.2 Navier-Stokes Flow in a 2D Forked Channel

In this section, we consider the modeling and control of a two-dimensional incompressible Navier-Stokes flow in a region Ω described in Fig. 5.1.

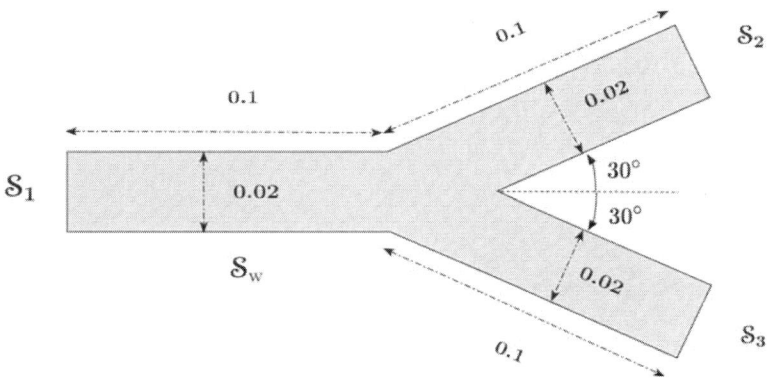

Fig. 5.1: Forked Channel Domain.

The domain Ω consists of a main channel of length 0.1 and altitude 0.02 which forks at the right extreme dividing into two equal branches of length 0.1 and altitude 0.02. The branches are inclined at $\pm30°$ with respect to the main channel. The channel is open on the left at $\mathcal{S}_1$ and on the right at $\mathcal{S}_2$ and $\mathcal{S}_3$. Walls are considered on the remaining part of the boundary, denoted by $\mathcal{S}_w$.

In the whole domain the incompressible Navier-Stokes system of equations is considered for some initial data $\boldsymbol{\varphi}$. There is a constant disturbance d entering through a normal stress on the boundary $\mathcal{S}_1$ and a boundary control u entering again through a normal stress on $\mathcal{S}_2$. Zero stress and zero velocity boundary conditions are considered on $\mathcal{S}_3$ and $\mathcal{S}_w$, respectively. We have a measured output on $\mathcal{S}_3$ given by the average velocity in the outward normal direction.

In mathematical terms, the system can be written

$$\frac{\partial \boldsymbol{v}}{\partial t} + (\boldsymbol{v} \cdot \nabla)\boldsymbol{v} = \nabla \cdot (\nu\,[(\nabla\boldsymbol{v}) + (\nabla\boldsymbol{v})^{\mathsf{T}}]) - \nabla p, \qquad (5.1)$$

$$\nabla \cdot \boldsymbol{v} = 0, \quad \boldsymbol{v}(x,0) = \boldsymbol{\varphi}(x), \qquad (5.2)$$

$$\boldsymbol{\tau}(z)\big|_{\mathcal{S}_1} = d\,\boldsymbol{n}_{\mathcal{S}_1}, \quad \boldsymbol{\tau}(z)\big|_{\mathcal{S}_2} = u\,\boldsymbol{n}_{\mathcal{S}_2}, \qquad (5.3)$$

$$\boldsymbol{\tau}(z)\big|_{\mathcal{S}_3} = \boldsymbol{0}, \qquad (5.4)$$

$$\boldsymbol{v}\big|_{\mathcal{S}_w} = \boldsymbol{0}, \qquad (5.5)$$

$$y = C(z) = \frac{1}{|\mathcal{S}_3|} \int_{\mathcal{S}_3} \boldsymbol{v} \cdot \boldsymbol{n}_{\mathcal{S}_3}\,ds.$$

Here $z = (\boldsymbol{v}, p)$ is the state variable, where $\boldsymbol{v} = [v_1,\, v_2]^\mathsf{T}$ denotes the velocity vector field and p the pressure. We also use the notations of ν for the kinematic viscosity and $\boldsymbol{\tau}$ for the surface stress on $\mathcal{S}$ given by

$$\boldsymbol{\tau}(z) = (pI - \nu\,[(\nabla\boldsymbol{v}) + (\nabla\boldsymbol{v})^\mathsf{T}]) \cdot \boldsymbol{n}_\mathcal{S},$$

with $\boldsymbol{n}_\mathcal{S} = [n_x,\, n_y]^\mathsf{T}$ the outward normal on the boundary $\mathcal{S}$.

In this example, our objective is to find a control input u so that the measured output $y(t)$ tracks a given reference signal y_r while rejecting the disturbance d.

The controller u is found by following the algorithm outlined in Section 4.3.1. First we solve the linear steady-state Stokes problem for $X = [\boldsymbol{V},\, P]^\mathsf{T}$, with homogeneous boundary condition everywhere except on $\mathcal{S}_2$, where a unit normal stress is considered

$$\boldsymbol{0} = \nabla \cdot (\nu\,[(\nabla\boldsymbol{V}) + (\nabla\boldsymbol{V})^\mathsf{T}]) - \nabla P,$$
$$\nabla \cdot \boldsymbol{V} = 0,$$
$$\boldsymbol{\tau}(X)\big|_{\mathcal{S}_1} = \boldsymbol{0}, \quad \boldsymbol{\tau}(X)\big|_{\mathcal{S}_3} = \boldsymbol{0}, \quad \boldsymbol{V}\big|_{\mathcal{S}_w} = \boldsymbol{0},$$
$$\boldsymbol{\tau}(X)\big|_{\mathcal{S}_2} = \boldsymbol{n}_{\mathcal{S}_2}.$$

We note that X is the response of the homogeneous linear Stokes operator to the unit normal stress input on $\mathcal{S}_2$ and produces the entry of the 1×1 matrix (i.e., a scalar in this case)

$$G = \big(C(X)\big).$$

With this we solve the coupled nonlinear steady-state system with unknowns $\overline{z} = [\overline{\boldsymbol{v}}, \overline{p}]$, $\widetilde{z} = [\widetilde{\boldsymbol{v}}, \widetilde{p}]$ and γ

$$\boldsymbol{0} = \nabla \cdot (\nu\,[(\nabla\overline{\boldsymbol{v}}) + (\nabla\overline{\boldsymbol{v}})^\mathsf{T}]) - (\overline{\boldsymbol{v}} \cdot \nabla)\overline{\boldsymbol{v}} - \nabla\overline{p},$$
$$\nabla \cdot \overline{\boldsymbol{v}} = 0,$$
$$\boldsymbol{\tau}(\overline{z})\big|_{\mathcal{S}_1} = d\,\boldsymbol{n}_{\mathcal{S}_1}, \quad \overline{\boldsymbol{v}}\big|_{\mathcal{S}_w} = \boldsymbol{0},$$
$$\boldsymbol{\tau}(\overline{z})\big|_{\mathcal{S}_2} = \gamma\,\boldsymbol{n}_{\mathcal{S}_2}, \quad \boldsymbol{\tau}(\overline{z})\big|_{\mathcal{S}_3} = \boldsymbol{0},$$
$$\boldsymbol{0} = \nabla \cdot (\nu\,[(\nabla\widetilde{\boldsymbol{v}}) + (\nabla\widetilde{\boldsymbol{v}})^\mathsf{T}]) - (\overline{\boldsymbol{v}} \cdot \nabla)\overline{\boldsymbol{v}} - \nabla\widetilde{p},$$
$$\nabla \cdot \widetilde{\boldsymbol{v}} = 0,$$
$$\boldsymbol{\tau}(\widetilde{z})\big|_{\mathcal{S}_1} = d\,\boldsymbol{n}_{\mathcal{S}_1}, \quad \widetilde{\boldsymbol{v}}\big|_{\mathcal{S}_w} = \boldsymbol{0},$$
$$\boldsymbol{\tau}(\widetilde{z})\big|_{\mathcal{S}_2} = \boldsymbol{0}, \quad \boldsymbol{\tau}(\widetilde{z})\big|_{\mathcal{S}_3} = \boldsymbol{0},$$
$$\gamma = G^{-1}(y_r - C(\widetilde{z})) = \frac{1}{C(X)}(y_r - C(\widetilde{z})).$$

Finally we set $u = \gamma$ and solve the IBVP (5.1)–(5.5), i.e., the closed loop system. Under Assumptions 4.3 and 4.4 we expect $z \to \overline{z}$, for $t \to \infty$, in such a way that the desired tracking $y(t) = C(z) \to C(\overline{z}) = y_r$ takes place.

For our specific numerical example we have chosen the following parameters: $\nu = 0.0001$, $d = 0.025$ and $y_r = 0.05$. Accordingly, the maximum

Reynolds number occurs in the lower branch and is

$$Re = \frac{y_r\, D}{\nu} = 400.$$

The initial data in our simulation is $\varphi(x) = \mathbf{0}$ and the transient solution $z(x,t)$ is evaluated between $t = 0$ and $t = 20$. The numerical solution of the weak formulation of the above three systems is obtained using the predefined Navier-Stokes incompressible model of the COMSOL Multiphysics package. Lagrange elements $\boldsymbol{P}2$-$P1$ are used for velocity and pressure in order to satisfy the LBB condition. No stabilization for the advection term is used.

In Fig. 5.2 a screenshot of the vector field $\boldsymbol{v}$ and its magnitude $\|\boldsymbol{v}\|$ is given at time $T = 20$.

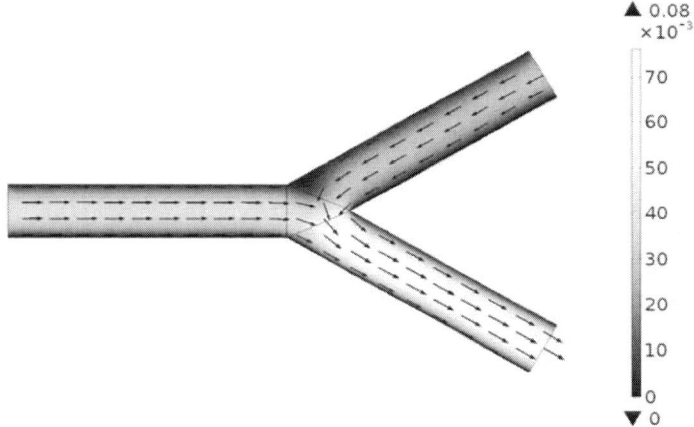

Fig. 5.2: Vector Field $\boldsymbol{v}$ and Its Magnitude $\|\boldsymbol{v}\|$ at $T = 5$.

The time evolutions of $y(t) = Cz$ and $y_r(t)$ are given in Fig. 5.3 for $0 \leq t \leq 20$. The asymptotic convergence of $e(t) = y_r(t) - y(t)$ to 0 is evident in Fig. 5.4.

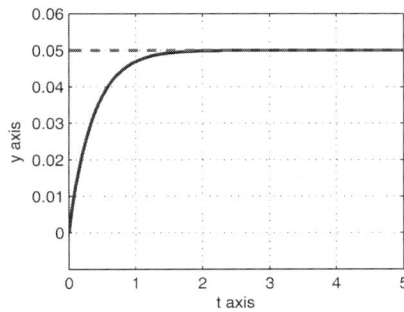

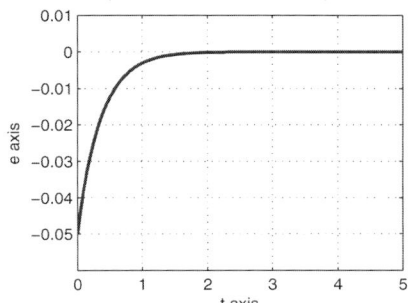

Fig. 5.3: $y(t)$, $y_r(t)$ for $0 \leq t \leq 5$.   **Fig. 5.4**: $e(t)$ for $0 \leq t \leq 5$.

5.3   Non-Isothermal Navier-Stokes Flow in a 2D Box

5.3.1   The Set-Point Case

In this section, we consider the modeling and control of a two-dimensional non-isothermal Navier-Stokes flow in the region Ω described in Fig. 5.5. The motivation for considering this example comes from discussions with researchers at Virginia Tech University actively engaged in research directed in part at control problems in the design of energy-efficient buildings.

The domain Ω consists of a main square box with side length 1. Inlet and outlet square regions, of side length 0.1, are located on the upper-right and lower-left sides of the main box, respectively. The boundary of the region Ω consists of an inflow boundary S_1, an outflow boundary S_2 and a hot wall S_3. Insulated walls are considered on the rest of the boundary S_w. An interior target region Ω_T is also considered.

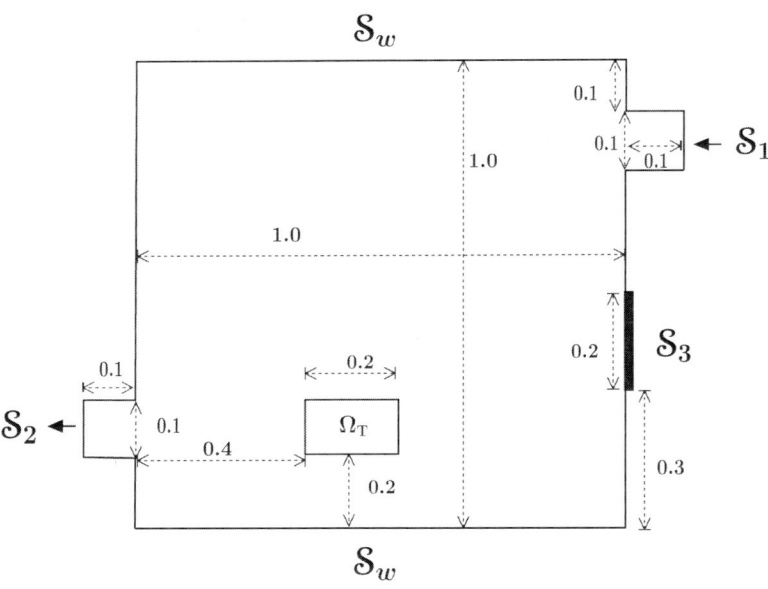

Fig. 5.5: Two-Dimensional Box Domain.

The physical model governing this problem consists of the non-isothermal incompressible Navier-Stokes equations in the entire domain. In the momentum equation we use the Boussinesq approximation to describe the buoyancy force. There is a boundary control u entering through a distributed heat flux on S_1, and a disturbance d entering on the wall S_3 through a constant given temperature. A parabolic inflow profile is considered for the fluid velocity on

S_1. Zero stress and zero flux boundary conditions are considered on S_2, while zero velocity and zero flux are considered on S_w. We have a measured output on Ω_T given by the average temperature.

Formulating the above within a mathematical framework we have,

$$\frac{\partial \boldsymbol{v}}{\partial t} + (\boldsymbol{v} \cdot \nabla)\boldsymbol{v} = \nabla \cdot (\nu\,[(\nabla\boldsymbol{v}) + (\nabla\boldsymbol{v})^\mathsf{T}]) - \nabla p + \langle 1, \beta \rangle T, \qquad (5.6)$$

$$\nabla \cdot \boldsymbol{v} = 0,$$

$$\frac{\partial T}{\partial t} + (\boldsymbol{v} \cdot \nabla)T = \alpha \Delta T,$$

with initial data

$$\boldsymbol{v}(x,0) = \boldsymbol{\varphi}(x), \quad T(x,0) = \psi(x),$$

boundary conditions

$$\boldsymbol{v} = \begin{bmatrix} f(s) \\ 0 \end{bmatrix} \quad \text{and} \quad \boldsymbol{q} = u\,\boldsymbol{n}_{S_1} \quad \text{on } S_1, \qquad (5.7)$$

$$\boldsymbol{\tau} = \boldsymbol{0} \quad \text{and} \quad \boldsymbol{q} = \boldsymbol{0} \quad \text{on } S_2, \qquad (5.8)$$

$$\boldsymbol{v} = \boldsymbol{0} \quad \text{and} \quad T = d \quad \text{on } S_3, \qquad (5.9)$$

$$\boldsymbol{v} = \boldsymbol{0} \quad \text{and} \quad \boldsymbol{q} = \boldsymbol{0} \quad \text{on } S_w \qquad (5.10)$$

and measured output

$$y = C(z) = \frac{1}{|\Omega_T|} \int_{\Omega_T} T\, dx. \qquad (5.11)$$

Here $z = (\boldsymbol{v}, p, T)$ is the state variable, where $\boldsymbol{v} = [v_1, v_2]^\mathsf{T}$ denotes the velocity vector field, p the pressure and T the temperature. We also use the notations of ν for the kinematic viscosity, α for the thermal diffusivity, β for the buoyancy coefficient and

$$f(s) = 4s(1-s)$$

for the parabolic inflow with maximum amplitude 1, where s is the arclength normalized between 0 and 1. The surface stress $\boldsymbol{\tau}$ and the heat flux $\boldsymbol{q}$ on S are given, respectively, by

$$\boldsymbol{\tau}(\boldsymbol{v}, p) = (pI - \nu\,[(\nabla\boldsymbol{v}) + (\nabla\boldsymbol{v})^\mathsf{T}]) \cdot \boldsymbol{n}_S,$$

$$\boldsymbol{q}(T) = -\alpha \nabla T \cdot \boldsymbol{n}_S,$$

with $\boldsymbol{n}_S = [n_x, n_y]^\mathsf{T}$ the outward normal on the boundary S.

Again, in this example, our objective is to find a control input u so that the measured output $y(t)$ tracks a given (constant) reference signal y_r while rejecting the disturbance d. The controller u is found by proceeding as described in the Section 4.3.1. We notice that the momentum equation is coupled to the temperature only through the buoyancy term. We first determine the steady-state velocity in absence of buoyancy force by solving the isothermal Navier-Stokes equation

$$(\boldsymbol{V} \cdot \nabla)\boldsymbol{V} = \nabla \cdot (\nu [(\nabla \boldsymbol{V}) + (\nabla \boldsymbol{V})^{\mathsf{T}}]) - \nabla P,$$

$$\nabla \cdot \boldsymbol{V} = 0,$$

$$\boldsymbol{V} = \begin{bmatrix} f(s) \\ 0 \end{bmatrix} \quad \text{on } \mathcal{S}_1,$$

$$\boldsymbol{\tau} = \boldsymbol{0} \quad \text{on } \mathcal{S}_2,$$

$$\boldsymbol{V} = \boldsymbol{0} \quad \text{on } \mathcal{S}_3 \cup \mathcal{S}_w$$

for the variables $\boldsymbol{V}$ and P. Then we solve the linear energy equation for X, with homogeneous boundary condition everywhere except on $\mathcal{S}_1$, where a unit heat flux is considered

$$0 = \alpha \Delta X - (\boldsymbol{V} \cdot \nabla)X, \qquad (5.12)$$

$$\boldsymbol{q} = 1\,\boldsymbol{n}_{\mathcal{S}_1} \quad \text{on } \mathcal{S}_1,$$

$$\boldsymbol{q} = \boldsymbol{0} \quad \text{on } \mathcal{S}_2 \cup \mathcal{S}_w,$$

$$X = 0 \quad \text{on } \mathcal{S}_3.$$

Here X is the response of the homogeneous energy equation to the unit input on $\mathcal{S}_1$, and produces the entry of the 1×1 matrix (i.e., a scalar in this case)

$$G = \big(C(X)\big).$$

Remark 5.1. Notice that the system Equation (5.12) is linear, since the advection velocity $\boldsymbol{V}$ is given. In the energy equation we then consider the linear homogeneous operator A applied to the state variable X to be given by $\alpha \Delta X - (\boldsymbol{V} \cdot \nabla)X$ rather than $\alpha \Delta X$, only. Again we assume $\boldsymbol{V}$ to be so that A is always invertible. Here, of course, the definition of the linear operator A is also supplemented by the homogeneous boundary conditions on the boundary of the domain Ω. The above choice of A is motivated by the fact that in the current example we expect the flow to be mostly driven by the inflow boundary condition rather than the buoyancy force. Under this assumption we included in A what we expect to be a good approximation of the advection term. Thus, considering both diffusion and convection in the linear operator gives a more realistic response than considering diffusion only.

With this we solve now the fully coupled nonlinear steady-state system

$$(\overline{\boldsymbol{v}} \cdot \nabla)\overline{\boldsymbol{v}} = \nabla \cdot (\nu [(\nabla \overline{\boldsymbol{v}}) + (\nabla \overline{\boldsymbol{v}})^{\mathsf{T}}]) - \nabla \overline{p} + \langle 1, \beta \rangle \overline{T},$$

$$\nabla \cdot \overline{\boldsymbol{v}} = 0,$$

$$(\overline{\boldsymbol{v}} \cdot \nabla)\overline{T} = \alpha \Delta \overline{T},$$

$$((\overline{\boldsymbol{v}} - \boldsymbol{V}) \cdot \nabla)\overline{T} = \alpha \Delta \widetilde{T} - (\boldsymbol{V} \cdot \nabla)\widetilde{T},$$

with boundary conditions

$$\overline{v} = \begin{bmatrix} f(s) \\ 0 \end{bmatrix} , \; q(\overline{T}) = \gamma\, n_{S_1} \text{ and } q(\widetilde{T}) = 0 \text{ on } S_1,$$

$$\tau(\overline{v}, \overline{p}) = 0 , \; q(\overline{T}) = 0 \text{ and } q(\widetilde{T}) = 0 \text{ on } S_2,$$

$$\overline{v} = 0 , \; \overline{T} = d \text{ and } \widetilde{T} = d \text{ on } S_3,$$

$$\overline{v} = 0 , \; q(\overline{T}) = 0 \text{ and } q(\widetilde{T}) = 0 \text{ on } S_w,$$

and control

$$\gamma = G^{-1}(y_r - C(\widetilde{T})) = \frac{1}{C(X)}(y_r - C(\widetilde{T})).$$

Finally we set $u = \gamma$ and solve the IBVP (5.6)–(5.10). Under Assumptions 4.5 and 4.6 we expect $T \to \overline{T}$, for $t \to \infty$, so that, in addition, the desired tracking $y(t) = C(T) \to C(\overline{T}) = y_r$ takes place.

For our specific numerical example we have chosen the following parameters: $\nu = 0.002$, $\alpha = 0.01$, $\beta = 1.$, $d = 1.$ and $y_r = 0.5$. Accordingly, the maximum Reynolds number occurs in the inlet region and is approximately $Re \simeq 250$. The Prandtl and Grashof numbers are

$$Pr = \frac{\nu}{\alpha} = 0.2, \qquad Gr = \frac{\beta y_r D^3}{\nu^2} = 1.25 \times 10^5,$$

respectively. The initial data in our simulation are $\varphi(x) = 0$ and $\phi(x) = 0$. The transient solution $z(x,t)$ is evaluated between $t = 0$ and $t = 2000$. The numerical solution of the weak formulation of the above systems is obtained again with the COMSOL Multiphysics package. The momentum and continuity equations are solved using the predefined Navier-Stokes incompressible model. The energy equation is solved by using the PDE coefficient model. In the Navier-Stokes model Lagrange elements $P2$-$P1$ are used for velocity and pressure in order to satisfy the LBB condition. In the PDE coefficient model Lagrange elements $P2$ are used for the Temperature. No stabilization is used for the advection term in any physics. In Fig. 5.6 a screenshot of the velocity vector field v and the temperature profile T are given. In Fig. 5.7 we show the velocity magnitude $\|v\|$ and streamlines at the final time $t = 2000$. The time evolution of CT is given in Fig. 5.8 for all time (solid line). The asymptotic convergence of $CT(t)$ to $y_r = 0.5$ is evident.

Remark 5.2. In this example we have used a flux boundary control on S_1. A Dirichlet boundary control can also be found to track the desired output y_r. For example, a constant temperature value u on the inlet boundary

$$T = u \text{ on } S_1,$$

can be obtained by proceeding as described in Section 4.3.1.

In particular the trace of $\overline{T}$ on the boundary S_1 can be used in the closed loop system as a Dirichlet controller

$$T = \overline{T} \text{ on } S_1. \tag{5.13}$$

The time evolution of $CT(t)$ in this case is given by the dashed line in Fig. 5.8. The two transients are different, but the asymptotic values are clearly the same.

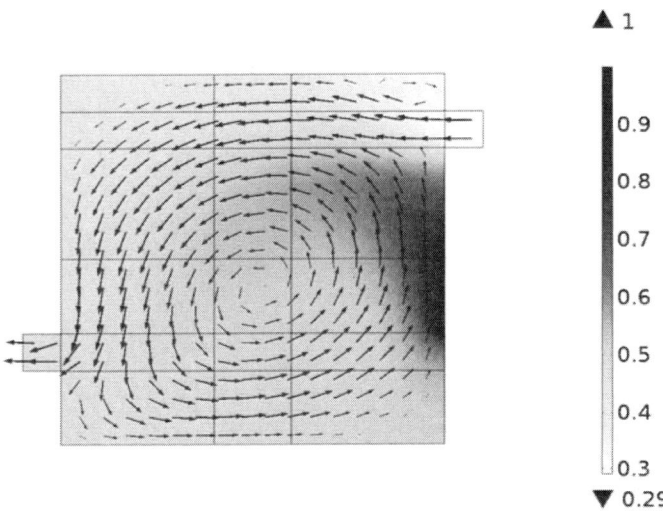

Fig. 5.6: Velocity Vector Field v and Temperature Profile $T = 2000$.

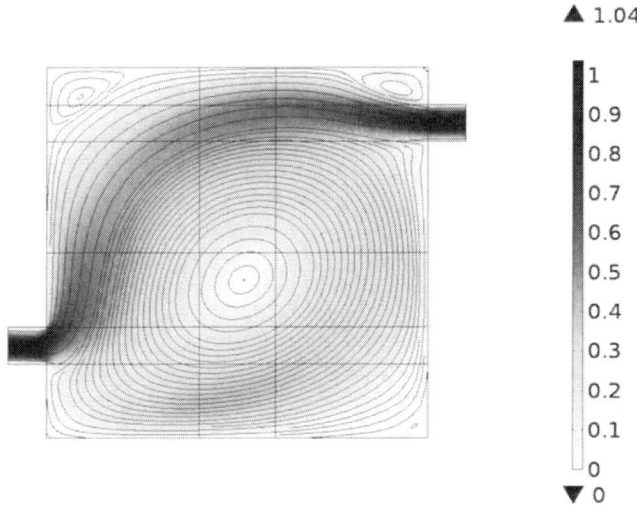

Fig. 5.7: Velocity Magnitude Profile $\|v\|$ and Velocity Streamlines at $t = 2000$.

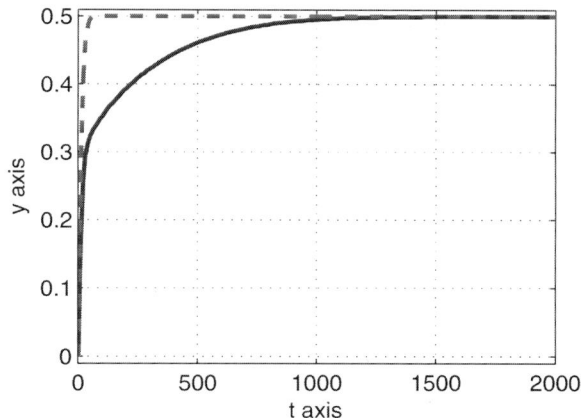

Fig. 5.8: Time Evolution of $C(T)$ for Flux Control (solid line) and Dirichlet Control (dashed line) for All Time.

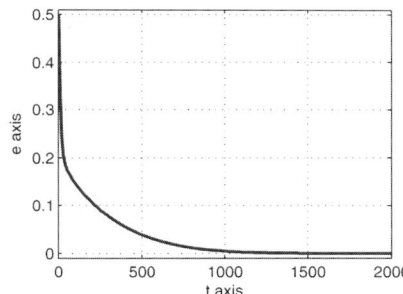

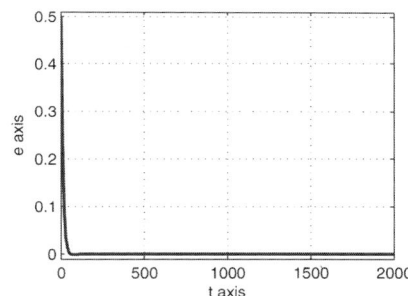

Fig. 5.9: $e(t)$ Flux Control.   **Fig. 5.10**: $e(t)$ Dirichlet Control.

5.3.2   The Piecewise Constant Case

In this Section we consider exactly the same thermal regulation problem discussed in the previous section (5.6)–(5.11), with the exception that now we seek a time-dependent control input $u(t)$ in Equation (5.7) so that the measured output $y(t)$ in (5.11) tracks a given periodic time-dependent piecewise constant reference signal $y_r(t)$ while rejecting a similar time-dependent piecewise constant disturbance $d(t)$. In particular, we consider

$$y_r(t) = \begin{cases} M_1, & 0 \leq t < t_{r1}, \\ M_2, & t_{r1} \leq t < t_r, \end{cases}$$

and $y_r(t + t_r) = y_r(t)$ for all $t > 0$. Similarly we consider a disturbance $d(t)$ in the form

$$d(t) = \begin{cases} d_1, & 0 \le t < t_{d1}, \\ d_2, & t_{d1} \le t < t_d, \end{cases}$$

and $d(t + t_d) = d(t)$ for all $t > 0$.

For simplicity, just as in [1], we will assume here that the step times for the reference signal and disturbance coincide. While this is not necessary it will shorten the discussion considerably. So we assume that $t_0 = 0$, $t_1 \equiv t_{r1} = t_{d1}$ and $t_2 \equiv t_r = t_d$ so the functions have the same jump point and period.

Our objective is to find a time-dependent control law $u(t) = \gamma(t)$, in order to track and reject piecewise constant time-dependent signals $y_r(t)$ and $d(t)$, respectively.

We proceed according to Section 4.4 to choose an initial condition for the new dynamic equation (see (4.56)–(4.58)). This is a particularly important point computationally. Initially we would expect to be able to choose any initial condition sufficiently close to the initial data z_0 in (1.2), but as it turns out it is computationally much more advantageous to begin by solving the following set-point control problem exactly as described in Section 4.4.

We briefly recall the general algorithm given in Section 4.4. First we solve the set-point control problem

$$\overline{Z}_t(x, t) = A\overline{Z}(x, t) + F(\overline{Z}(x, t))$$
$$+ B_d \, d_0 + B_{in} \, \gamma_0,$$
$$\overline{Z}(x, 0) = z_0(x),$$
$$y(t) = (C\overline{Z})(t),$$

where the disturbances and the signals to be tracked are

$$d_0 = d_1 \, , \quad y_{r_0} = M_1,$$

respectively. Thus they are time independent. Notice that to find $\overline{Z}(x)$, we follow the strategy described in Section 4.3.2. First we solve

$$0 = AX(x) + B_{in}.$$

This equation produces

$$X = -A^{-1}B_{in}. \tag{5.14}$$

From X we define

$$G = CX, \tag{5.15}$$

which in linear systems terminology is the transfer function $G(s) = C(sI - A)^{-1}B_{in}$ (for the linearized control system) evaluated at $s = 0$. In this way we obtain G and the constant set-point control gain $u_0 = \gamma_0$.

Next, we seek a control $u(t) = \gamma(t)$ and solution (state variable) $\overline{z}(x,t)$ satisfying

$$\overline{z}_t(x,t) = A\overline{z}(x,t) + F(\overline{z}(x,t))$$
$$+ B_{\mathrm{d}}(x)\, d(t) + B_{\mathrm{in}}(x)\, \gamma(t), \tag{5.16}$$
$$\overline{z}(x,0) = \overline{Z}(x),$$
$$y_r(t) = (C\overline{z})(t) \tag{5.17}$$

for all time. To find $\gamma(t)$ we proceed similar to Section 4.4 and rewrite Equation (5.16) as

$$\overline{z}(x,t) = -A^{-1}\big[-\overline{z}_t(x,t) + F(\overline{z}(x,t)) + B_{\mathrm{d}}\, d(t)\big]$$
$$+ \big(-A^{-1}B_{\mathrm{in}}(x)\, \gamma(t)\big). \tag{5.18}$$

Let $\widetilde{z}(x,t)$ be the auxiliary state variable obtained as the solution of

$$\overline{z}_t(x,t) = A\widetilde{z}(x,t) + F(\overline{z}(x,t)) + B_{\mathrm{d}}(x)d(t).$$

Here, $\widetilde{z}$ is the response of the linear operator A to the sum of the nonlinear term $F(\overline{z}(x))$, the inertial term $-\overline{z}_t(x,t)$ and the disturbance B_{d}, namely

$$\widetilde{z}(x,t) = -A^{-1}\big[F(\overline{z}(x,t)) - \overline{z}_t(x,t) + B_{\mathrm{d}}(x)d(t)\big]. \tag{5.19}$$

Notice that although $\widetilde{z}$ is time-dependent the solution of (5.19) does not require any initial data. Substituting Equations (5.14) and (5.19) in Equation (5.18) yields

$$\overline{z}(x,t) = \widetilde{z}(x,t) + X\gamma(t).$$

Applying the output operator C to each side of the above equation, and substituting Equations (5.17) and (5.15) it follows that

$$y_r(t) = (C\overline{z})(t) = (C\widetilde{z})(t) + G\gamma(t).$$

The above equation can be solved for the desired time-dependent control

$$\gamma(t) = G^{-1}\big[y_r(t) - (C\widetilde{z})(t)\big]. \tag{5.20}$$

Notice that the initial boundary value problem (5.16)–(5.17), Equation (5.19), and the coupling condition (5.20) should be solved together for all time.

In each example considered by the authors, it is observed that introducing the inertial term in Equation (5.19) leads to numerical instabilities. To overcome this difficulty an approximation of Equation (5.19) is introduced by reducing by 95% the inertial effects of $\overline{z}_t(x,t)$. Namely, instead of (5.19), we consider

$$\widetilde{z}(x,t) = -A^{-1}\big[F(\overline{z}(x,t)) - 0.95\,\overline{z}_t(x,t) + B_{\mathrm{d}}(x)d(t)\big]. \tag{5.21}$$

Although this leads to an approximate solution for $\overline{z}(x,t)$, and therefore to

$$y_r(t) \simeq (C\widetilde{z})(t),$$

in most of the cases the error is very small, and is immediately reabsorbed.

Example 5.1 (Control for the 2D Thermal Fluid). For our example we notice that the momentum equation is coupled to the temperature only through the buoyancy term. We first determine the steady-state velocity in the absence of buoyancy force by solving the isothermal Navier-Stokes equation

$$(\boldsymbol{V} \cdot \nabla)\boldsymbol{V} = \nabla \cdot (\nu \left[(\nabla \boldsymbol{V}) + (\nabla \boldsymbol{V})^\mathsf{T}\right]) - \nabla P,$$
$$\nabla \cdot \boldsymbol{V} = 0,$$

$$\boldsymbol{V} = \begin{bmatrix} f(s) \\ 0 \end{bmatrix} \quad \text{on} \ \ \mathcal{S}_1,$$

$$\boldsymbol{\tau} = \boldsymbol{0} \ \ \text{on} \ \ \mathcal{S}_2,$$
$$\boldsymbol{V} = \boldsymbol{0} \ \ \text{on} \ \ \mathcal{S}_3 \cup \mathcal{S}_w,$$

for the variables $\boldsymbol{V}$ and P. Then we solve the linear energy equation for X, with homogeneous boundary conditions everywhere except on $\mathcal{S}_1$, where a unit heat flux is considered

$$0 = \alpha \Delta X - (\boldsymbol{V} \cdot \nabla)X, \qquad (5.22)$$
$$\boldsymbol{q} = 1\,\boldsymbol{n}_{\mathcal{S}_1} \ \ \text{on} \ \ \mathcal{S}_1,$$
$$\boldsymbol{q} = \boldsymbol{0} \ \ \text{on} \ \ \mathcal{S}_2 \cup \mathcal{S}_w,$$
$$X = 0 \ \ \text{on} \ \ \mathcal{S}_3.$$

Here X is the response of the homogeneous energy equation to the unit input on $\mathcal{S}_1$, and produces the entry of the 1×1 matrix (i.e., a scalar in this case)

$$G = \big(C(X)\big).$$

Remark 5.3. The system Equation (5.22) is linear, since the advection velocity $\boldsymbol{V}$ is given. In the energy equation we then consider the linear homogeneous operator A applied to the state variable X to be $\alpha \Delta X - (\boldsymbol{V} \cdot \nabla)X$ rather than $\alpha \Delta X$, only. Again we assume $\boldsymbol{V}$ to be so that A is always invertible. Here, of course, the definition of the linear operator A is also supplemented by the homogeneous boundary conditions on the boundary of the domain Ω.

With this we solve now the fully coupled nonlinear steady-state system

$$(\overline{\boldsymbol{V}} \cdot \nabla)\overline{\boldsymbol{V}} = \nabla \cdot (\nu \left[(\nabla \overline{\boldsymbol{V}}) + (\nabla \overline{\boldsymbol{V}})^\mathsf{T}\right])$$
$$- \nabla \overline{P} + \langle 1, \beta \rangle \overline{Z},$$
$$\nabla \cdot \overline{\boldsymbol{V}} = 0,$$
$$(\overline{\boldsymbol{V}} \cdot \nabla)\overline{Z} = \alpha \Delta \overline{Z},$$
$$((\overline{\boldsymbol{V}} - \boldsymbol{V}) \cdot \nabla)\overline{Z} = \alpha \Delta \widetilde{Z} - (\boldsymbol{V} \cdot \nabla)\widetilde{Z},$$

with boundary conditions

$$\overline{\boldsymbol{V}} = \begin{bmatrix} f(s) \\ 0 \end{bmatrix}, \ \ \boldsymbol{q}(\overline{Z}) = \gamma_0\,\boldsymbol{n}_{\mathcal{S}_1}, \ \ \boldsymbol{q}(\widetilde{Z}) = \boldsymbol{0} \ \ \text{on} \ \ \mathcal{S}_1,$$

$$\boldsymbol{\tau}(\overline{\boldsymbol{V}}, \overline{P}) = \boldsymbol{0} \,, \boldsymbol{q}(\overline{Z}) = \boldsymbol{0}, \ \boldsymbol{q}(\widetilde{Z}) = \boldsymbol{0} \text{ on } \mathcal{S}_2,$$

$$\overline{\boldsymbol{V}} = \boldsymbol{0} \,, \ \overline{Z} = d_0 \ \text{ and } \ \widetilde{Z} = d_0 \ \text{ on } \mathcal{S}_3,$$

$$\overline{\boldsymbol{V}} = \boldsymbol{0} \,, \ \boldsymbol{q}(\overline{Z}) = \boldsymbol{0} \ \text{ and } \ \boldsymbol{q}(\widetilde{Z}) = \boldsymbol{0} \ \text{ on } \mathcal{S}_w,$$

and control

$$\gamma_0 = G^{-1}(y_{r_0} - C(\widetilde{Z})) = \frac{1}{C(X)}(y_{r_0} - C(\widetilde{Z})). \tag{5.23}$$

Then, we solve the fully coupled nonlinear time-dependent system

$$\overline{\boldsymbol{v}}_t + (\overline{\boldsymbol{v}} \cdot \nabla)\overline{\boldsymbol{v}} = \nabla \cdot (\nu \left[(\nabla \overline{\boldsymbol{v}}) + (\nabla \overline{\boldsymbol{v}})^{\mathsf{T}} \right])$$
$$- \nabla \overline{p} + \langle 1, \beta \rangle \overline{T},$$

$$\nabla \cdot \overline{\boldsymbol{v}} = 0,$$

$$\overline{T}_t + (\overline{\boldsymbol{v}} \cdot \nabla)\overline{T} = \alpha \Delta \overline{T},$$

$$0.95\overline{T}_t + \big((\overline{\boldsymbol{v}} - \boldsymbol{V}) \cdot \nabla\big)\overline{T} = \alpha \Delta \widetilde{z} - (\boldsymbol{V} \cdot \nabla)\widetilde{T},$$

with initial conditions

$$\overline{\boldsymbol{v}}(x, 0) = \overline{\boldsymbol{V}}(x), \quad \overline{T}(x, 0) = \overline{Z}(x),$$

boundary conditions

$$\overline{\boldsymbol{v}} = \begin{bmatrix} f(s) \\ 0 \end{bmatrix}, \ \boldsymbol{q}(\overline{T}) = \gamma(t)\, \boldsymbol{n}_{\mathcal{S}_1}, \ \boldsymbol{q}(\widetilde{T}) = \boldsymbol{0} \ \text{ on } \mathcal{S}_1,$$

$$\boldsymbol{\tau}(\overline{\boldsymbol{v}}, \overline{p}) = \boldsymbol{0} \,, \boldsymbol{q}(\overline{T}) = \boldsymbol{0}, \ \boldsymbol{q}(\widetilde{T}) = \boldsymbol{0} \text{ on } \mathcal{S}_2,$$

$$\overline{\boldsymbol{v}} = \boldsymbol{0} \,, \ \overline{T} = d(t) \ \text{ and } \ \widetilde{T} = d(t) \ \text{ on } \mathcal{S}_3,$$

$$\overline{\boldsymbol{v}} = \boldsymbol{0} \,, \ \boldsymbol{q}(\overline{T}) = \boldsymbol{0} \ \text{ and } \ \boldsymbol{q}(\widetilde{T}) = \boldsymbol{0} \ \text{ on } \mathcal{S}_w,$$

and control

$$\gamma(t) = G^{-1}(y_r - C(\widetilde{T})) = \frac{1}{C(X)}(y_r(t) - C(\widetilde{T})).$$

Finally, we set $u(t) = \gamma(t)$ and solve the IBVP (5.6)–(5.11) and observe that the desired tracking $y(t) = C(T) \to C(\overline{T}) = y_r$ takes place.

In our numerical simulation we have set the following parameter values:

$$\alpha = 0.01, \quad \nu = 0.002, \quad \beta = 1.0, \quad t_1 = 750s, \quad t_2 = 2250s,$$
$$M_1 = 0.25, \quad M_2 = 0.5, \quad d_1 = 0.75, \quad d_2 = 0.5.$$

The transient solution is evaluated between $t = 0$ and $t = 3000$. The numerical solution of the weak formulation of the above systems is obtained again with the COMSOL Multiphysics package. The momentum and continuity equations are solved using the predefined Navier-Stokes incompressible model. The energy equation is solved by using the PDE coefficient model. In the Navier-Stokes model Lagrange elements $P2$-$P1$ are used for velocity and pressure in order to satisfy the LBB condition. In the PDE coefficient model Lagrange elements $P2$ are used for the Temperature. No stabilization is used for the advection term in any physics.

The time evolution of CT is given in Fig. 5.11 for all time (solid line). Once there is a change in the value of the reference signal or the disturbance a spike will occur in CT. These spikes are due to the inertial approximation in Equation (5.21) and are immediately reabsorbed. Fig. 5.12 contains a plot of the disturbance $e(t)$, Fig. 5.13 contains a plot of the disturbance $d(t)$, and finally Fig. 5.14 contains a plot of the resulting control $u(t)$.

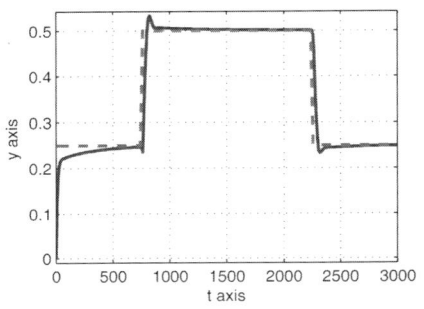

Fig. 5.11: $y(t)$ and $y_r(t)$.

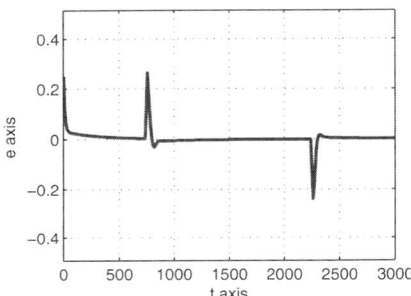

Fig. 5.12: Plot of $e(t)$.

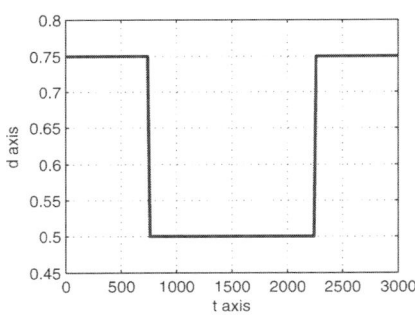

Fig. 5.13: Plot of $d(t)$.

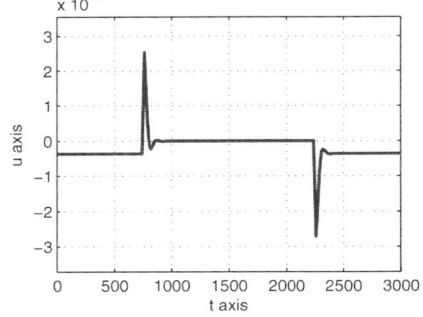

Fig. 5.14: Plot of $u(t)$.

5.3.3   The Harmonic Case

In this subsection we consider exactly the same thermal regulation problem discussed in the previous two sections, see Equations (5.6)–(5.11), where now we seek a time-dependent control input $u(t)$ in Equation (5.7) so that the measured output $y(t)$ in (5.11) tracks a given periodic time-dependent reference signal $y_r(t)$ while rejecting a smooth time-dependent periodic harmonic disturbance $d(t)$. In particular, for the reference signal we consider two cases:

Case A (piecewise continuous): $y_r(t) = \begin{cases} M_1, & 0 \le t < 750, \\ M_2, & 750 \le t < 2250, \\ M_1, & 2250 \le t \le 3000. \end{cases}$

Case B (smooth): $y_r(t) = M_1 + (M_2 - M_1) \sin\left(\dfrac{2\pi t}{1500}\right)$.

For the disturbance we always take

$$d(t) = d_1 + (d_2 - d_1) \sin\left(\frac{2\pi t}{1500}\right).$$

In order to solve these control problems we refer to Section 4.5 where we introduced an iterative scheme, called the β−iteration method, which aims to reduce the error generated by introducing the undervalued inertial term in the regulator equation (5.21). In that section we described a general setup to evaluate $\overline{z}$, $\widetilde{z}$ and γ as finite sums obtained by solving a sequence of regulator equations. In this specific example we stop our iterative procedure after three steps so the necessary expressions are given by sums of three terms

$$\overline{T} = \sum_{i=1}^{3} \overline{T}_i, \ \widetilde{T} = \sum_{i=1}^{3} \widetilde{T}_i, \ \gamma = \sum_{i=1}^{3} \gamma_i.$$

We proceed exactly as described in the previous section up to equation (5.23), and then we solve the fully coupled nonlinear time-dependent system

$$\overline{\boldsymbol{v}}_t + (\overline{\boldsymbol{v}} \cdot \nabla)\overline{\boldsymbol{v}} = \nabla \cdot (\nu \left[(\nabla \overline{\boldsymbol{v}}) + (\nabla \overline{\boldsymbol{v}})^\mathsf{T} \right]) - \nabla \overline{p} + \langle 1, \beta \rangle \overline{T},$$

$$\nabla \cdot \overline{\boldsymbol{v}} = 0,$$

$$\overline{T}_{1t} + (\overline{\boldsymbol{v}} \cdot \nabla)\overline{T}_1 = \alpha \Delta \overline{T}_1,$$

$$0.95\overline{T}_{1t} + ((\overline{\boldsymbol{v}} - \boldsymbol{V}) \cdot \nabla)\overline{T}_1 = \alpha \Delta \widetilde{T}_1 - (\boldsymbol{V} \cdot \nabla)\widetilde{T}_1,$$

$$\overline{T}_{2t} + (\overline{\boldsymbol{v}} \cdot \nabla)\overline{T}_2 = \alpha \Delta \overline{T}_2,$$

$$0.95\overline{T}_{2t} + ((\overline{\boldsymbol{v}} - \boldsymbol{V}) \cdot \nabla)\overline{T}_2 = \alpha \Delta \widetilde{T}_2 - (\boldsymbol{V} \cdot \nabla)\widetilde{T}_2,$$

$$\overline{T}_{3t} + (\overline{\boldsymbol{v}} \cdot \nabla)\overline{T}_3 = \alpha \Delta \overline{T}_3,$$

$$0.95\overline{T}_{3t} + ((\overline{\boldsymbol{v}} - \boldsymbol{V}) \cdot \nabla)\overline{T}_3 = \alpha \Delta \widetilde{T}_3 - (\boldsymbol{V} \cdot \nabla)\widetilde{T}_3,$$

with initial conditions

$$\overline{\boldsymbol{v}}(x,0) = \overline{\boldsymbol{V}}(x), \ \overline{T}_1(x,0) = \overline{Z}(x), \ \overline{T}_2(x,0) = 0, \ \overline{T}_3(x,0) = 0,$$

boundary conditions

$$\overline{v} = \begin{bmatrix} f(s) \\ 0 \end{bmatrix}, \ q(\overline{T}_1) = \gamma_1(t)\, \boldsymbol{n}_{\mathcal{S}_1}, \ q(\widetilde{T}_1) = \boldsymbol{0},$$

$$q(\overline{T}_2) = \gamma_2(t)\, \boldsymbol{n}_{\mathcal{S}_2}, \ q(\widetilde{T}_2) = \boldsymbol{0},$$

$$q(\overline{T}_3) = \gamma_3(t)\, \boldsymbol{n}_{\mathcal{S}_3}, \ q(\widetilde{T}_3) = \boldsymbol{0}, \qquad \text{on } \mathcal{S}_1,$$

$$\boldsymbol{\tau}(\overline{v}, \overline{p}) = \boldsymbol{0}, \ q(\overline{T}_1) = \boldsymbol{0}, \ q(\widetilde{T}_1) = \boldsymbol{0},$$

$$q(\overline{T}_2) = \boldsymbol{0}, \ q(\widetilde{T}_2) = \boldsymbol{0},$$

$$q(\overline{T}_3) = \boldsymbol{0}, \ q(\widetilde{T}_3) = \boldsymbol{0}, \qquad \text{on } \mathcal{S}_2,$$

$$\overline{v} = \boldsymbol{0}, \ \overline{T}_1 = d(t), \ \widetilde{T}_1 = d(t),$$

$$\overline{T}_2 = 0, \ \widetilde{T}_2 = 0,$$

$$\overline{T}_3 = 0, \ \widetilde{T}_3 = 0, \qquad \text{on } \mathcal{S}_3,$$

$$\overline{v} = \boldsymbol{0}, \ q(\overline{T}_1) = \boldsymbol{0}, \ q(\widetilde{T}_1) = \boldsymbol{0},$$

$$q(\overline{T}_2) = \boldsymbol{0}, \ q(\widetilde{T}_2) = \boldsymbol{0},$$

$$q(\overline{T}_3) = \boldsymbol{0}, \ q(\widetilde{T}_3) = \boldsymbol{0}, \qquad \text{on } \mathcal{S}_w,$$

and controls

$$\gamma_1(t) = \frac{1}{C(X)} (y_r(t) - C(\widetilde{T}_1)),$$

$$\gamma_2(t) = \frac{1}{C(X)} \left[(y_r(t) - C(\overline{T}_1)) - C(\widetilde{T}_2) \right],$$

$$\gamma_3(t) = \frac{1}{C(X)} \left[(y_r(t) - C(\overline{T}_1 + \overline{T}_2)) - C(\widetilde{T}_3) \right].$$

Finally, we set $u = \gamma = \gamma_1 + \gamma_2 + \gamma_3$ and solve the IBVP (5.6)–(5.11) and observe that a very accurate approximate tracking $y(t) = C(T) \to C(\overline{T}) \simeq y_r$ as $t \to \infty$ takes place.

In our numerical simulation we have set the following parameter values:

$$\alpha = 0.01, \quad \nu = 0.002, \quad \beta = 1.0,$$

$$M_1 = 0.25, \quad M_2 = 0.5, \quad d_1 = 0.75, \quad d_2 = 0.5.$$

The transient solution is evaluated between $t = 0$ and $t = 3000$.

We consider two simulation cases. In the first case we consider a reference signal y_r which is piecewise constant periodic and a smooth periodic disturbance d. In the second case we consider both y_r and d to be smooth periodic signals. We note that in the first case, as would be expected, there is a jump in the output when a transition to a different constant value occurs. This is expected since our theoretical estimates require smoothness of the reference signals. On the other hand we notice that following the jump in y_r, the output immediately responds to quickly track the new constant value. In both cases we have applied the β-iteration algorithm to enhance the tracking regulation.

Example 5.2 (Case A: Piecewise Constant Periodic Reference Signal). In Figs. 5.15 and 5.16 we have plotted the numerical solutions obtained after three iterations of the β-iteration method. In the first figure we have plotted the output y (solid line) and the reference signal y_r (dashed line). In the second figure we have plotted the resulting error $e(t)$. In both figures the simulation is computed for t ranging from $t = 0$ to $t = 3000$.

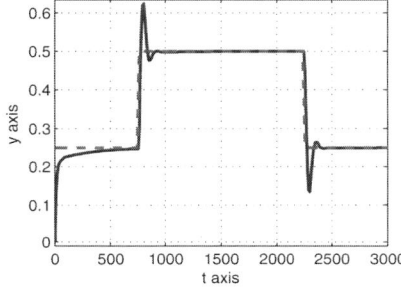
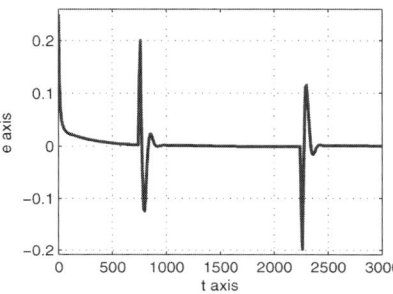

Fig. 5.15: $y(t)$ and $y_r(t)$.   **Fig. 5.16**: Plot of $e(t)$.

In Figs. 5.17 and 5.18 we have plotted the smooth periodic disturbance d and the control $u = \gamma$ used in the closed loop system.

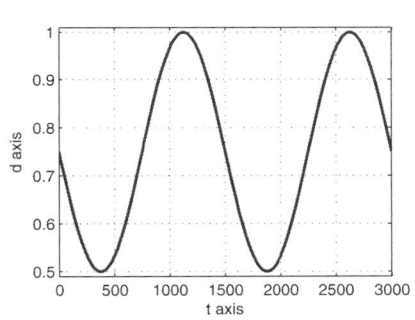
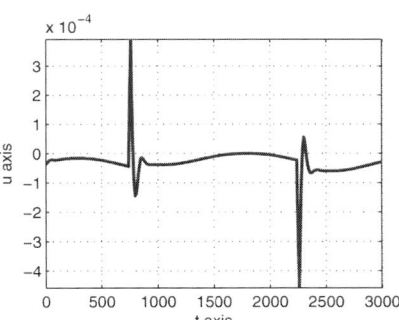

Fig. 5.17: $d(t)$.   **Fig. 5.18**: Plot of $u = \gamma(t)$.

In the following figures we have plotted the outputs of the first and second β-iterations, $C(\overline{T}_1)$ (dashed line), $C(\overline{T}_1 + \overline{T}_2)$ (solid line) in Fig. 5.19 and the corresponding errors $e_1(t)$ (dashed line) and $e_2(t)$ (solid line) in Fig. 5.20. Notice how the solid line gives better results far away from the transient regions.

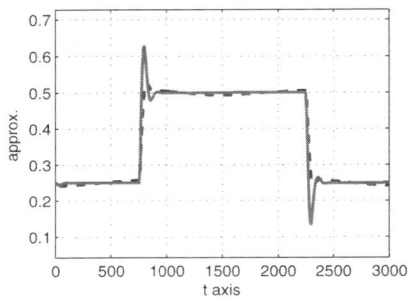

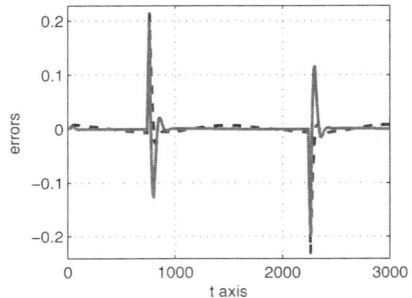

Fig. 5.19: β-Iterations.   **Fig. 5.20**: $e_1(t)$, $e_2(t)$.

Example 5.3 (Case B: Smooth Periodic Reference Signal). Here we assume that both y_r and d are smooth periodic functions. Notice that in this case the β-iteration produces very accurate tracking after only 3 iterations.

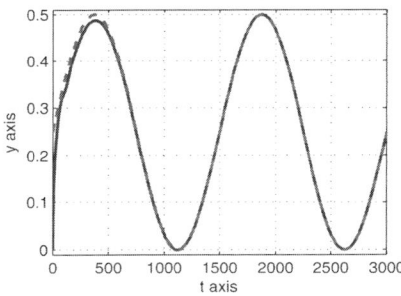

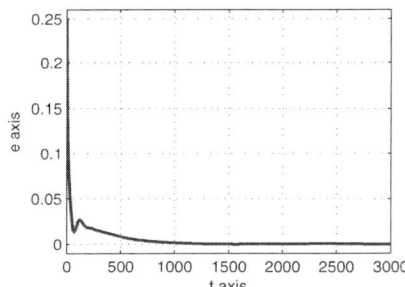

Fig. 5.21: $y(t)$ and $y_r(t)$.   **Fig. 5.22**: Plot of $e(t)$.

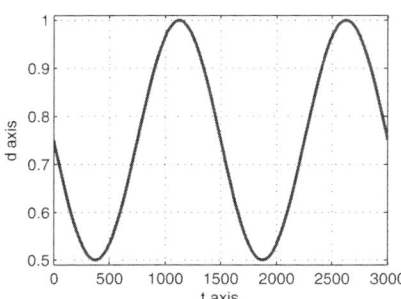

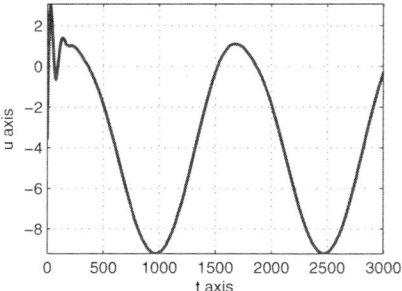

Fig. 5.23: $d(t)$.   **Fig. 5.24**: Plot of $u = \gamma(t)$.

In Figs. 5.21 and 5.22 we have plotted the numerical values obtained after 3 iterations of the β-iteration method. In the first figure we have plotted the measured output y (solid line) and the reference signal y_r (dashed line). In the second figure we have plotted the resulting error $e(t)$. In both figures the simulation is computed for t ranging from $t = 0$ to $t = 3000$.

In Figs. 5.23 and 5.24 we have plotted the smooth periodic disturbance d and the control $u = \gamma$ used in the closed loop system.

In the following figures we have depicted the outputs of the first, $C(\overline{T}_1)$ (solid line), and second, $C(\overline{T}_1 + \overline{T}_2)$ (dashed line), β-iterations in Fig. 5.25, and the corresponding errors $e_1(t)$ and $e_2(t)$ in Fig. 5.26.

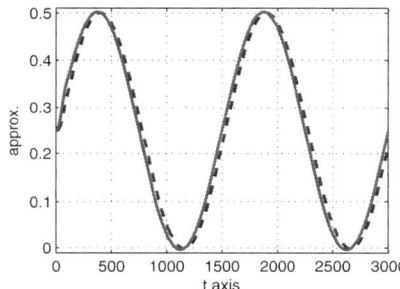
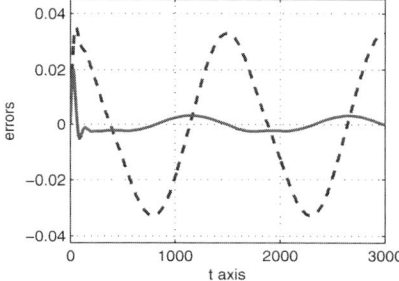

Fig. 5.25: β-Iterations.   **Fig. 5.26**: $e_1(t)$, $e_2(t)$.

5.4   2D Chafee-Infante with Time-Dependent Regulation

In this example we discuss the β-iteration method presented in Section 4.5 for a 2D boundary control problem nonlinear Chafee-Infante equation with state variable $z = z(x,t)$ on the rectangular region Ω depicted in Fig. 5.27. Here $x = (x_1, x_2) \in (0, 2L) \times (0, L) \subset \mathbb{R}^2$.

$$z_t - c\Delta z = \lambda(z - z^3)0, \text{ for } t > 0, \tag{5.24}$$
$$z(x,0) = \varphi(x),$$

We impose boundary conditions

$$\mathcal{B}_{\text{in}}(z) = (c\nabla z \cdot \boldsymbol{n} + kz) = u(t), \text{ on } \mathcal{S}_1,$$
$$\mathcal{B}_{\text{d}}(z) = z(x,t) = d(t), \text{ on } \mathcal{S}_4,$$
$$\mathcal{B}_0(z) = \nabla z \cdot \boldsymbol{n} = 0, \text{ on } \mathcal{S}_2 \cup \mathcal{S}_3, \tag{5.25}$$

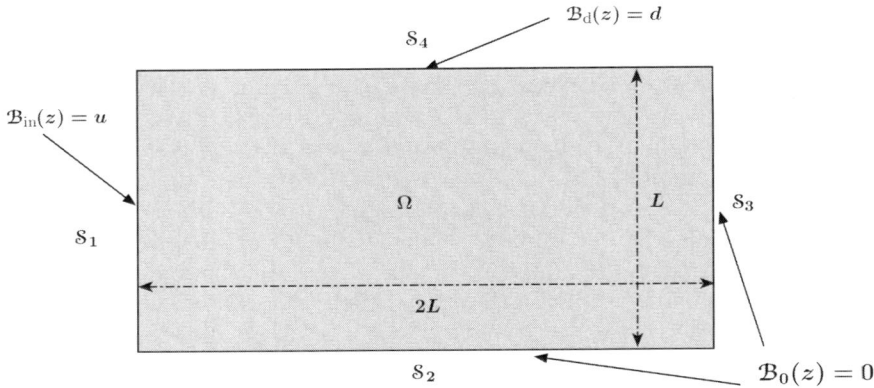

Fig. 5.27: Two-Dimensional Rectangle.

where $u(t)$ is the control, entering throughout a mixed boundary condition on the left of the domain, and $d(t)$ is some disturbance, entering throughout a Dirichlet boundary condition on the top of the domain. Neumann homogeneous boundary conditions are considered on the right and bottom part of the domain.

We also consider the output to be the average of the solution on the right side of the domain

$$y(t) = C(z) = \frac{1}{|\mathcal{S}_3|} \int_{\mathcal{S}_3} z \, ds.$$

Our design objective is to have the measured output $y(t)$ track a prescribed trajectory $y_r(t)$, while rejecting a disturbance $d(t)$.

To solve the regulation problem we follow the procedure described in Section 4.5. We synthesize it in the following steps:

1. Find G, i.e. solve

$$-c\Delta X = \lambda X,$$
$$\mathcal{B}_{\text{in}}(X) = 1, \quad \mathcal{B}_{\text{d}}(X) = 0, \quad \mathcal{B}_0(X) = 0,$$
$$G = C(X).$$

2. Find an appropriate initial data $\overline{Z}_0$ that satisfies the set-point problem with $y_{r0} = y_r(0)$ and $d_0 = d(0)$, i.e., solve

$$-c\Delta \overline{Z}_0 = \lambda(\overline{Z}_0 - \overline{Z}_0^3), \qquad -c\Delta \widetilde{Z}_0 = \lambda(\widetilde{Z}_0 - \overline{Z}_0^3),$$
$$\mathcal{B}_{\text{in}}(\overline{Z}_0) = \gamma_0, \qquad\qquad \mathcal{B}_{\text{in}}(\widetilde{Z}_0) = 0,$$
$$\mathcal{B}_{\text{d}}(\overline{Z}_0) = d_0, \qquad\qquad \mathcal{B}_{\text{d}}(\widetilde{Z}_0) = d_0,$$
$$\mathcal{B}_0(\overline{Z}_0) = 0, \qquad\qquad \mathcal{B}_0(\widetilde{Z}_0) = 0,$$
$$\gamma_0 = G^{-1}(y_{r0} - C(\widetilde{Z}_0)).$$

3. Build the β-iterative procedure to approximate the solution of the time-dependent regulator equations, i.e., for $i = 1$ solve

$$\overline{Z}_{1t} - c\Delta\overline{Z}_1 = \lambda(\overline{Z}_1 - \overline{Z}_1^3), \quad (1-\beta)\widetilde{Z}_{1t} - c\Delta\widetilde{Z}_1 = \lambda(\widetilde{Z}_1 - \overline{Z}_1^3),$$
$$\overline{Z}_1(x,0) = \overline{Z}_0, \qquad\qquad \widetilde{Z}_1(x,0) = \widetilde{Z}_0,$$
$$\mathcal{B}_{\mathrm{in}}(\overline{Z}_1) = \gamma_0, \qquad\qquad \mathcal{B}_{\mathrm{in}}(\widetilde{Z}_1) = 0,$$
$$\mathcal{B}_{\mathrm{d}}(\overline{Z}_1) = d(t), \qquad\qquad \mathcal{B}_{\mathrm{d}}(\widetilde{Z}_1) = d(t),$$
$$\mathcal{B}_0(\overline{Z}_1) = 0, \qquad\qquad \mathcal{B}_0(\widetilde{Z}_1) = 0,$$
$$\gamma_1 = G^{-1}(y_r(t) - C(\widetilde{Z}_1)), \quad e_1(t) = y_r(t) - C(\overline{Z}_1),$$

and for $i = 2, 3, \ldots, n$, set $\overline{Z}_i = \overline{Z}_{i-1} + \overline{z}_i$,

$$\overline{H}_i = (\overline{z}_i - \overline{Z}_i^3 + \overline{Z}_{i-1}^3), \quad \widetilde{H}_i = (\widetilde{z}_i - \overline{Z}_i^3 + \overline{Z}_{i-1}^3)$$

and solve

$$\overline{z}_{it} - c\Delta\overline{z}_i = \lambda\,\overline{H}_i, \qquad\qquad (1-\beta)\widetilde{z}_{it} - c\Delta\widetilde{z}_i = \lambda\,\widetilde{H}_i,$$
$$\overline{z}_i(x,0) = 0, \qquad\qquad \widetilde{z}_i(x,0) = 0,$$
$$\mathcal{B}_{\mathrm{in}}(\overline{z}_i) = \gamma_i, \qquad\qquad \mathcal{B}_{\mathrm{in}}(\widetilde{z}_i) = 0,$$
$$\mathcal{B}_{\mathrm{d}}(\overline{z}_i) = 0, \qquad\qquad \mathcal{B}_{\mathrm{d}}(\widetilde{z}_i) = 0,$$
$$\mathcal{B}_0(\overline{z}_i) = 0, \qquad\qquad \mathcal{B}_0(\widetilde{z}_i) = 0,$$

$$\gamma_i = \begin{cases} 0 & \text{for } t < T_i, \\[2mm] \dfrac{e_{i-1}(t) - C(\widetilde{z}_i)}{G} & \text{for } t \geq T_i, \end{cases} \qquad e_i(t) = y_r(t) - C(\overline{Z}_i).$$

4. Finally set $u(t) = \gamma(t) = \displaystyle\sum_{i=1}^{n} \gamma_i(t)$ and solve the closed loop system (5.24)–(5.25).

Notice that we have introduced time switches for γ_i for $i \geq 2$. As a result the solutions $\overline{z}_i$ are identically zero until $t = T_i$. As already pointed out in Example 4.4 this is an expedient used for stability reasons in the numerical computation, and it is needed to dump out the oscillations introduced by using the approximate initial data $\overline{Z}_0$.

In our numerical simulation we track a polynomial reference signal and reject a polynomial disturbance. Notice that such signals are not generated by a finite-dimensional neutrally stable exosystem. For that reason our example is more general than what has been considered earlier in the book.

$$y_r(t) = \left(\frac{10-t}{10}\right)\left(\frac{40-t}{40}\right)\left(\frac{80-t}{80}\right)\left(\frac{100-t}{100}\right), \quad d(t) = \frac{t^2}{100^2}.$$

We also choose the following parameters

$$c = 1, \ L = 1, \ k = 0.1, \ \lambda = 0.4, \ \beta = 0.5, \ n = 3, \ T_2 = 5, \ T_3 = 10 \,.$$

The closed loop system is then solved for $t \in [0, 100]$ with initial condition $\varphi(x) = 0$.

In Fig. 5.28 we have plotted the output $y(t)$ (solid line) and the reference signal $y_r(t)$ to be tracked (dashed line) for the whole time interval $0 < t < 100$. The convergence of $y(t)$ to $y_r(t)$ as t increases is evident. In Fig. 5.29 we have also plotted the time evolution of the resulting error $e(t) = y_r(t) - y(t)$.

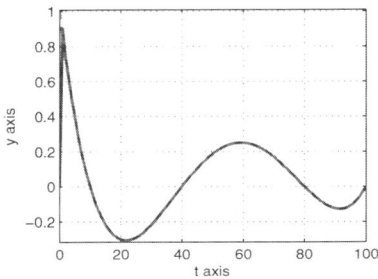

Fig. 5.28: $y(t)$ and $y_r(t)$.   **Fig. 5.29**: $e(t) = y_r(t) - y(t)$.

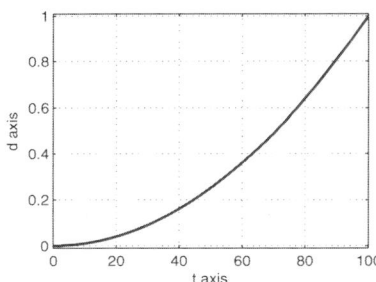

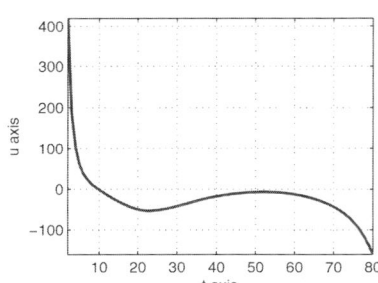

Fig. 5.30: $d(t)$.   **Fig. 5.31**: $u = \gamma(t)$ for $2 \leq t \leq 80$.

In Figs. 5.30 and 5.31 we have plotted the smooth disturbance $d(t)$ and the control $u = \gamma(t)$ used in the closed loop system. Notice that since neither y_r nor d are periodic functions, the control $\gamma(t)$ is not periodic as well.

We note that the values of the control are very large for t near $t = 0$ and $t = 100$. For this reason we have plotted $u(t)$ only on the interval $2 \leq t \leq 80$ since otherwise the control appears to be zero on most of the interval. For example near $t = 0$, $u(t)$ is on the order 10^4 and near $t = 100$ it is on the order 10^3. In Figs. 5.32 and 5.33 we have plotted the errors $e_1(t)$ and $e_2(t)$ and the errors $e_2(t)$ and $e_3(t)$, respectively, in the time interval $20 < t < 100$, when the error due to the approximate initial data $\overline{Z}_0$ has been reabsorbed. Notice that $e_2(t)$ is consistently smaller than $e_1(t)$ and that $e_3(t)$ is consistently smaller than $e_2(t)$ as expected by using the β-iteration method.

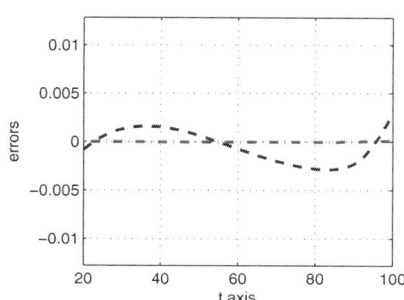
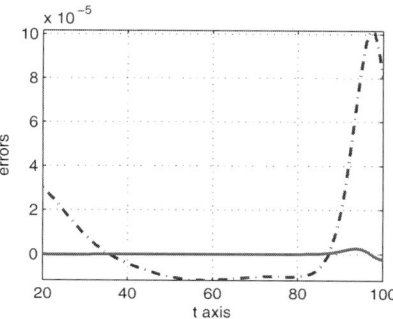

Fig. 5.32: $e_1(t)$ and $e_2(t)$ for $20 < t < 100$.

Fig. 5.33: $e_2(t)$ and $e_3(t)$ for $20 < t < 100$.

5.5 Regulation of 2D Burgers' Using Fourier Series

In this example we use the Fourier series decomposition method presented in Section 4.6 applied to a boundary controlled 2D nonlinear viscous Burgers's system on a rectangular region Ω depicted in Fig. 5.34. The system is given in terms of the state variable $z = \begin{bmatrix} u , v \end{bmatrix}$ where $z = z(x,t)$, and $x = (x_1, x_2) \in (0, 2L) \times (0, L) = \Omega \subset \mathbb{R}^2$,

$$z_t + z \cdot \nabla z - \nu \Delta z = 0, \text{ for } t > 0, \qquad (5.26)$$
$$z(x,0) = \varphi(x),$$

We consider the system (5.26) together with boundary conditions

$$\mathcal{B}_{\text{in}}(z) = (\nu \nabla z \cdot n + kz) = \begin{bmatrix} \gamma , 0 \end{bmatrix}, \text{ on } \mathcal{S}_1,$$
$$\mathcal{B}_{\text{d}}(z) = z(x,t) = \begin{bmatrix} 0 , d \end{bmatrix}, \text{ on } \mathcal{S}_2,$$
$$\mathcal{B}_0(z) = \nabla z \cdot n = 0, \text{ on } \mathcal{S}_3 \cup \mathcal{S}_4,$$

where $\gamma(t)$ is the control, entering through a mixed boundary condition in the u-component on the left of the domain, and $d(t)$ is a disturbance, entering throughout a Dirichlet boundary condition for the v-component on the bottom of the domain. Homogeneous Neumann boundary conditions are considered on the right and top part of the domain.

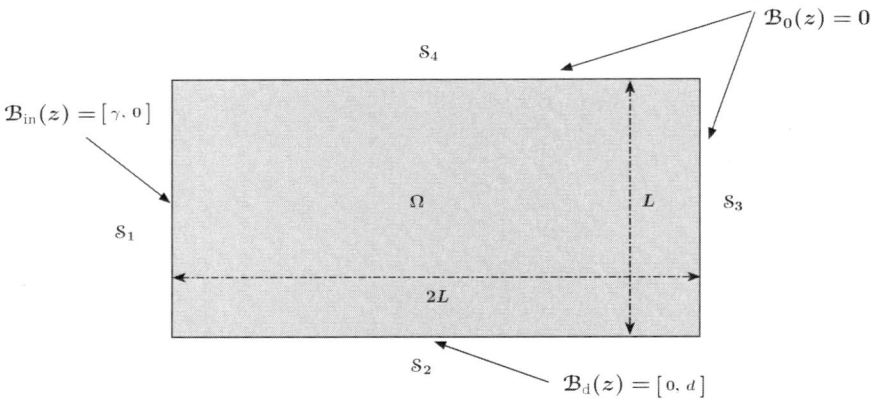

Fig. 5.34: Two-Dimensional Rectangle.

In (5.26) the nonlinearity is given by the convective term

$$z \cdot \nabla z = \left[uu_x + vu_y \ , \ uv_x + vv_y \right].$$

For this example we consider the output to be the average of the u-component on the right side of the domain

$$y(t) = C(z) = \frac{1}{|S_3|} \int_{S_3} u \, ds.$$

Our design objective is to have the measured output $y(t)$ track a prescribed periodic trajectory $y_r(t)$, while rejecting a periodic disturbance $d(t)$.

As pointed out in Assumption 4.7, we require the reference signal and the disturbance to have commensurate periods. Let

$$P = \frac{2\pi}{\alpha}$$

be the smallest common period between y_r and d. Let $\alpha_i = i\,\alpha$ and proceed formally as described in Section 4.6, taking into account that now we have 2 components. Truncating the Fourier series decomposition at the N-th term, and defining the following quantities

$$\mathbf{\Pi} = \begin{bmatrix} \boldsymbol{a}_0^\pi \\ \boldsymbol{a}_1^\pi \\ \boldsymbol{b}_1^\pi \\ \vdots \\ \boldsymbol{a}_N^\pi \\ \boldsymbol{b}_N^\pi \end{bmatrix} = \begin{bmatrix} a_{0,u}^\pi \ , \ a_{0,v}^\pi \\ a_{1,u}^\pi \ , \ a_{1,v}^\pi \\ b_{1,u}^\pi \ , \ b_{1,v}^\pi \\ \vdots \\ a_{N,u}^\pi \ , \ a_{N,v}^\pi \\ b_{N,u}^\pi \ , \ b_{N,v}^\pi \end{bmatrix}, \quad \mathcal{S}\mathbf{\Pi} = \begin{bmatrix} 0 \ , & 0 \\ \alpha_1 \, b_{1,u}^\pi \ , & \alpha_1 \, b_{1,v}^\pi \\ -\alpha_1 \, a_{1,u}^\pi \ , & -\alpha_1 \, a_{1,v}^\pi \\ \vdots & \vdots \\ \alpha_N \, b_{N,u}^\pi \ , & \alpha_N \, b_{N,v}^\pi \\ -\alpha_N \, a_{N,u}^\pi \ , & -\alpha_N \, a_{N,v}^\pi \end{bmatrix},$$

$$
D = \begin{bmatrix} \mathbf{0} , \ \mathcal{D} \end{bmatrix} = \begin{bmatrix} 0 , & a_0^d \\ 0 , & a_1^d \\ 0 , & b_1^d \\ & \vdots \\ 0 , & a_N^d \\ 0 , & b_N^d \end{bmatrix}, \quad
F(\mathbf{\Pi}) = \begin{bmatrix} \boldsymbol{f}_0 \\ \boldsymbol{f}_1^c \\ \boldsymbol{f}_1^s \\ \vdots \\ \boldsymbol{f}_N^c \\ \boldsymbol{f}_N^s \end{bmatrix} = \begin{bmatrix} f_{0,u} , & f_{0,v} \\ f_{1,u}^c , & f_{1,v}^c \\ f_{1,u}^s , & f_{1,v}^s \\ \vdots & \vdots \\ f_{N,u}^c , & f_{N,v}^c \\ f_{N,u}^s , & f_{N,v}^s \end{bmatrix},
$$

$$
\mathbf{\Gamma} = \begin{bmatrix} \mathcal{G} , \ \mathbf{0} \end{bmatrix} = \begin{bmatrix} a_0^\gamma , & 0 \\ a_1^\gamma , & 0 \\ b_1^\gamma , & 0 \\ \vdots & \vdots \\ a_N^\gamma , & 0 \\ b_N^\gamma , & 0 \end{bmatrix},
$$

the first regulator equation can be written in matrix form as

$$
\mathcal{S}\mathbf{\Pi} = A\mathbf{\Pi} + F(\mathbf{\Pi}) + B_{\mathrm{d}}D + B_{\mathrm{in}}\mathbf{\Gamma},
$$

where the contributions in $F(\mathbf{\Pi})$ depend on the particular structure of the nonlinear term. For the multi-dimensional Burgers' equation here considered, the nonlinear term, given by $\boldsymbol{z} \cdot \nabla \boldsymbol{z}$, yields

$$
\boldsymbol{f}_0 = -\frac{1}{2}\left(\boldsymbol{a}_0^\pi \cdot \nabla \boldsymbol{a}_0^\pi + 2\sum_{i=1}^{N}[\boldsymbol{a}_i^\pi \cdot \nabla \boldsymbol{a}_i^\pi + \boldsymbol{b}_i^\pi \cdot \nabla \boldsymbol{b}_i^\pi] \right),
$$

For $n = 1, \cdots, N$, $1 \le i \le N$ and $1 \le j \le N$ set

$$
\begin{aligned}
\delta_1(i,j,n) &= (\delta_{i,j+n} + \delta_{j,i+n} + \delta_{n,i+j}), \\
\delta_2(i,j,n) &= (\delta_{i,j+n} + \delta_{j,i+n} - \delta_{n,i+j}), \\
\delta_3(i,j,n) &= (-\delta_{i,j+n} + \delta_{j,i+n} + \delta_{n,i+j})
\end{aligned} \tag{5.27}
$$

where $\delta_{l,m}$ is the standard Kronecker delta. Then we can write the remaining terms $\boldsymbol{f}_n^c$ and $\boldsymbol{f}_n^s$ in $F(\mathbf{\Pi})$ as follows:

$$
\begin{aligned}
\boldsymbol{f}_n^c = -\frac{1}{2}\Big(& \boldsymbol{a}_0^\pi \cdot \nabla \boldsymbol{a}_n^\pi + \boldsymbol{a}_n^\pi \cdot \nabla \boldsymbol{a}_0^\pi \\
& + \frac{1}{2}\sum_{i=1}^{N}\sum_{j=1}^{N}\big[(\boldsymbol{a}_i^\pi \cdot \nabla \boldsymbol{a}_j^\pi + \boldsymbol{a}_j^\pi \cdot \nabla \boldsymbol{a}_i^\pi)\,\delta_1(i,j,n) \\
& \qquad\qquad + (\boldsymbol{b}_i^\pi \cdot \nabla \boldsymbol{b}_j^\pi + \boldsymbol{b}_j^\pi \cdot \nabla \boldsymbol{b}_i^\pi)\,\delta_2(i,j,n)\big] \Big),
\end{aligned}
$$

$$
\begin{aligned}
\boldsymbol{f}_n^s = -\frac{1}{2}\Big(& \boldsymbol{a}_0^\pi \cdot \nabla \boldsymbol{b}_n^\pi + \boldsymbol{b}_n^\pi \cdot \nabla \boldsymbol{a}_0^\pi \\
& + \sum_{i=1}^{N}\sum_{j=1}^{N}\big[(\boldsymbol{a}_i^\pi \cdot \nabla \boldsymbol{b}_j^\pi + \boldsymbol{b}_j^\pi \cdot \nabla \boldsymbol{a}_i^\pi)\,\delta_3(i,j,n)\big] \Big),
\end{aligned}
$$

The first regulator equation is supplemented with boundary conditions and constraints due to the second regulator equation. Let $\mathcal{Y}_r$ be given by

$$\mathcal{Y}_r = \begin{bmatrix} a_0^r, & a_1^r, & b_1^r, & \cdots, & a_N^r, & b_N^r \end{bmatrix}^{\mathsf{T}},$$

then we proceed by solving the system

$$0 = A\mathbf{\Pi} - \mathcal{S}\mathbf{\Pi} + \boldsymbol{F}(\mathbf{\Pi}) + B_{\mathrm{d}}\boldsymbol{D} + B_{\mathrm{in}}\boldsymbol{\Gamma},$$

$$0 = A\widetilde{\mathbf{\Pi}} - \mathcal{S}\mathbf{\Pi} + \boldsymbol{F}(\mathbf{\Pi}) + B_{\mathrm{d}}\boldsymbol{D},$$

$$\boldsymbol{\Gamma} = \begin{bmatrix} \mathcal{G}, \mathbf{0} \end{bmatrix} = \begin{bmatrix} G^{-1}(\mathcal{Y}_r - C(\widetilde{\mathbf{\Pi}})), \mathbf{0} \end{bmatrix},$$

where, as usual, $G = C(-A^{-1})B_{\mathrm{in}}$. With this we then solve the closed loop system where the control γ is given by

$$\gamma = \frac{1}{2}a_0^\gamma + \sum_{i=0}^{N} a_i^\gamma \cos(\alpha_i t) + b_i^\gamma \sin(\alpha_i t).$$

In our numerical simulation we have set

$$y_r(t) = M_1 + A_1 \sin(\alpha t),$$
$$d(t) = M_2 + A_2 \sin(\alpha t).$$

Given this choice the corresponding Fourier coefficients for $N = 5$ are given by

$$\mathcal{Y}_r = \begin{bmatrix} 2M_1, 0, A_1, 0, 0, 0, 0, 0, 0, 0, 0 \end{bmatrix}^{\mathsf{T}},$$

$$\mathcal{D} = \begin{bmatrix} 2M_2, 0, A_2, 0, 0, 0, 0, 0, 0, 0, 0 \end{bmatrix}^{\mathsf{T}}.$$

We have also chosen the following parameters

$$\nu = 1, \ L = 1, \ k = 5, \ \alpha = 1,$$

$$M_1 = 0.1, \ A_1 = 0.2, \ M_2 = 0.2, \ A_2 = 0.1.$$

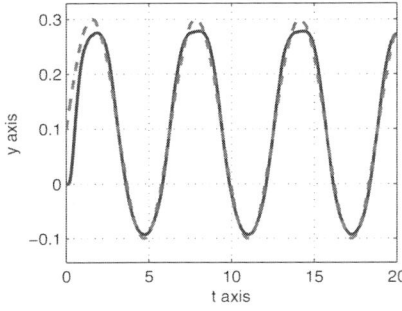

Fig. 5.35: $y(t)$ with $N = 2$.

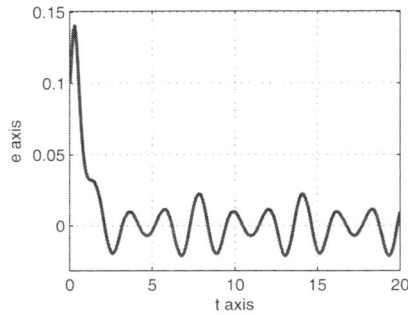

Fig. 5.36: $e(t)$ with $N = 2$.

The closed loop system is solved for $t \in [0, 20]$ with initial condition $\varphi(x) = \mathbf{0}$. First we have considered the case of Fourier series truncated at $N = 2$. In Fig. 5.35 we have plotted the output of the system and the reference signal and in Fig. 5.36 we have plotted the corresponding error for this case. We notice that already the error is quite small.

Next we considered the case of Fourier series truncated at $N = 3$. In Fig. 5.37 we have plotted the output of the system and the reference signal and in Fig. 5.38 we have plotted the corresponding error for this case. In this case we note that the error is much smaller.

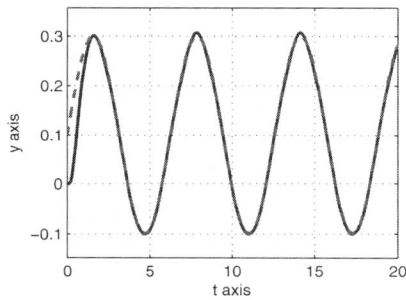

Fig. 5.37: $y(t)$ with $N = 3$.

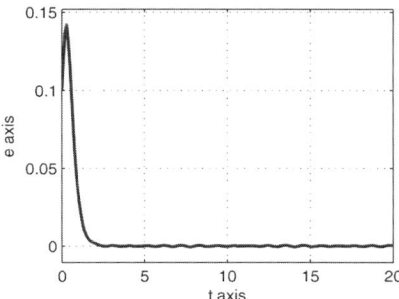

Fig. 5.38: $e(t)$ with $N = 3$.

Finally, in our last case we have considered the case of Fourier series truncated at $N = 5$. In Fig. 5.39 we have plotted the output of the system and the reference signal and in Fig. 5.40 we have plotted the corresponding error for this case. In this case we note that the error is extremely small.

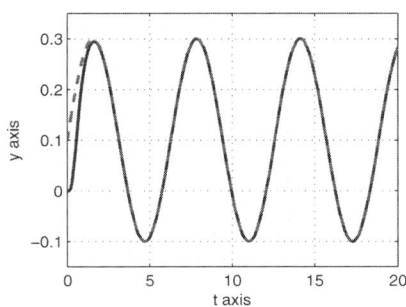
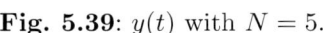

Fig. 5.39: $y(t)$ with $N = 5$.

Fig. 5.40: $e(t)$ with $N = 5$.

5.6   Back-Step Navier-Stokes Flow

In this example we are interested in a tracking regulation problem for a back-step incompressible Navier-Stokes flow whose geometry is given in Fig. 5.41. An input flow enters from the left side of the domain, S_{in}, and exits out the right side, S_{out}. Wall boundary conditions are considered elsewhere on the boundary denoted by S_w.

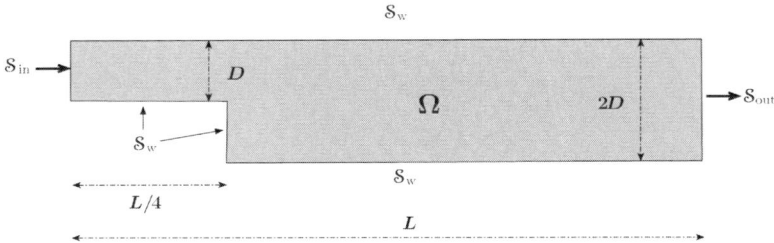

Fig. 5.41: Back-Step Navier-Stokes Geometry.

The mathematical formulation of the above problem for the state variable

$$z = \begin{bmatrix} v, \ p \end{bmatrix} = \begin{bmatrix} u \ , \ v \ , \ p \end{bmatrix}$$

is considered for $x = (x_1, x_2) \in \Omega \subset \mathbb{R}^2$, and $t > 0$ via the system

$$v_t + v \cdot \nabla v - \nu \Delta v + \nabla p = 0, \text{ for } t > 0, \tag{5.28}$$

$$\nabla \cdot v = 0, \text{ for } t > 0, \tag{5.29}$$

$$v(x, 0) = \varphi(x), \tag{5.30}$$

with boundary conditions

$$\mathcal{B}_{in}(z) = \tau(z)\big|_{S_{in}} = \gamma \, n_{S_{in}}, \text{ on } S_{in}, \tag{5.31}$$

$$\mathcal{B}_{out}(z) = \tau(z)\big|_{S_{out}} = 0, \text{ on } S_{out}, \tag{5.32}$$

$$\mathcal{B}_w(z) = v = 0, \text{ on } S_w, \tag{5.33}$$

where

$$\tau(z) = (pI - \nu\,[(\nabla v) + (\nabla v)^{\mathsf{T}}]) \cdot n_S, \tag{5.34}$$

and n_S the outward normal on a boundary surface S.

In the momentum equation (5.28) the nonlinearity is given by the convective term

$$v \cdot \nabla v = \begin{bmatrix} uu_x + vu_y \ , \ uv_x + vv_y \end{bmatrix}.$$

In equations (5.28) and (5.34), ν represents the kinematic viscosity and Equation (5.28) is the incompressibility constraint. The control input corresponds to the normal stress to the surface S_{in}.

We consider the output to be the total flux across the right side boundary of the domain

$$y(t) = C(z) = \int_{S_{out}} \boldsymbol{v} \cdot \boldsymbol{n}_{S_{out}} \, ds.$$

Our design objective is to have the measured output $y(t)$ track a prescribed periodic trajectory $y_r(t)$, and in particular we choose the reference trajectory to be the periodic signal $y_r(t) = M_1 + A_1 \sin(\alpha t)$.

We solve this problem combining the features of the two main methodologies described and used throughout the book. Namely, we first use a truncated Fourier series decomposition approach to get a reasonably accurate description of the projection of the reference trajectory on the "center manifold trajectory". Then we apply a single β−iteration to further correct and better approximate the attractive trajectory on the center manifold.

5.6.1 Fourier Series Decomposition

We proceed as discussed in the previous examples. Let $\alpha_i = i\,\alpha$. Truncating the Fourier series decomposition at the N-*th* term, and defining the quantities

$$
\boldsymbol{\pi} =
\begin{bmatrix}
\boldsymbol{a}_0^\pi \\
\boldsymbol{a}_1^\pi \\
\boldsymbol{b}_1^\pi \\
\vdots \\
\boldsymbol{a}_N^\pi \\
\boldsymbol{b}_N^\pi
\end{bmatrix}
=
\begin{bmatrix}
a_{0,u}^\pi \ , \ a_{0,v}^\pi \\
a_{1,u}^\pi \ , \ \alpha_{1,v}^\pi \\
b_{1,u}^\pi \ , \ b_{1,v}^\pi \\
\vdots \\
a_{N,u}^\pi \ , \ a_{N,v}^\pi \\
b_{N,u}^\pi \ , \ b_{N,v}^\pi
\end{bmatrix}
, \
\boldsymbol{p} =
\begin{bmatrix}
a_0^p \\
a_1^p \\
b_1^p \\
\vdots \\
a_N^p \\
b_N^p
\end{bmatrix}
, \
\mathcal{S}\boldsymbol{\pi} =
\begin{bmatrix}
0 \ , & 0 \\
\alpha_1 b_{1,u}^\pi \ , & \alpha_1 b_{1,v}^\pi \\
-\alpha_1 a_{1,u}^\pi \ , & -\alpha_1 a_{1,v}^\pi \\
\vdots & \vdots \\
\alpha_N b_{N,u}^\pi \ , & \alpha_N b_{N,v}^\pi \\
-\alpha_N a_{N,u}^\pi \ , & -\alpha_N a_{N,v}^\pi
\end{bmatrix}
,
$$

$$
\boldsymbol{F}(\boldsymbol{\pi}) =
\begin{bmatrix}
\boldsymbol{f}_0 \\
\boldsymbol{f}_1^c \\
\boldsymbol{f}_1^s \\
\vdots \\
\boldsymbol{f}_N^c \\
\boldsymbol{f}_N^s
\end{bmatrix}
=
\begin{bmatrix}
f_{0,u} \ , \ f_{0,v} \\
f_{1,u}^c \ , \ f_{1,v}^c \\
f_{1,u}^s \ , \ f_{1,v}^s \\
\vdots \quad \vdots \\
f_{N,u}^c \ , \ f_{N,v}^c \\
f_{N,u}^s \ , \ f_{N,v}^s
\end{bmatrix}
, \
\boldsymbol{\gamma} = [\mathcal{G} \ , \boldsymbol{0}] =
\begin{bmatrix}
a_0^\gamma \ , \ 0 \\
a_1^\gamma \ , \ 0 \\
b_1^\gamma \ , \ 0 \\
\vdots \quad \vdots \\
a_N^\gamma \ , \ 0 \\
b_N^\gamma \ , \ 0
\end{bmatrix}
,
$$

allows us to write the first regulator equation in matrix form as

$$\mathcal{S}\boldsymbol{\pi} = A\boldsymbol{\pi} - \nabla \boldsymbol{p} + \boldsymbol{F}(\boldsymbol{\pi}) + B_{\text{in}}\boldsymbol{\gamma},$$
$$\nabla \cdot \boldsymbol{\pi} = \boldsymbol{0},$$

where the nonlinear contributions in $\boldsymbol{F}(\boldsymbol{\pi})$ derive from the structure of the nonlinear term. As in the Burger's example discussed in Section 5.5 they are given by

$$\boldsymbol{f}_0 = -\frac{1}{2}\left(\boldsymbol{a}_0^\pi \cdot \nabla \boldsymbol{a}_0^\pi + 2 \sum_{i=1}^{N} [\boldsymbol{a}_i^\pi \cdot \nabla \boldsymbol{a}_i^\pi + \boldsymbol{b}_i^\pi \cdot \nabla \boldsymbol{b}_i^\pi] \right),$$

and for $n = 1, \cdots, N$ using the notain in (5.27) we have

$$
\begin{aligned}
\boldsymbol{f}_n^c = -\frac{1}{2} \Big(& \boldsymbol{a}_0^\pi \cdot \nabla \boldsymbol{a}_n^\pi + \boldsymbol{a}_n^\pi \cdot \nabla \boldsymbol{a}_0^\pi \\
& + \frac{1}{2} \sum_{i=1}^N \sum_{j=1}^N \big[\big(\boldsymbol{a}_i^\pi \cdot \nabla \boldsymbol{a}_j^\pi + \boldsymbol{a}_j^\pi \cdot \nabla \boldsymbol{a}_i^\pi \big) \delta_1(i, j, n) \\
& \qquad + \big(\boldsymbol{b}_i^\pi \cdot \nabla \boldsymbol{b}_j^\pi + \boldsymbol{b}_j^\pi \cdot \nabla \boldsymbol{b}_i^\pi \big) \delta_2(i, j, k) \big] \Big), \\
\boldsymbol{f}_n^s = -\frac{1}{2} \Big(& \boldsymbol{a}_0^\pi \cdot \nabla \boldsymbol{b}_n^\pi + \boldsymbol{b}_n^\pi \cdot \nabla \boldsymbol{a}_0^\pi \\
& + \sum_{i=1}^N \sum_{j=1}^N \big[\big(\boldsymbol{a}_i^\pi \cdot \nabla \boldsymbol{b}_j^\pi + \boldsymbol{b}_j^\pi \cdot \nabla \boldsymbol{a}_i^\pi \big) \delta_3(i, j, n) \big] \Big),
\end{aligned}
$$

where $\delta_{l,m}$ denotes the standard Kronecker delta.

The first regulator equation is supplemented with boundary conditions and constraints due to the second regulator equation. Given our choice for $y_r(t)$, we define $\mathcal{Y}_r$ to be

$$
\mathcal{Y}_r = \big[2M_1, \, 0, \, A_1, \, 0, \, \ldots, \, 0 \big]^\mathsf{T},
$$

then we proceed by solving the system

$$
\begin{aligned}
0 &= A\boldsymbol{\pi} - \mathcal{S}\boldsymbol{\pi} + \boldsymbol{F}(\boldsymbol{\pi}) + B_{\mathrm{in}}\boldsymbol{\gamma}, \\
0 &= A\widetilde{\boldsymbol{\pi}} - \mathcal{S}\boldsymbol{\pi} + \boldsymbol{F}(\boldsymbol{\pi}), \\
\boldsymbol{\gamma} &= \big[\mathcal{G}, \boldsymbol{0} \big] = \big[G^{-1}(\mathcal{Y}_r - C(\widetilde{\boldsymbol{\pi}})), \boldsymbol{0} \big].
\end{aligned}
$$

5.6.2   The β–Iteration Step

We first use the solution of the Fourier Decomposition Method to reconstruct the approximation

$$
\overline{\boldsymbol{Z}}_1 = \big[\overline{\boldsymbol{V}}_1, \, \overline{P}_1 \big] = \big[\overline{U}_1, \, \overline{V}_1, \, \overline{P}_1 \big]
$$

to the periodic motion on the center manifold. Then we solve a single β–iteration step for

$$
\begin{aligned}
\overline{\boldsymbol{z}}_2 &= \big[\overline{\boldsymbol{v}}_2, \, \overline{p}_2 \big] = \big[\overline{u}_2, \, \overline{v}_2, \, \overline{p}_2 \big], \\
\widetilde{\boldsymbol{z}}_2 &= \big[\widetilde{\boldsymbol{v}}_2, \, \widetilde{p}_2 \big] = \big[\widetilde{u}_2, \, \widetilde{v}_2, \, \widetilde{p}_2 \big],
\end{aligned}
$$

to further refine the approximation of the solution. Namely we proceed as follows:

- Set

$$
\begin{aligned}
\gamma_1 &= \tfrac{1}{2} a_0^\gamma + \textstyle\sum_{i=0}^N a_i^\gamma \cos(\alpha_i t) + b_i^\gamma \sin(\alpha_i t), \\
\boldsymbol{V}_0 &= \tfrac{1}{2} \boldsymbol{a}_0^\pi + \textstyle\sum_{i=0}^N \boldsymbol{a}_i^\pi.
\end{aligned}
$$

Solve the IBVP

$$\overline{\boldsymbol{V}}_{1t} + \overline{\boldsymbol{V}}_1 \cdot \nabla \overline{\boldsymbol{V}}_1 - \nu \Delta \overline{\boldsymbol{V}}_1 + \nabla \overline{P}_1 = 0,$$

$$\nabla \cdot \overline{\boldsymbol{V}}_1 = 0,$$

$$\overline{\boldsymbol{V}}_1(x,0) = \boldsymbol{V}_0,$$

$$\mathcal{B}_{\text{in}}(\overline{\boldsymbol{Z}}_1) = \gamma_1, \ \mathcal{B}_{out}(\overline{\boldsymbol{Z}}_1) = \boldsymbol{0}, \ \mathcal{B}_w(\overline{\boldsymbol{Z}}_1) = \boldsymbol{0}.$$

- Define

$$e_1 = y_r(t) - C(\overline{\boldsymbol{Z}}_1),$$

and solve the following differential algebraic IBVP:

$$\overline{\boldsymbol{v}}_{2t} + \overline{\boldsymbol{V}}_2 \cdot \nabla \overline{\boldsymbol{V}}_2 - \overline{\boldsymbol{V}}_1 \cdot \nabla \overline{\boldsymbol{V}}_1 - \nu \Delta \overline{\boldsymbol{v}}_2 + \nabla \overline{p}_2 = 0,$$

$$\nabla \cdot \overline{\boldsymbol{v}}_2 = 0,$$

$$\overline{\boldsymbol{v}}_2(x,0) = \boldsymbol{0},$$

$$\mathcal{B}_{\text{in}}(\overline{\boldsymbol{z}}_2) = \gamma_2, \ \mathcal{B}_{out}(\overline{\boldsymbol{z}}_2) = \boldsymbol{0}, \ \mathcal{B}_w(\overline{\boldsymbol{z}}_2) = \boldsymbol{0},$$

$$(1-\beta)\overline{\boldsymbol{v}}_{2t} + \overline{\boldsymbol{V}}_2 \cdot \nabla \overline{\boldsymbol{V}}_2 - \overline{\boldsymbol{V}}_1 \cdot \nabla \overline{\boldsymbol{V}}_1 - \nu \Delta \widetilde{\boldsymbol{v}}_2 + \nabla \widetilde{p}_2 = 0,$$

$$\nabla \cdot \widetilde{\boldsymbol{v}}_2 = 0,$$

$$\mathcal{B}_{\text{in}}(\widetilde{\boldsymbol{z}}_2) = \gamma_2, \ \mathcal{B}_{out}(\widetilde{\boldsymbol{z}}_2) = \boldsymbol{0}, \ \mathcal{B}_w(\widetilde{\boldsymbol{z}}_2) = \boldsymbol{0},$$

$$\gamma_2 = G^{-1}(e_1 - C(\widetilde{\boldsymbol{z}}_2)),$$

where we set

$$\overline{\boldsymbol{V}}_2 = \overline{\boldsymbol{V}}_1 + \overline{\boldsymbol{v}}_2.$$

- Set

$$\gamma = \gamma_1 + \gamma_2,$$

as the desired control in (5.31) and solve the closed loop system (5.28)–(5.33).

In our numerical simulation we have set the following parameters

$$\nu = 0.005, \ L = 1, \ D = 0.1, \ \alpha = 1,$$

$$M_1 = 0.1, \ A_1 = 0.1, \ N = 3.$$

The closed loop system is solved for $t \in [0, 20]$ with initial condition $\boldsymbol{\varphi}(x) = \boldsymbol{0}$.

In Fig. 5.42 we have plotted $y = C(\boldsymbol{z})$, $C(U_1)$ and $C(U_2)$. We notice that only y and y_r are distinguishable as is clear from Fig. 5.43 where we have plotted y, $C(U_1)$ and $C(U_2)$ which essentially lay on top of each other. Indeed, in Fig. 5.44 we have plotted $e(t) = y_r - C(\boldsymbol{z})$, $e_1(t) = y_r - C(U_1)$ and $e_2(t) = y_r - C(U_2)$. Notice the scale on the vertical axis is 10^{-4} which explains why the curves in Figs. 5.42 and 5.43 appear to be indistinguishable. In Fig. 5.45 we have plotted the control $u = \gamma$ applied in the closed loop system.

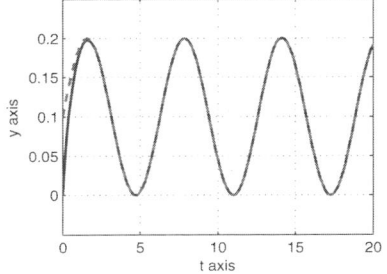

Fig. 5.42: $y = C(z)$ and y_r.

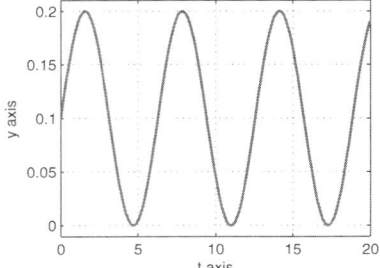

Fig. 5.43: $C(U1)$ and $C(U2)$.

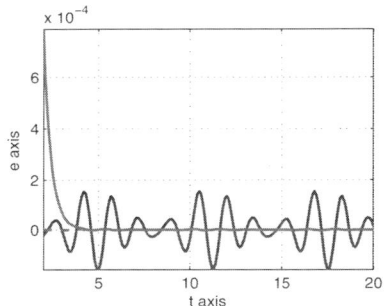

Fig. 5.44: e, e_1 and e_2.

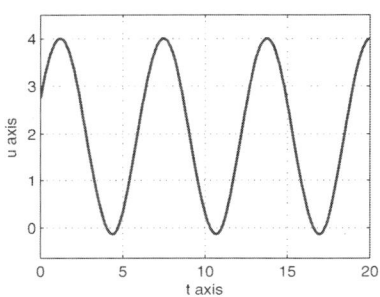

Fig. 5.45: $C(U1)$ and $C(U2)$.

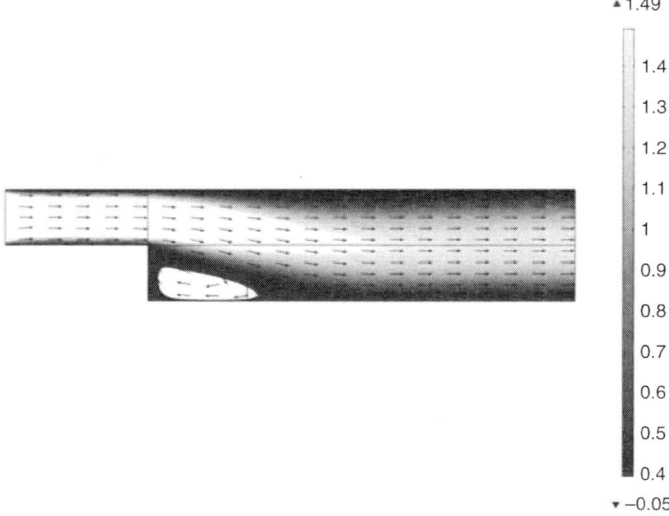

Fig. 5.46: Velocity Field at $t = 20$ Seconds.

Notice in Fig. 5.44 we have plotted $e(t)$, e_1 and e_2 on $2 \leq t \leq 20$ since following a short transient $e(t)$ and the others are on the order of 10^{-4} so if the three were plotted from time $t = 0$ the large transient near time 0 would dominate the figure and we would miss the more precise behavior displayed in the figure.

In our final Fig. 5.46 we have depicted the velocity flow field z at time $t = 20$ for the closed loop system.

5.7 Nonlinear Regulation Using Zero Dynamics Design

5.7.1 Vibrating Nonlinear Beam

Consider the following SISO boundary controlled nonlinear beam equation defined on the region $x \in (0, L) \subset \mathbb{R}$, depicted in Fig. 5.47:

$$u_{tt} - D_1 \left(u_x + \frac{1}{2} w_x^2 \right)_x = 0, \tag{5.35}$$

$$w_{tt} + \beta_1 w_t - \beta_2 w_{xxt} - D_1 \left(w_x \left(u_x + \frac{1}{2} w_x^2 \right) \right)_x + D_2 w_{xxxx} = 0, \tag{5.36}$$

$$u(x, 0) = \varphi, \quad w(x, 0) = \phi, \quad u_t(x, 0) = \psi, \quad w_t(x, 0) = \eta, \tag{5.37}$$

$$u(0, t) = 0, \quad w(0, t) = 0, \quad w_{xx}(0, t) = 0, \tag{5.38}$$

$$u_x(L, t) = 0, \quad w(L, t) = 0, \quad M = D_2 w_{xx}(L, t) = u(t). \tag{5.39}$$

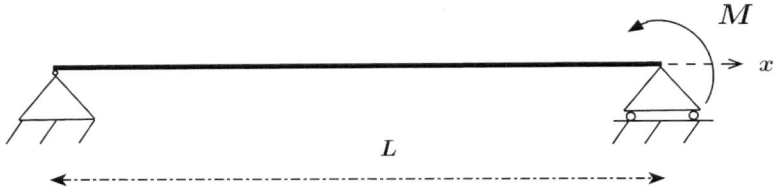

Fig. 5.47: Beam Geometry.

Here the solution vector $z = (u, w)$ corresponds to the horizontal end vertical displacements; $D_1 = E/\rho$ and $D_2 = EI/A\rho$ are the stiffness coefficients, where E is the Young's modulus, ρ the density, A and I the cross section area and the momentum of inertia; β_1 and β_2 are damping coefficients. On the left end of the domain the beam is simply supported in both directions ($u = w = 0$) and no loads are applied ($w_{xx} = 0$). On the right end, the beam is simply supported in the vertical direction ($w = 0$), is free to move in the horizontal direction ($u_x = 0$), and a point torque boundary control $\left(M = D_2 w_{xx} = u(t) \right)$ is applied.

We consider a co-located output given in terms of the trace of the solution spatial derivative on the right end

$$y(t) = C(z(t)) = w_x(L, t),$$

which corresponds to the rotation of the beam end.

Our design objective is to force the measured output $y(t)$ to track a pre-scribed reference trajectory $y_r(t)$.

As discussed in Section 4.7, we obtain the controls from the zero dynamics system obtained from (5.35)–(5.39) by replacing the boundary control term with the constraint $e(t) = y_r(t) - y(t) \equiv 0$ for all time. Thus we obtain the zero dynamics system

$$p_{tt} - D_1 \left(p_x + \frac{1}{2} q_x^2 \right)_x = 0, \tag{5.40}$$

$$q_{tt} + \beta_1 q_t - \beta_2 q_{xxt} - D_1 \left(q_x \left(p_x + \frac{1}{2} q_x^2 \right) \right)_x + D_2 q_{xxxx} = 0, \tag{5.41}$$

$$p(x, 0) = 0, \quad q(x, 0) = 0, \quad p_t(x, 0) = 0, \quad q_t(x, 0) = 0, \tag{5.42}$$

$$p(0, t) = 0, \quad q(0, t) = 0, \quad q_{xx}(0, t) = 0, \tag{5.43}$$

$$p_x(L, t) = 0, \quad q(L, t) = 0, \quad q_x = y_r(t), \tag{5.44}$$

where we have arbitrarily chosen all the initial data to be zero. We note that other choices would simply give control gains with different transient effects.

The desired control is finally obtained from (5.40)–(5.44) by setting in (5.39)

$$u(t) = D_2 q_{xx}(L, t).$$

We choose the parameters to be $L = 5$, $D_1 = 33333$, $D_2 = 27778$, $\beta_1 = 0$, $\beta_2 = D_2$. For the reference signal we choose $y_r(t) = 0.01 \sin(t)$, and set the initial data

$$u(x, 0) = 0, \quad w(x, 0) = 0.01 \frac{x^2(L - x)}{L^3}, \quad u_t(x, 0) = 0, \quad w_t(x, 0) = 0.$$

In our numerical simulation the closed loop system is solved for $t \in (0, 2.5)$.

In Fig. 5.48 we present the reference signals $y_r(t)$ and the resulting measured outputs $y(t) = C(z(t))$. In Fig. 5.49 we present the corresponding error $e(t) = y_r(t) - y(t)$. It is clear from the above figures that the goal of harmonic tracking for $x = L$ is achieved very quickly.

In Figs. 5.50, 5.51, 5.52 and 5.53 we depict the numerical solution surfaces respectively for $u(x, t)$, $w(x, t)$, $w_x(x, t)$ and $w_{xx}(x, t)$, for $0 \le x \le 5$ and for $0 \le t \le 2.5$.

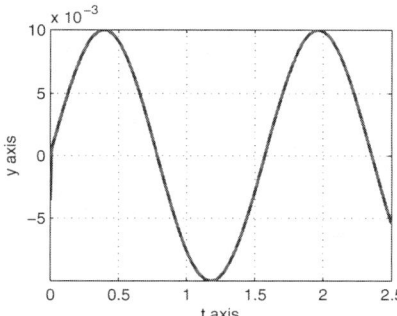

Fig. 5.48: $y_r(t)$ and $y(t)$.

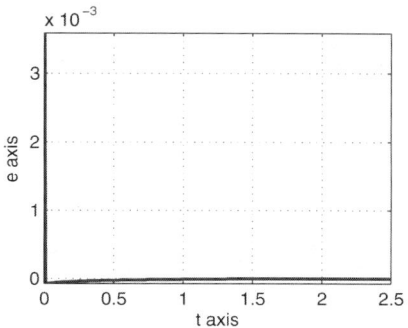

Fig. 5.49: $e(t) = y_r(t) - y(t)$.

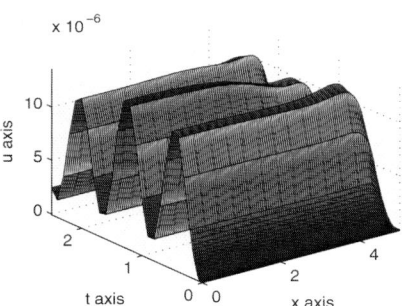

Fig. 5.50: $u(x,t)$.

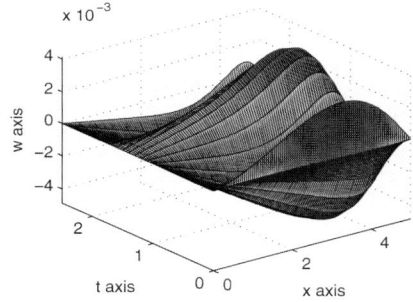

Fig. 5.51: $w(x,t)$.

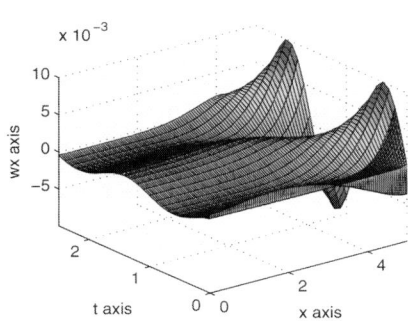

Fig. 5.52: $w_x(x,t)$.

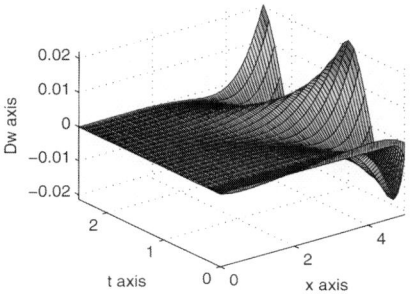

Fig. 5.53: $w_{xx}(x,t)$.

5.7.2   2D Chafee-Infante Equation MIMO

Consider the following MIMO boundary controlled Chafee-Infante equation on the finite rectangular region, for $x = (x_1, x_2) \in (0, a) \times (0, b) \subset \mathbb{R}^2$, as depicted in Fig. 5.54.

$$\frac{\partial z}{\partial t}(x, t) = \Delta z(x, t) + \lambda \left(z(x, t) - z^3(x, t) \right), \ t > 0, \tag{5.45}$$

$$\mathcal{B}_1(z)(x_2, t) = \left(\frac{\partial z}{\partial x_1} - k_1 z \right)(0, x_2, t) = d_1(x_2, t), \tag{5.46}$$

$$\mathcal{B}_2(z)(x_1, t) = \left(\frac{\partial z}{\partial x_2} - k_2 z \right)(x_1, 0, t) = d_2(x_1, t), \tag{5.47}$$

$$\mathcal{B}_3(z)(x_2, t) = \left(\frac{\partial z}{\partial x_1} + k_3 z \right)(a, x_2, t) = u_3(x_2, t), \tag{5.48}$$

$$\mathcal{B}_4(z)(x_1, t) = \left(\frac{\partial z}{\partial x_2} + k_4 z \right)(x_1, b, t) = u_4(x_1, t), \tag{5.49}$$

$$z(x, 0) = \varphi(x),$$

where $d_1(x_2, t)$ and $d_2(x_1, t)$ are distributed disturbances entering through the left and bottom boundary and $u_3(x_2, t)$ and $u_4(x_1, t)$ are distributed controls entering on the right and upper parts of the boundary. Notice, for this problem we have controls u_3 and u_4 entering through mixed boundary conditions on the right and top sides of the rectangle, and again mixed boundary disturbances d_1 and d_2 entering on the left and bottom sides.

In keeping with the zero dynamics design methodology we consider co-located outputs given in terms of the trace of the solution on the right and top boundaries, respectively,

$$y_3(x_2, t) = (\mathcal{C}_3 z)(t) = z(a, x_2, t), \quad y_4(x_1, t) = (\mathcal{C}_4 z)(t) = z(x_1, b, t).$$

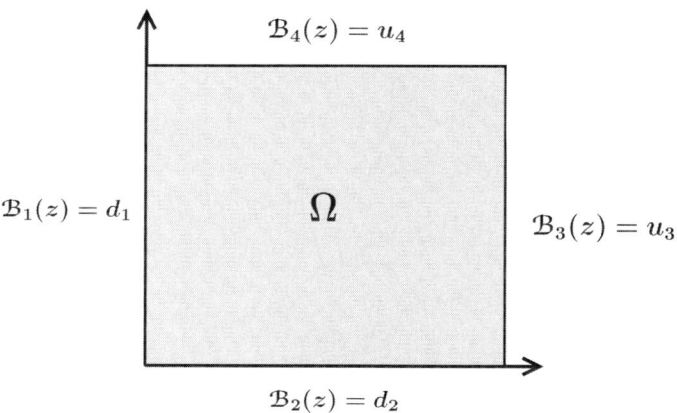

Fig. 5.54: Two-Dimensional Rectangle.

Our design objective is to have the measured outputs y_i, $i = 3, 4$, track prescribed reference trajectories on the right and top sides of the rectangle, i.e., on $x_1 = a$ and $x_2 = b$. More specifically, we assume there are reference signals (to be tracked) in the form

$$y_{r,3}(x_2, t) = w_3(t)q_3(x_2), \quad y_{r,4}(x_1, t) = w_4(t)q_4(x_1),$$
$$w_3, w_4 \in C_b^\infty([0, \infty]), \quad q_3 \in C_c^2(0, b), \quad q_4 \in C_c^2(0, a).$$

Here $C_b^\infty([0, \infty])$ is the class of infinitely differentiable functions with w and dw/dt bounded on $[0, \infty)$ and $C_c^2(I)$ denotes C^2 functions with compact support in the interval I.

We define the error vector by

$$e(t) = \begin{bmatrix} e_1(t) \\ e_2(t) \end{bmatrix} = \begin{bmatrix} \left(\int_0^b (y_{r,3} - y_3)^2 \, dx_2 \right)^{1/2} \\ \left(\int_0^a (y_{r,4} - y_4)^2 \, dx_1 \right)^{1/2} \end{bmatrix}. \tag{5.50}$$

Our main regulation problem consists of finding control inputs u_3, u_4 in (5.48), (5.49), so that the error defined in (5.50) satisfies

$$\|e(t)\| \xrightarrow{t \to \infty} 0.$$

As discussed in Section 4.7, we obtain the controls from the zero dynamics system obtained from (5.45)–(5.48) by replacing the boundary control terms with the constraint $e(t) \equiv 0$ for all time which is equivalent to the conditions (5.54) and (5.55) below. Thus we obtain the system on the rectangular region depicted in Fig. 5.55

$$\frac{\partial \xi}{\partial t}(x, t) = \Delta \xi(x, t) + \lambda \left(\xi(x, t) - \xi^3(x, t) \right), \tag{5.51}$$

$$\mathcal{B}_1(\xi)(x_2, t) = \left(\frac{\partial \xi}{\partial x_1} - k_1 \xi \right) (0, x_2, t) = d_1(x_2, t), \tag{5.52}$$

$$\mathcal{B}_2(\xi)(x_1, t) = \left(\frac{\partial \xi}{\partial x_2} - k_2 \xi \right) (x_1, 0, t) = d_2(x_1, t), \tag{5.53}$$

$$\mathcal{C}_3(\xi)(x_2, t) = (\xi)(a, x_2, t) = y_{3,r}(x_2, t), \tag{5.54}$$

$$\mathcal{C}_4(\xi)(x_1, t) = (\xi)(x_1, b, t) = y_{4,r}(x_1, t), \tag{5.55}$$

$$\xi(x, 0) = \psi(x). \tag{5.56}$$

The controls are finally obtained from (5.51)–(5.56) by setting

$$u_3(x_2, t) = \mathcal{B}_3(z)(x_2, t) = \left(\frac{\partial \xi}{\partial x_1} + k_3 \xi \right) (a, x_2, t),$$

$$u_4(x_1, t) = \mathcal{B}_4(z)(x_1, t) = \left(\frac{\partial \xi}{\partial x_2} + k_4 \xi \right) (x_1, b, t).$$

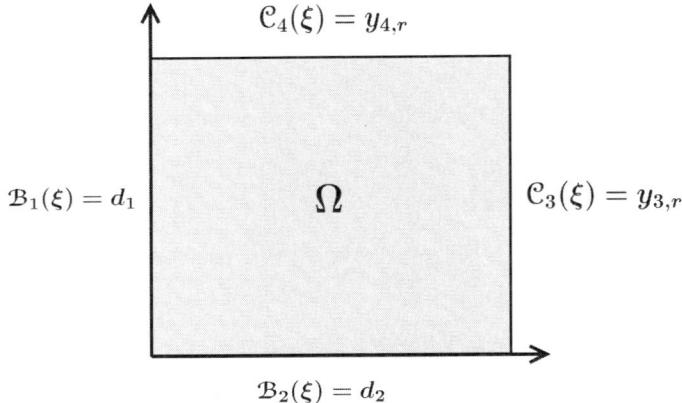

Fig. 5.55: Two-Dimensional Rectangle for Zero Dynamics.

For our numerical example we have considered the MIMO control problem with set-point signals to be tracked given by

$$y_{3,r} = Q(x_2), \quad y_{4,r} = -2Q(x_1),$$

where $Q(s)$ is a smooth approximation to the characteristic function on the intervals $[0.2, 0.55]$. The disturbances to be rejected are given by

$$d_1(x_2, t) = 5\cos\left(2\pi(x_2 - t)\right),$$

$$d_2(x_1, t) = 10\cos\left(2\pi(x_1 - t)\right).$$

We also choose the coefficients and the initial data to be

$$\lambda = 3, \ a = b = 1, \ k_1 = 1, \ k_2 = 2, \ k_3 = 0.5, \ k_4 = 0.25,$$

$$\varphi = 3\cos(2\pi x_1 x_2) \text{ and } \psi = 0.$$

Notice that for this example the boundary disturbances $d_1(x_2, t)$ and $d_2(x_1, t)$ are rather strong disturbances as depicted in Figs. 5.56 and 5.57.

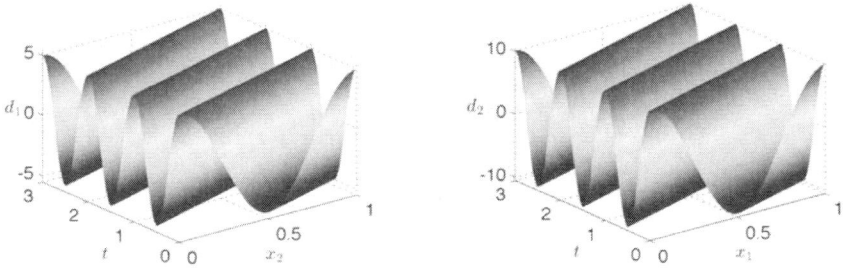

Fig. 5.56: Plot of $d_1(t)$.   **Fig. 5.57**: Plot of $d_2(t)$.

We also note that the convergence of the errors to zero takes place very rapidly. Indeed, by time $t = .75$ the order of the errors are both 10^{-3}. We have displayed the numerical values of the errors in Figs. 5.58 and 5.59.

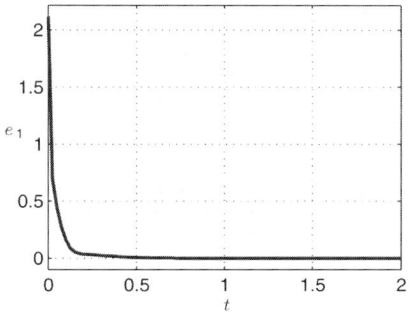 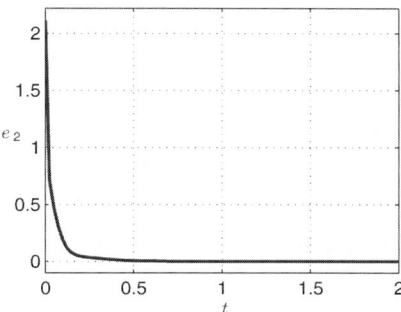

Fig. 5.58: Plot of $e_1(t)$.   **Fig. 5.59**: Plot of $e_2(t)$.

In order to display the convergence of $z(1, x_2, t)$ to the required shape $y_{3,r}(x_2)$, in Fig. 5.60 we have plotted $y_{3,r}(x_2)$ using a dashed line and the curves $z(1, x_2, t)$ for $t = 0.125$, $t = 0.5$ and we notice that by time $t = .75$ the curve lies almost completely on top of the desired target shape so we have not labeled that curve.

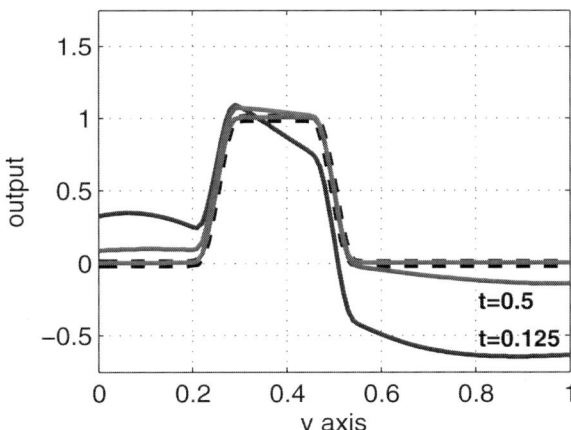

Fig. 5.60: Convergence of Outputs to Desired Shape $y_{3,r}$.

In Fig. 5.61, just as in Fig.5.60, we have displayed the convergence of $z(x_1, 1, t)$ to the required shape $y_{4,r}(x_1)$. Once again we have plotted $y_{4,r}(x_2)$ using a dashed line and the curves $z(x_1, 1, t)$ for $t = 0.125$, $t = 0.5$ and we notice that by time $t = .75$ the curve lies virtually on top of the desired target shape so we have not labeled that curve.

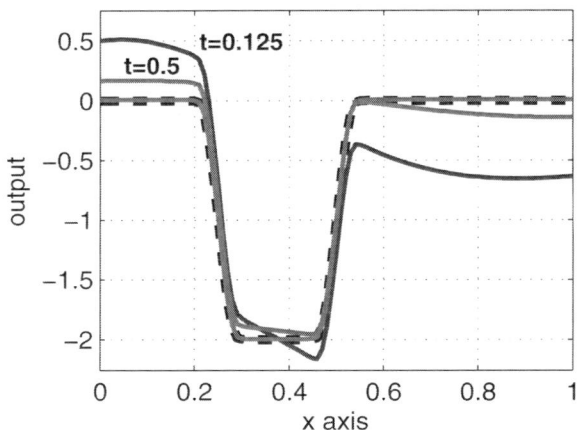

Fig. 5.61: Convergence of Outputs to Desired Shape $y_{4,r}$.

5.7.3 2D Burgers' Equation MIMO

Consider the following MIMO boundary controlled 2D Burgers' equation on the finite rectangular region $x = (x_1, x_2) \in (0, a) \times (0, b) \subset \mathbb{R}^2$, as depicted in Fig. 5.62,

$$z_t = \Delta z - z \nabla \cdot z, \quad z = \begin{bmatrix} z_1 \\ z_2 \end{bmatrix}, \text{ for } t > 0, \tag{5.57}$$

with external controls and disturbances entering through boundary conditions given by

$$\mathcal{B}_1(z)(x_2, t) = \left(\frac{\partial z}{\partial x_1} - k_1 z \right)(0, x_2, t) = d_1(x_2, t), \tag{5.58}$$

$$\mathcal{B}_2(z)(x_1, t) = \left(\frac{\partial z}{\partial x_2} - k_2 z \right)(x_1, 0, t) = d_2(x_1, t), \tag{5.59}$$

$$\mathcal{B}_3(z)(x_2, t) = \left(\frac{\partial z}{\partial x_1} + k_3 z \right)(a, x_2, t) = u_3(x_2, t), \tag{5.60}$$

$$\mathcal{B}_4(z)(x_1, t) = \left(\frac{\partial z}{\partial x_2} + k_4 z \right)(x_1, b, t) = u_4(x_1, t), \tag{5.61}$$

$$z(x, 0) = \varphi(x),$$

where

$$d_1 = \begin{bmatrix} d_{1,1} \\ d_{1,2} \end{bmatrix}, \quad d_2 = \begin{bmatrix} d_{2,1} \\ d_{2,2} \end{bmatrix}, \quad u_3 = \begin{bmatrix} u_{3,1} \\ u_{3,2} \end{bmatrix}, \quad u_4 = \begin{bmatrix} u_{4,1} \\ u_{4,2} \end{bmatrix}.$$

Similarly to the previous example, we seek two-dimensional controls u_3 and u_4 entering through mixed boundary conditions on the right and top sides of the rectangle, while rejecting mixed boundary disturbances $\boldsymbol{d}_1$ and $\boldsymbol{d}_2$ on the left and bottom sides.

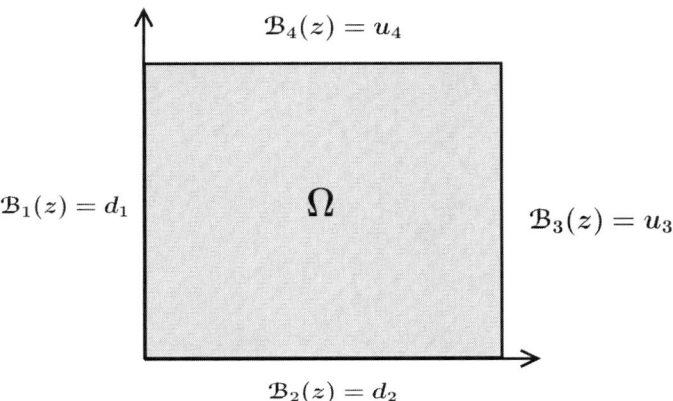

Fig. 5.62: Two-Dimensional Rectangle.

As co-located outputs we take the trace of the solution on the right and top boundaries

$$\boldsymbol{y}_3(x_2, t) = (\mathcal{C}_3 \boldsymbol{z})(t) = \boldsymbol{z}(a, x_2, t), \quad \boldsymbol{y}_4(x_1, t) = (\mathcal{C}_4 \boldsymbol{z})(t) = \boldsymbol{z}(x_1, b, t),$$

respectively.

Our design objective is to have the measured outputs $\boldsymbol{y}_i$, $i = 3, 4$, to track prescribed reference trajectories on the right and top sides of the rectangle, i.e., on $x_1 = a$ and $x_2 = b$.

More specifically, we assume there are reference signals (to be tracked) in the form

$$\boldsymbol{y}_{r3}(x_2, t) = \begin{bmatrix} y_{r3,1}(x_2, t) \\ y_{r3,2}(x_2, t) \end{bmatrix} = \begin{bmatrix} w_{3,1}(t) q_{3,1}(x_2) \\ w_{3,2}(t) q_{3,2}(x_2) \end{bmatrix},$$

$$\boldsymbol{y}_{r4}(x_1, t) = \begin{bmatrix} y_{r4,1}(x_1, t) \\ y_{r4,2}(x_1, t) \end{bmatrix} = \begin{bmatrix} w_{4,1}(t) q_{4,1}(x_1) \\ w_{4,2}(t) q_{4,2}(x_1) \end{bmatrix},$$

where

$$w_{3,i}, w_{4,i} \in C_b^\infty([0, \infty]), \quad q_{3,i} \in C_c^2(0, b), \quad q_{4,i} \in C_c^2(0, a), \text{ for } i = 1, 2.$$

Here $C_b^\infty([0, \infty])$ is the class of infinitely differentiable functions with w and dw/dt bounded on $[0, \infty)$ and $C_c^2(I)$ denotes C^2 functions with compact support in the interval I.

We define the error vector by

$$
e(t) = \begin{bmatrix} e_{1,1}(t) \\ e_{1,2}(t) \\ e_{2,1}(t) \\ e_{2,2}(t) \end{bmatrix} = \begin{bmatrix} \left(\displaystyle\int_0^b (y_{r3,1} - y_{3,1})^2 \, dx_2 \right)^{1/2} \\ \left(\displaystyle\int_0^b (y_{r3,2} - y_{3,2})^2 \, dx_2 \right)^{1/2} \\ \left(\displaystyle\int_0^b (y_{r4,1} - y_{4,1})^2 \, dx_2 \right)^{1/2} \\ \left(\displaystyle\int_0^b (y_{r4,2} - y_{4,2})^2 \, dx_2 \right)^{1/2} \end{bmatrix}. \tag{5.62}
$$

Our main problem is finding vectors of control inputs $\boldsymbol{u}_3$, $\boldsymbol{u}_4$ in (5.60), (5.61), so that the error defined in (5.62) satisfies

$$
\|e(t)\| \xrightarrow{t \to \infty} 0.
$$

Following the discussion in Section 4.7, we obtain the controls from the zero dynamics system obtained from (5.57)–(5.61) by replacing the boundary control term with the constraint $e(t) \equiv 0$ for all time. These conditions imply the boundary constraints imposed in (5.66)–(5.67). Thus we obtain the zero dynamics system acting in the domain in Fig. 5.63:

$$
\frac{\partial \boldsymbol{\xi}}{\partial t}(x, t) = \Delta \boldsymbol{\xi} - \boldsymbol{\xi} \cdot \nabla \boldsymbol{\xi}, \ \boldsymbol{\xi} = \begin{bmatrix} \xi_1 \\ \xi_2 \end{bmatrix}, \ \text{for } t > 0, \tag{5.63}
$$

$$
\mathcal{B}_1(\boldsymbol{\xi})(x_2, t) = \left(\frac{\partial \boldsymbol{\xi}}{\partial x_1} - k_1 \boldsymbol{\xi} \right)(0, x_2, t) = \boldsymbol{d}_1(x_2, t), \tag{5.64}
$$

$$
\mathcal{B}_2(\boldsymbol{\xi})(x_1, t) = \left(\frac{\partial \boldsymbol{\xi}}{\partial x_2} - k_2 \boldsymbol{\xi} \right)(x_1, 0, t) = \boldsymbol{d}_2(x_1, t), \tag{5.65}
$$

$$
\mathcal{C}_3(\boldsymbol{\xi})(x_2, t) = (\boldsymbol{\xi})(a, x_2, t) = \boldsymbol{y}_{3,r}(x_2, t), \tag{5.66}
$$

$$
\mathcal{C}_4(\boldsymbol{\xi})(x_1, t) = (\boldsymbol{\xi})(x_1, b, t) = \boldsymbol{y}_{4,r}(x_1, t), \tag{5.67}
$$

$$
\boldsymbol{\xi}(x, 0) = \boldsymbol{\psi}(x). \tag{5.68}
$$

The controls are finally obtained from (5.63)–(5.68) by setting

$$
\boldsymbol{u}_3(x_2, t) = \mathcal{B}_3(z)(x_2, t) = \left(\frac{\partial \boldsymbol{\xi}}{\partial x_1} + k_3 \boldsymbol{\xi} \right)(a, x_2, t),
$$

$$
\boldsymbol{u}_4(x_1, t) = \mathcal{B}_4(z)(x_1, t) = \left(\frac{\partial \boldsymbol{\xi}}{\partial x_2} + k_4 \boldsymbol{\xi} \right)(x_1, b, t).
$$

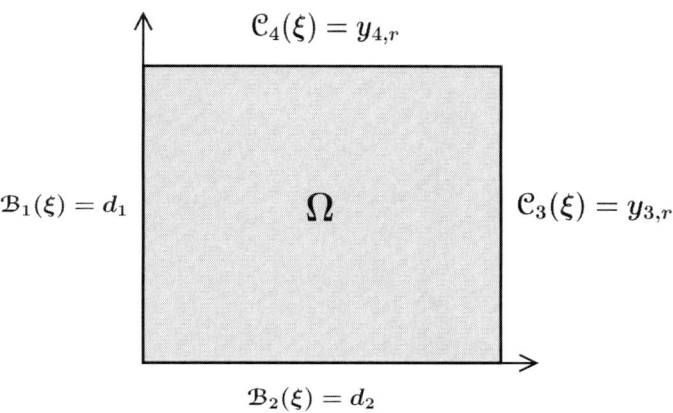

$\mathcal{C}_4(\xi) = y_{4,r}$

$\mathcal{B}_1(\xi) = d_1$

Ω

$\mathcal{C}_3(\xi) = y_{3,r}$

$\mathcal{B}_2(\xi) = d_2$

Fig. 5.63: Two-Dimensional Rectangle for Zero Dynamics.

In our numerical example we have considered the MIMO control problem with set-point signals to be tracked given by

$$\boldsymbol{y}_{3,r} = \begin{bmatrix} Q(x_2) \\ -Q(x_2) \end{bmatrix}, \quad \boldsymbol{y}_{4,r} = \begin{bmatrix} 2Q(x_1) \\ -2Q(x_1) \end{bmatrix},$$

where $Q(s)$ is a smooth approximation to the characteristic function on the intervals $[0.2, 0.55]$ as considered in the previous example. The disturbances to be rejected are given by

$$\boldsymbol{d}_1(x_2, t) = \begin{bmatrix} 5\cos\left(2\pi(x_2 - t)\right) \\ 5\cos\left(2\pi(x_2 - t)\right) \end{bmatrix},$$

$$\boldsymbol{d}_2(x_1, t) = \begin{bmatrix} 10\cos\left(2\pi(x_1 - t)\right) \\ 10\cos\left(2\pi(x_1 - t)\right) \end{bmatrix}.$$

We also choose the coefficients and the initial data to be

$$a = b = 1, \ k_1 = 1, \ k_2 = 2, \ k_3 = 0.5, \ k_4 = 0.25,$$

$$\varphi = \begin{bmatrix} 3\cos(2\pi x_1 x_2) \\ 3\sin(2\pi x_1 x_2) \end{bmatrix} \text{ and } \psi = \begin{bmatrix} 0 \\ 0 \end{bmatrix}.$$

For this example the boundary disturbances $d_1(x_2, t)$ and $d_2(x_1, t)$ are exactly those used in the previous example. The only difference is that in this case the disturbances enter in both components of the state variable through the left boundary $(0, x_2)$ for $0 \le x_2 \le 1$ and the lower boundary at $(x_1, 0)$ for $0 \le x_1 \le 1$.

Due to the vector nature of 2D Burgers we depict several outputs and the associated errors.

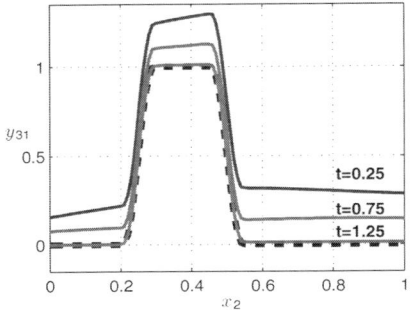

Fig. 5.64: $z_1(1, x_2, t)$, $y_{3,1}$.

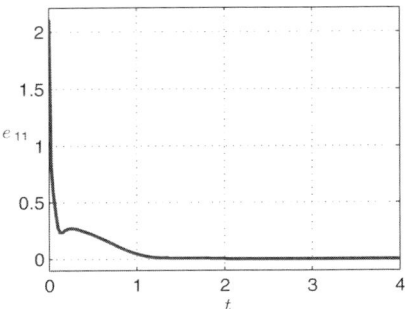

Fig. 5.65: Plot of $e_{11}(t)$.

In Fig. 5.64 we have plotted the values of $z_1(1, x_2, t)$ at the times $t = 0.125, 0.50, 1.25$. It is evident that the trajectories are approaching the desired reference shape function which is depicted using a dashed line. The error $e_{11}(t)$ is plotted in Fig. 5.65. We note that the error $e_{11}(1.25) = 0.0109$ and so the a plot of $z_1(1, x_2, 1.25)$ is nearly indistinguishable from the reference signal.

In Fig. 5.66 we have plotted the values of $z_2(1, x_2, t)$ at the times $t = 0.25, 0.50, 0.75$. It is evident that the trajectories are approaching the desired reference shape function which is depicted using a dashed line. The error $e_{21}(t)$ is plotted in Fig. 5.67.

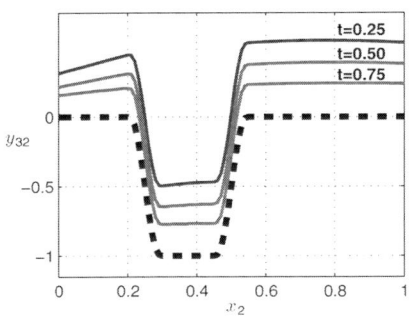
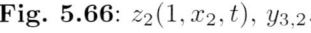

Fig. 5.66: $z_2(1, x_2, t)$, $y_{3,2}$.

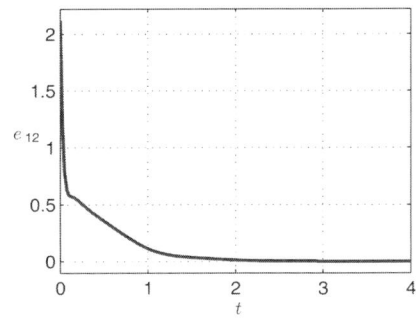

Fig. 5.67: Plot of $e_{21}(t)$.

In Fig. 5.68 we have plotted the values of $z_1(x_1, 1, t)$ at the times $t = 0.25, 1$. It is evident that the trajectories are approaching the desired reference shape function which is depicted using a dashed line. The error $e_{12}(t)$ is plotted in Fig. 5.69. We note that the error $e_{12}(1.25) = 0.0092$ and so a plot of $z_1(x_1, 1, 1.25)$ would be indistinguishable from the reference signal.

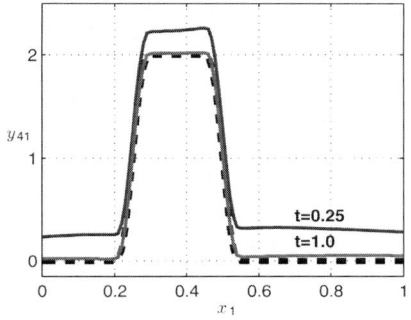

Fig. 5.68: $z_1(x_1, 1, t)$, $y_{4,1}$.

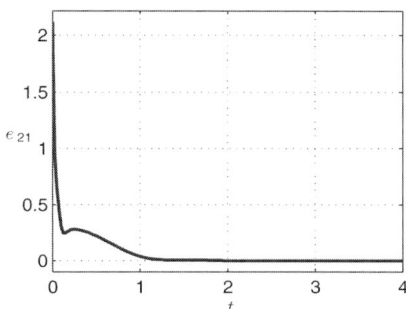

Fig. 5.69: Plot of $e_{12}(t)$.

In Fig. 5.70 we have plotted the values of $z_2(x_1, 1, t)$ at the times $t = 0.25, 1$. It is evident that the trajectories are approaching the desired reference shape function which is depicted using a dashed line. The error $e_{22}(t)$ is plotted in Fig. 5.71.

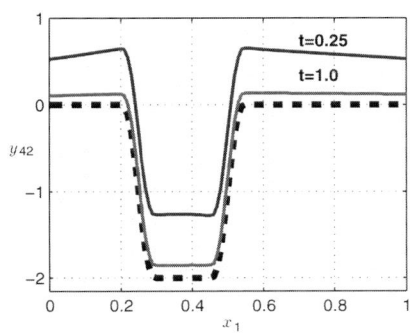

Fig. 5.70: $z_2(x_1, 1, t)$, $y_{4,2}$.

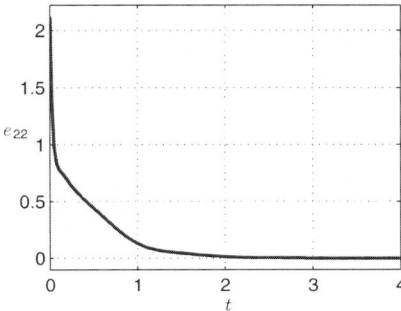

Fig. 5.71: Plot of $e_{22}(t)$.

Bibliography

[1] E. Aulisa, J. A. Burns, and D. S. Gilliam. An example of thermal regulation of a two dimensional non-isothermal incompressible flow. In *Decision and Control (CDC), 2012 IEEE 51st Annual Conference on*, pages 1578–1583. IEEE, Los Alamitos, CA, 2012.

[2] E. Aulisa and D. S. Gilliam. A numerical algorithm for set-point regulation of non-linear parabolic control systems. *International Journal of Numerical Analysis and Modeling*, 11(1):54–85, 2014.

[3] E. Aulisa, D. S. Gilliam, and T. W. Pathiranage. Analysis of the error in an iterative algorithm for solution of the regulator equations for linear distributed parameter control systems. Preprint, Texas Tech University, 2015.

[4] H. T. Banks. *A functional analysis framework for modeling, estimation and control in science and engineering*. CRC Press, Boca Raton, FL, 2012.

[5] H. T. Banks, W. Fang, and R. C. Smith. Active noise control: piezoceramic actuators in fluid/structure interaction models. In *Decision and Control, 1991, Proceedings of the 30th IEEE Conference on*, pages 2328–2333. IEEE, Los Alamitos, CA, 1991.

[6] Alain Bensoussan, Giuseppe Da Prato, Michel C. Delfour, and Sanjoy K. Mitter. *Representation and control of infinite-dimensional systems. Vol. 1*. Systems & Control: Foundations & Applications. Birkhäuser Boston Inc., Boston, MA, 1992.

[7] Alain Bensoussan, Giuseppe Da Prato, Michel C. Delfour, and Sanjoy K. Mitter. *Representation and control of infinite-dimensional systems. Vol. 2*. Systems & Control: Foundations & Applications. Birkhäuser Boston Inc., Boston, MA, 1993.

[8] C. I. Byrnes, F. Delli Priscoli, A. Isidori, and W. Kang. Structurally stable output regulation of nonlinear systems. *Automatica J. IFAC*, 33(3):369–385, 1997.

[9] C. I. Byrnes and D. S. Gilliam. Asymptotic properties of root locus for distributed parameter systems. In *Decision and Control, 1988, Proceed-*

ings of the 27th IEEE Conference on, pages 45–51. IEEE, Los Alamitos, CA, 1988.

[10] C. I. Byrnes and D. S. Gilliam. Approximate solutions of the regulator equations for nonlinear dps. In *Decision and Control, 2007 46th IEEE Conference on*, pages 854–859. IEEE, Los Alamitos, CA, 2007.

[11] C. I. Byrnes and D. S. Gilliam. Geometric output regulation for a class of nonlinear distributed parameter systems. In *American Control Conference, 2008*, pages 254–259. IEEE, Los Alamitos, CA, 2008.

[12] C. I. Byrnes, D. S. Gilliam, and C. Hu. Set-point boundary control for a kuramoto-sivashinsky equation. In *Decision and Control, 2006 45th IEEE Conference on*, pages 75–80. IEEE, Los Alamitos, CA, 2006.

[13] C. I. Byrnes, D. S. Gilliam, C. Hu, and V. I. Shubov. Zero dynamics boundary control for regulation of the Kuramoto-Sivashinsky equation. *Math. Comput. Modelling*, 52(5-6):875–891, 2010.

[14] C. I. Byrnes, D. S. Gilliam, C. Hu, and V. I. Shubov. Asymptotic regulation for distributed parameter systems via zero dynamics inverse design. *Internat. J. Robust Nonlinear Control*, 23(3):305–333, 2013.

[15] C. I. Byrnes, D. S. Gilliam, A. Isidori, and V. I. Shubov. Static and dynamic controllers for boundary controlled distributed parameter systems. In *Decision and Control, 2004 43rd IEEE Conference on*, volume 3, pages 3324–3325. IEEE, Los Alamitos, CA, 2004.

[16] C. I. Byrnes, D. S. Gilliam, A. Isidori, and V. I. Shubov. Set-point boundary control for a viscous Burgers equation. In *New directions and applications in control theory*, volume 321 of *Lecture Notes in Control and Inform. Sci.*, pages 43–60. Springer, Berlin, 2005.

[17] C. I. Byrnes, D. S. Gilliam, A. Isidori, and V. I. Shubov. Interior point control of a heat equation using zero dynamics design. In *American Control Conference, 2006*, pages 1138–1143. IEEE, Los Alamitos, CA, 2006.

[18] C. I. Byrnes, D. S. Gilliam, A. Isidori, and V. I. Shubov. Zero dynamics modeling and boundary feedback design for parabolic systems. *Math. Comput. Modelling*, 44(9-10):857–869, 2006.

[19] C. I. Byrnes, D. S. Gilliam, and V. I. Shubov. On the global dynamics of a controlled viscous Burgers' equation. *J. Dynam. Control Systems*, 4(4):457–519, 1998.

[20] C. I. Byrnes, D. S. Gilliam, and V. I. Shubov. Examples of regular linear systems governed by partial differential equations. In *Decision and Control, 2001. Proceedings of the 40th IEEE Conference on*, volume 1, pages 129–130. IEEE, Los Alamitos, CA, 2001.

[21] C. I. Byrnes, D. S. Gilliam, V. I. Shubov, and G. Weiss. Regular linear systems governed by a boundary controlled heat equation. *J. Dynam. Control Systems*, 8(3):341–370, 2002.

[22] C. I. Byrnes and A. Isidori. Nonlinear internal models for output regulation. *IEEE Trans. Automat. Control*, 49(12):2244–2247, 2004.

[23] Christopher I. Byrnes, David S. Gilliam, and Jianqiu He. A root locus methodology for parabolic distributed parameter feedback systems. In *Computation and control, II (Bozeman, MT, 1990)*, volume 11 of *Progr. Systems Control Theory*, pages 63–83. Birkhäuser Boston, Boston, MA, 1991.

[24] Christopher I. Byrnes, David S. Gilliam, and Jianqiu He. Root-locus and boundary feedback design for a class of distributed parameter systems. *SIAM Journal on Control and Optimization*, 32(5):1364–1427, 1994.

[25] Christopher I. Byrnes, David S. Gilliam, and Victor I. Shubov. High gain limits of trajectories and attractors for a boundary controlled viscous Burgers' equation. *J. Math. Systems Estim. Control*, 6(4):40 pp. (electronic), 1996.

[26] Christopher I. Byrnes, David S. Gilliam, and Victor I. Shubov. Boundary control, stabilization and zero-pole dynamics for a non-linear distributed parameter system. *Internat. J. Robust Nonlinear Control*, 9(11):737–768, 1999.

[27] Christopher I. Byrnes and Alberto Isidori. Asymptotic stabilization of minimum phase nonlinear systems. *IEEE Trans. Automat. Control*, 36(10):1122–1137, 1991.

[28] Christopher I. Byrnes and Alberto Isidori. Limit sets, zero dynamics, and internal models in the problem of nonlinear output regulation. *IEEE Trans. Automat. Control*, 48(10):1712–1723, 2003.

[29] Christopher I. Byrnes, Istvan G. Lauko, D. S. Gilliam, and Victor I. Shubov. Zero dynamics for relative degree one siso distributed parameter systems. In *Decision and Control, 1998, Proceedings of the 37th IEEE Conference on*, volume 3, pages 2390–2391. IEEE, Los Alamitos, CA, 1998.

[30] Christopher I. Byrnes, István G. Laukó, David S. Gilliam, and Victor I. Shubov. Output regulation for linear distributed parameter systems. *IEEE Trans. Automat. Control*, 45(12):2236–2252, 2000.

[31] Jack Carr. *Applications of centre manifold theory*, volume 35 of *Applied Mathematical Sciences*. Springer-Verlag, New York, 1981.

[32] Ada Cheng and Kirsten Morris. Well-posedness of boundary control systems. *SIAM J. Control Optim.*, 42(4):1244–1265, 2003.

[33] Ruth F. Curtain and George Weiss. Well posedness of triples of operators (in the sense of linear systems theory). In *Control and estimation of distributed parameter systems (Vorau, 1988)*, volume 91 of *Internat. Ser. Numer. Math.*, pages 41–59. Birkhäuser, Basel, 1989.

[34] Ruth F. Curtain and Hans Zwart. *An introduction to infinite-dimensional linear systems theory*, volume 21 of *Texts in Applied Mathematics*. Springer-Verlag, New York, 1995.

[35] Edward J. Davison. The robust control of a servomechanism problem for linear time-invariant multivariable systems. *IEEE Trans. Automatic Control*, AC-21(1):25–34, 1976.

[36] Nelson Dunford and Jacob T. Schwartz. *Linear Operators, Part III: Spectral Operators*. Wiley Classics Library. John Wiley & Sons, Inc., New York, 1988.

[37] Nelson Dunford and Jacob T. Schwartz. *Linear Operators, Spectral Theory, Self Adjoint Operators in Hilbert Space, Part II*. Wiley Classics Library. John Wiley & Sons, Inc., New York, 1988.

[38] Klaus-Jochen Engel and Rainer Nagel. *One-parameter semigroups for linear evolution equations*, volume 194 of *Graduate Texts in Mathematics*. Springer-Verlag, New York, 2000.

[39] Lawrence C. Evans. *Partial Differential Equations*. American Mathematical Society, Providence, RI, 1998.

[40] B. A. Francis and W. M. Wonham. The internal model principle of control theory. *Automatica–J. IFAC*, 12(5):457–465, 1976.

[41] Bruce A. Francis. The linear multivariable regulator problem. *SIAM J. Control Optimization*, 15(3):486–505, 1977.

[42] Bruce A. Francis and William M. Wonham. The role of transmission zeros in linear multivariable regulators. *International Journal of Control*, 22(5):657–681, 1975.

[43] Timo Hämäläinen and Seppo Pohjolainen. Robust control and tuning problem for distributed parameter systems. *International Journal of Robust and Nonlinear Control*, 6(5):479–500, 1996.

[44] Timo Hämäläinen and Seppo Pohjolainen. Robust regulation of distributed parameter systems with infinite-dimensional exosystems. *SIAM J. Control Optim.*, 48(8):4846–4873, 2010.

[45] Timo Hämäläinen and Seppo Pohjolainen. A self-tuning robust regulator for infinite-dimensional systems. *IEEE Trans. Automat. Control*, 56(9):2116–2127, 2011.

[46] M. Hautus. Linear matrix equations with applications to the regulator problem. *Outils et modèles mathématiques pour lAutomatique*, pages 399–412, 1983.

[47] Daniel Henry. *Geometric theory of semilinear parabolic equations*, volume 840 of *Lecture Notes in Mathematics*. Springer-Verlag, Berlin, 1981.

[48] L. Herrmann. Vibration of the euler-bernoulli beam with allowance for dampings. *Proceedings of the World Congress on Engineering*, II, July 2008.

[49] Jie Huang. *Nonlinear Output Regulation: Theory and Applications*, volume 8 of *Advances in Design and Control*. Society for Industrial and Applied Mathematics (SIAM), Philadelphia, PA, 2004.

[50] Eero Immonen. A feedforward-feedback controller for infinite-dimensional systems and regulation of bounded uniformly continuous signals. *Internat. J. Robust Nonlinear Control*, 16(5):259–280, 2006.

[51] Eero Immonen. Some properties of infinite-dimensional systems capable of asymptotically tracking bounded uniformly continuous signals. *Mathematics of Control, Signals and Systems*, 18(4):323–344, 2006.

[52] Eero Immonen and Seppo Pohjolainen. Output regulation of periodic signals for dps: an infinite-dimensional signal generator. *Automatic Control, IEEE Transactions on*, 50(11):1799–1804, 2005.

[53] Eero Immonen and Seppo Pohjolainen. Feedback and feedforward output regulation of bounded uniformly continuous signals for infinite-dimensional systems. *SIAM J. Control Optim.*, 45(5):1714–1735 (electronic), 2006.

[54] A. Isidori and C. H. Moog. On the nonlinear equivalent of the notion of transmission zeros. In *Modelling and Adaptive Control*, pages 146–158. Springer, New York, 1988.

[55] Alberto Isidori. *Nonlinear control systems*. Communications and Control Engineering Series. Springer-Verlag, Berlin, third edition, 1995.

[56] Alberto Isidori and Christopher I. Byrnes. Output regulation of nonlinear systems. *IEEE Trans. Automat. Control*, 35(2):131–140, 1990.

[57] Tosio Kato. *Perturbation theory for linear operators*. Die Grundlehren der mathematischen Wissenschaften, Band 132. Springer, New York, 1966.

[58] Hans-W. Knobloch, Alberto Isidori, and Dietrich Flockerzi. *Topics in control theory*, volume 22 of *DMV Seminar*. Birkhäuser Verlag, Basel, 1993.

[59] A. J. Krener and A. Isidori. Nonlinear zero distributions. In *Decision and Control including the Symposium on Adaptive Processes, 1980 19th IEEE Conference on*, volume 19, pages 665–668. IEEE, Los Alamitos, CA, 1980.

[60] Olga Aleksandrovna Ladyženskaâ. *The Boundary Value Problems of Mathematical Physics*, volume 3. Springer-Verlag, Berlin, 1985.

[61] Olga Aleksandrovna Ladyženskaâ, Vsevolod Alekseevič Solonnikov, and Nina N Ural'tseva. *Linear and quasi-linear equations of parabolic type*, volume 23. American Mathematical Society, Providence, RI, 1988.

[62] Jacques Louis Lions. *Optimal control of systems governed by partial differential equations problèmes aux limites*. Springer, Berlin, 1971.

[63] Kirsten Morris and Richard Rebarber. Zeros of siso infinite-dimensional systems. In David S. Gilliam and Joachim Rosenthal, editors, *MTNS 2002*, volume Electronic Proceedings of 15th International Symposium on the Mathematical Theory of Networks and Systems, pages 3–10, 2002.

[64] Kirsten Morris and Richard Rebarber. Invariant zeros of SISO infinite-dimensional systems. *Internat. J. Control*, 83(12):2573–2579, 2010.

[65] COMSOL Multiphysics. 5.1 user's guide, 2015.

[66] Vivek Natarajan, David S. Gilliam, and George Weiss. The state feedback regulator problem for regular linear systems. *Automatic Control, IEEE Transactions on*, 59(10):2708–2723, 2014.

[67] T. W. Pathiranage. Error analysis for harmonic tracking algorithm using geometric control. Master's thesis, Texas Tech University, 2013.

[68] T. W. Pathiranage. *Error Analysis for a Tracking Algorithm Using Geometric Control*. PhD thesis, Texas Tech University, 2015.

[69] L. Paunonen and S. Pohjolainen. Internal model theory for distributed parameter systems. *SIAM J. Control Optim.*, 48(7):4753–4775, 2010.

[70] Lassi Paunonen and Seppo Pohjolainen. Periodic output regulation for distributed parameter systems. *Mathematics of Control, Signals, and Systems*, 24(4):403–441, 2012.

[71] Aleksej Pavlov, Nathan van de Wouw, and Hendrik Nijmeijer. *Uniform output regulation of nonlinear systems: a convergent dynamics approach*. Springer, New York, 2006.

[72] A. Pazy. *Semigroups of linear operators and applications to partial differential equations*, volume 44 of *Applied Mathematical Sciences*. Springer-Verlag, New York, 1983.

[73] Sulanie Perera. Comparison of optimal and geometric control methods for regulation of distributed parameter systems. Master's thesis, Texas Tech University, Department of Mathematics and Statistics, 2013.

[74] S. A. Pohjolainen. On the asymptotic regulation problem for distributed parameter systems. In *Control of distributed parameter systems, 1982 (Toulouse, 1982)*, pages 197–201. IFAC, Laxenburg, 1983.

[75] Seppo Pohjolainen. Computation of transmission zeros for distributed parameter systems. *International journal of control*, 33(2):199–212, 1981.

[76] Seppo A. Pohjolainen. Robust multivariable PI-controller for infinite-dimensional systems. *IEEE Trans. Automat. Control*, 27(1):17–30, 1982.

[77] Jean-Pierre Raymond. Feedback boundary stabilization of the two-dimensional Navier-Stokes equations. *SIAM J. Control Optim.*, 45(3):790–828 (electronic), 2006.

[78] Dietmar Salamon. Infinite-dimensional linear systems with unbounded control and observation: a functional analytic approach. *Trans. Amer. Math. Soc.*, 300(2):383–431, 1987.

[79] J. M. Schumacher. *Dynamic feedback in finite- and infinite-dimensional linear systems*, volume 143 of *Mathematical Centre Tracts*. Mathematisch Centrum, Amsterdam, 1981.

[80] J. M. Schumacher. A direct approach to compensator design for distributed parameter systems. *SIAM J. Control Optim.*, 21(6):823–836, 1983.

[81] J. M. Schumacher. Finite-dimensional regulators for a class of infinite-dimensional systems. *Systems Control Lett.*, 3(1):7–12, 1983.

[82] Olof Staffans and George Weiss. Transfer functions of regular linear systems part ii: The system operator and the lax–phillips semigroup. *Transactions of the American Mathematical Society*, 354(8):3229–3262, 2002.

[83] Ivar Stakgold. *Green's functions and boundary value problems*. John Wiley & Sons, New York-Chichester-Brisbane, 1979.

[84] Marius Tucsnak and George Weiss. *Observation and control for operator semigroups*. Birkhäuser Advanced Texts: Basler Lehrbücher. [Birkhäuser Advanced Texts: Basel Textbooks]. Birkhäuser Verlag, Basel, 2009.

[85] George Weiss. Admissibility of unbounded control operators. *SIAM J. Control Optim.*, 27(3):527–545, 1989.

[86] George Weiss. Admissible observation operators for linear semigroups. *Israel J. Math.*, 65(1):17–43, 1989.

[87] George Weiss. Regular linear systems with feedback. *Mathematics of Control, Signals and Systems*, 7(1):23–57, 1994.

[88] George Weiss. Transfer functions of regular linear systems. i. characterizations of regularity. *Transactions of the American Mathematical Society*, 342(2):827–854, 1994.

[89] George Weiss and Ruth F. Curtain. Exponential stabilization of a Rayleigh beam using collocated control. *IEEE Trans. Automat. Control*, 53(3):643–654, 2008.

[90] J. Wloka. *Partial Differential Equations*. Cambridge University Press, Cambridge, UK, 1987.

[91] W. Murray Wonham. *Linear multivariable control: a geometric approach*, volume 10 of *Applications of Mathematics*. Springer-Verlag, New York, second edition, 1979.

[92] T. I. Zelenyak, M. M. Lavrentiev, Jr., and M. P. Vishnevskii. *Qualitative theory of parabolic equations. Part 1*. VSP, Utrecht, 1997.

[93] H. Zwart and M. B. Hof. Zeros of infinite-dimensional systems. *IMA J. Math. Control Inform.*, 14(1):85–94, 1997.

[94] H. J. Zwart. *Geometric theory for infinite-dimensional systems*, volume 115 of *Lecture Notes in Control and Information Sciences*. Springer-Verlag, Berlin, 1989.

Index

List of Symbols